Energy Harvesting Properties of Electrospun Nanofibers

Energy Harvesting Properties of Electrospun Nanofibers

Edited by
Jian Fang and Tong Lin
Institute for Frontier Materials, Deakin University, Geelong, Australia

IOP Publishing, Bristol, UK

ISBN 978-0-7503-2005-4 (ebook)
ISBN 978-0-7503-2003-0 (print)
ISBN 978-0-7503-2006-1 (myPrint)
ISBN 978-0-7503-2004-7 (mobi)

DOI 10.1088/978-0-7503-2005-4

Version: 20191201

IOP ebooks

British Library Cataloguing-in-Publication Data: A catalogue record for this book is available from the British Library.

Published by IOP Publishing, wholly owned by The Institute of Physics, London

IOP Publishing, Temple Circus, Temple Way, Bristol, BS1 6HG, UK

US Office: IOP Publishing, Inc., 190 North Independence Mall West, Suite 601, Philadelphia, PA 19106, USA

Contents

Preface

Harvesting small kinetic energies for electric power using piezoelectric or the triboelectric principle is an emerging technology attracting broad attention in materials and technology fields. It offers a favorable solution to powering microelectronic devices and the development of self-powered sensors, robot systems, and personal electronics, thus contributing to solving our over-dependence on fossil energy.

With the rapid progress in nanotechnology and nanomaterials, research on piezoelectric and triboelectric energy harvesting has made amazing progress over recent decades. In particular, polymeric nanofibers produced by an electrostatic spinning technology, as active substances, have shown enormous potential in energy conversion. They offer many advantages, including a wide range of selection in materials, controllability of components, fiber diameter, fibrous structure and surface features, high flexibility, light weight, and large-scale production, and thus are superior to most other active materials.

This book provides a state-of-the-art review about the applications of electrospun nanofibers for piezoelectric and triboelectric energy harvesting. It comprises eight chapters covering the basis of electrospinning, nanofibers, piezoelectricity, triboelectricity, and the emerging research results about the use of nanofibers in energy harvesting devices. In chapter 1, we provide a general introduction about electrospinning technology and electrospun nanofibers. Chapter 2 introduces fundamentals about piezoelectricity and piezoelectric materials fabricated by electrospinning. Chapter 3 introduces triboelectricity and triboelectric energy harvesting. Chapter 4 reviews the current progress in using electrospun nanofibers in piezoelectric energy harvesters. Chapter 5 discusses the approaches to control the β crystal phase in polyvinylidene fluoride nanofibers. Chapter 6 describes the applications of electrospun nanofibers in acoustoelectric conversions. Chapter 7 presents the development of flexible and stretchable energy harvesting devices using electrospun nanofibers. Finally, chapter 8 reviews the applications of electrospun nanofibers in triboelectric energy harvesting.

We expect that this book is useful for graduate students, researchers, and scientists who are working in the fields of energy harvesting or nanofibers. It might benefit engineers and many others who work on or are interested in energy materials.

We sincerely appreciate the invaluable contributions from all the contributors to this book. We couldn't complete it without their great support. We are very grateful to the IOP Publishing editors Ms Caroline Mitchell and Mr Daniel Heatley for their fantastic assistance. Their professional instructions, great patience, and tireless communications are much appreciated. Last but not least, we want to express our sincere gratitude to our family members for their love and support throughout our research careers.

Jian Fang and Tong Lin

Acknowledgments

The Editors acknowledge the great support from the Australian Research Council for its funding support through a Discovery project (Grant No. DP180101161).

Editor biographies

Jian Fang

Jian Fang received his PhD degree in Materials Engineering from Deakin University in 2009. He worked as a Senior Research Fellow in the Institute for Frontier Materials (IFM) at Deakin University. In 2019, he took a full professor position in the College of Textile and Clothing Engineering at Soochow University in China. His research focuses on functional fibrous materials for energy storage and harvesting devices, biosensing and environment protection applications.

Tong Lin

Professor Tong Lin received his PhD degree in Physical Chemistry in 1998. He is a Professor and Personal Chair in Fibrous Materials at Deakin University, Australia. His research covers functional fibers, electrospinning and polymers.

List of contributors

Haitao Niu
Institute for Frontier Materials, Deakin University, Victoria 3216, Australia

Hua Zhou
Institute for Frontier Materials, Deakin University, Victoria 3216, Australia

Hongxia Wang
Institute for Frontier Materials, Deakin University, Victoria 3216, Australia

T Rodrigues-Marinho
Centre of Physics, University of Minho, 4710-057, Braga, Portugal
IB-S Institute of Science and Innovation for Sustainability, Universidade do Minho, 4710-057, Braga, Portugal

A C Lima
Centre of Physics, University of Minho, 4710-057, Braga, Portugal
IB-S Institute of Science and Innovation for Sustainability, Universidade do Minho, 4710-057, Braga, Portugal

P Martins
Centre of Physics, University of Minho, 4710-057, Braga, Portugal
IB-S Institute of Science and Innovation for Sustainability, Universidade do Minho, 4710-057, Braga, Portugal

P Costa
Centre of Physics, University of Minho, 4710-057, Braga, Portugal
Institute for Polymers and Composites IPC/I3N, University of Minho, 4800-058 Guimarães, Portugal

S Lanceros-Mendez
B C Materials, Basque Center for Materials, Applications and Nanostructures, UPV/EHU Science Park, 48940 Leioa, Spain
IKERBASQUE, Basque Foundation for Science, 48013, Bilbao, Spain

Yuying Cao
Institute for Frontier Materials, Deakin University, Geelong, Victoria 3216, Australia

Hao Shao
Institute for Frontier Materials, Deakin University, Geelong, Victoria 3216, Australia

Fatemeh Mokhtari
Intelligent Polymer Research Institute, ARC Centre of Excellence for Electromaterials Science, University of Wollongong, Wollongong NSW, 2522, Australia
Textile Engineering Department, Textile Excellence and Research Centers, Amirkabir University of Technology, Tehran, Iran

Javad Foroughi
Intelligent Polymer Research Institute, ARC Centre of Excellence for Electromaterials Science, University of Wollongong, Wollongong NSW, 2522, Australia

Masoud Latifi
Textile Engineering Department, Textile Excellence and Research Centers, Amirkabir University of Technology, Tehran, Iran

Chenhong Lang
College of Textiles and Apparel, Quanzhou Normal University, Quanzhou, Fujian 362000, China

Roohollah Bagherzadeh
Institute for Advanced Textile Materials and Technologies, Textile Engineering Department, Amirkabir University of Technology, Tehran, Iran

M S Sorayani
Textile Engineering Department, Amirkabir University of Technology, Textile Excellence & Research Centers, Tehran, Iran

Nikoo Saveh Shemshaki
Department of Biomedical Engineering, University of Connecticut, Storrs, CT 06269, USA

Z Moaref
Textile Engineering Department, Amirkabir University of Technology, Textile Excellence & Research Centers, Tehran, Iran

S Ghasemi-Nezhad
Textile Engineering Department, Amirkabir University of Technology, Textile Excellence & Research Centers, Tehran, Iran

F Maleki
Textile Engineering Department, Amirkabir University of Technology, Textile Excellence & Research Centers, Tehran, Iran

Parisa Fakhri
Textile Engineering Department, Amirkabir University of Technology, Textile Excellence & Research Centers, Tehran, Iran
Instrumentation Research group, Niroo Research Institute (NRI), Tehran, Iran

Zhaoling Li
Key Laboratory of Textile Science and Technology, Ministry of Education, College of Textiles, Donghua University, Shanghai 201620, China
Innovation Center for Textile Science and Technology, Donghua University, Shanghai 200051, China

Ahmed Salim
State Key Laboratory for Modification of Chemical Fibers and Polymer Materials, College of Materials Science and Engineering, Donghua University, Shanghai 201620, China

Ibrahim Abdalla
State Key Laboratory for Modification of Chemical Fibers and Polymer Materials, College of Materials Science and Engineering, Donghua University, Shanghai 201620, China

Miaomiao Zhu
State Key Laboratory for Modification of Chemical Fibers and Polymer Materials, College of Materials Science and Engineering, Donghua University, Shanghai 201620, China

Jianyong Yu
Innovation Center for Textile Science and Technology, Donghua University, Shanghai 200051, China

Bin Ding
Innovation Center for Textile Science and Technology, Donghua University, Shanghai 200051, China

IOP Publishing

Energy Harvesting Properties of Electrospun Nanofibers

Jian Fang and Tong Lin

Chapter 1

Electrospinning: an advanced nanofiber production technology

Haitao Niu, Hua Zhou and Hongxia Wang

The electrospinning process has been regarded as one of the most facile and versatile techniques to prepare nanoscale fiber materials. Electrospinning technology has undergone enormous progress since its appearance in the 1930s. This chapter briefly reviews the recent improvements on electrospinning technologies and summarizes the state-of-the-art electrospinning setups. Thus, this chapter will be a valuable resource for scientists in the electrospinning field and engineers in related areas.

1.1 Introduction to electrospinning

Nanomaterials are materials that have at least one dimension below 100 nm, e.g. nanoparticles, nanorods, nanowires, nanotubes, and nanosheets. Nanomaterials have attracted considerable attention in the past decades owing to their excellent properties, outstanding performances, and wide applications. As the dimension of a material decreases, the percentage of atoms on the surface and the surface-to-volume ratio increase remarkably [1]. For a 100 nm nanoparticle, less than 0.2% of atoms are on the surface; meanwhile, 10% of the atoms are on the surface of a 10 nm nanoparticle and around 90% atoms distribute on the surface of a 2 nm nanoparticle [2]. The atoms on the nanomaterial surface have more dangling bonds, which make them very active and tend to bond with adjacent molecules. As a result, nanomaterials, in comparison with their bulk counterparts, often exhibit higher chemical activity, lower melting points, and higher phase transition pressure and solubility [3].

In the fiber and textile industry, fibers with a diameter less than 1 μm (1000 nm) are normally defined as nanofibers. In comparison with conventional fibers, these nanofibers have an enlarged surface area; for instance, the surface area of electrospun nanofibers are dozens of times larger than that of conventional fibers (figure 1.1). Accordingly, the pore size reduces and pore volume increases in nanofibrous materials. As a result, nanofibers show excellent performance in

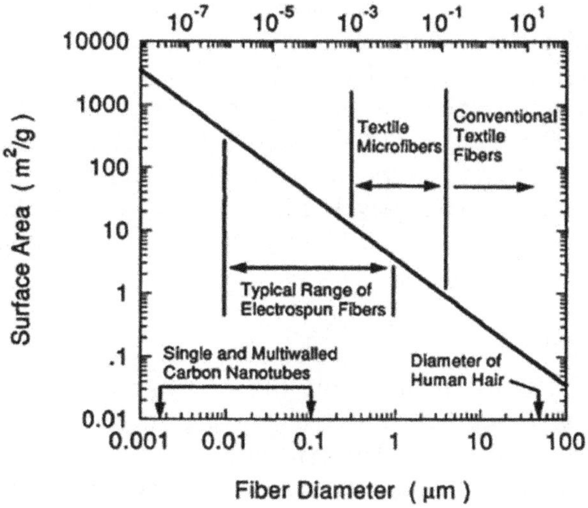

Figure 1.1. Relationship between specific surface area and diameter of different fibers. Reproduced with permission from Elsevier [11].

many different areas; for example, air/liquid filtration, energy generation and storage, biomedical and tissue engineering, sensors and catalysts, drug delivery, and nanocomposites [4–10].

Nanofibers can be fabricated by numerous approaches, such as phase separation [12], self-assembly [13, 14], template synthesis [15], melt-blowing [16], flash spinning [17], bicomponent spinning [18], and electrospinning [19, 20]. Among all the existing nanofiber-fabricating approaches, electrospinning technology is the most investigated and widely used because of its high efficiency, cost-effectiveness, and great adaptability. Especially, electrospinning technology is easy to scale up and has high commercialization potential.

1.1.1 Electrospinning history

In as early as 1600, William Gilbert first reported the observation about a spherical water drop on a dry surface being drawn and deformed into a cone under the influence of electrostatic force [21]. Nearly 300 years after Gilbert's observation, Formhal patented in the 1930s a setup to produce continuous fine fibers [22, 23]; this is regarded as the real beginning of electrospinning technology. In the 1960s, Geffrey Taylor investigated the shape of a cone formed by a fluid droplet under the action of an electric field, and reported the existence of a conical angle of 49.3° [24]. This angle was later named the 'Taylor cone' and it has been widely used to explain electrospinning and electrospraying. In addition, Taylor [25] proposed that a jet in a parallel electric field experiences two critical instabilities: Rayleigh instability and bending instability. His works greatly improved the understandings of electrospinning and promoted its development.

Henceforth, there were no noteworthy improvements on electrospinning technology, until Doshi and Reneker [19, 26] reported their work on using this technique to fabricate nanostructured materials. Since then, electrospinning technology and electrospun nanofibers have begun to draw more and more interest from both academic and industrial fields. Tremendous efforts have been devoted to the development of electrospinning technology, finding the fiber-making mechanism, characterization of nanofibers, and exploring novel applications. Large-scale production and commercialization of electrospun nanofibers have been accomplished. Electrospinning technology and processes have made fast advancement in recent years to meet the ever-increasing demands on nanofibers for various applications.

1.1.2 Basic apparatus

The conventional electrospinning setup (figure 1.2) comprises a capillary nozzle connected to a high-voltage DC power supply, a grounded collector, and a solution reservoir to supply solution [27]. This type of electrospinning setup works based on the capillary effect: the solution is transported to the tip of a thin nozzle and jet initiation happens at the nozzle tip. In the past two decades, many advancements have been made to control nanofiber collection, and various fiber-generating designs including needleless, near-field, melt electrospinning, yarn electrospinning, and multicomponent electrospinning designs, have been developed. An overview of this advanced fiber-making technology and state-of-the-art progress are reviewed in this chapter.

An electrospinning process can be briefly described as follows. Spinning fluid is fed to the capillary tube (usually a syringe needle) from a fluid reservoir. When a high voltage is applied to the needle (typically around 10 kV–20 kV), a high electric field is formed between the needle and collector (grounded or oppositely charged), which applies electrostatic force on the fluid droplet. Electrospinning fluid is described as a 'leaky dielectric' that has sufficient conductivity for the induced charges to quickly accumulate on the free surface in a short time scale or acts as a dielectric [28]. The repulsion between charges on the free surface of fluid droplet works against surface tension and fluid viscosity to deform the droplet into a cone

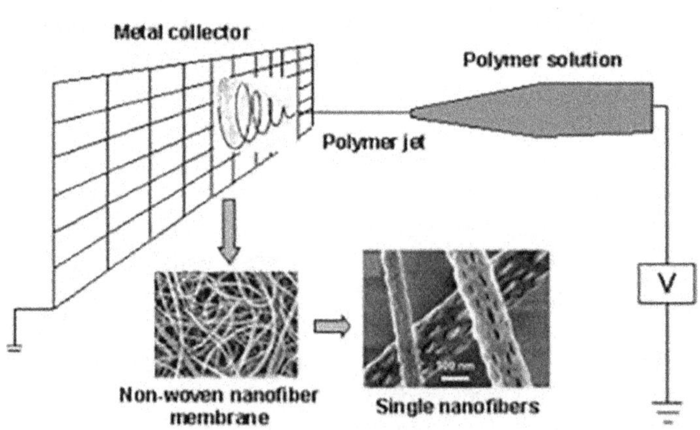

Figure 1.2. Illustration of typical electrospinning process. Reproduced with permission from Elsevier [27].

shape (Taylor cone) with increasing applied voltage (higher electric field intensity) [24]. When the applied voltage exceeds a critical value, the electrostatic force can overcome fluid surface tension, and jet initiation proceeds from the vertex of the Taylor cone. The generated fluid jets fly to the grounded collector under the action of electrostatic force. During the flying process, solvent evaporation from the fluid jet results in dry nanofibers depositing on the collector. Although the electrospinning process is relatively easy to implement, it is very complex considering the co-existence and combined action of Coulombic force, gravity, fluid surface tension, and viscosity during fiber formation.

1.2 Electrospinning basis

1.2.1 Mechanism of electrospinning process

The fiber formation process during electrospinning can be divided into three stages: jet initiation, jet whipping instability, and fiber deposition [29]. As the first stage of electrospinning, jet initiation has been widely studied [30–32]. In the first milli-seconds, the solution droplet begins to transform into a conical shape by the high electric potential. The round solution droplet tip becomes more and more sharp. Finally, a jet is emitted from the tip of the cone. Then the cone gradually changes back to the rounded shape, indicating the system transits from the jet initiation stage to the electrospinning stage. As long as the solution taken away by the solution jet is replenished promptly, the electrospinning process can last.

After ejection, the solution jet is straight for a short section (figure 1.3), which may extend from a few millimeters to several centimeters away from the nozzle tip along its axis direction. This ejected solution jet carries away electrical charges in the

Figure 1.3. Digital photo of an electrospinning solution jet, including a stable jet and jet whipping stage. Reproduced with permission from Elsevier [39].

form of uncompensated ions from the nozzle. Attributed to the effect of charge repulsion in the solution jet, free charges migrate radially onto the jet surface to satisfy the equilibrium condition. As a result, electrostatic force induced by the electric field is applied on the jet surface. The charged fluid jet accelerates under the action of the electric field, accompanied by the thinning of the fluid jet [31].

After the initial stable stage, the jet enters an instable stage (whipping) under the influence of the charges carried by the jet, which may involve bending, winding, spiraling, and looping movements. Many theoretical models have been proposed to describe this jet instability, and there is a prevalent belief that the jet is continuously elongated and becomes longer and thinner with continual stretching [33–37]. The existence of jet instability is due to the action of axisymmetric and nonaxisymmetric instabilities caused by the perturbations of surface charges. The axisymmetric instability derives from perturbation of the surface charges along the jet axis direction, which makes different segments of the jet be under different strengths of electrostatic force, resulting in an uneven jet. The nonaxisymmetric instability derives from perturbation of surface charges around the circumference of the jet, which induces a localized torque around the jet that accounts for the whipping motion. Under the condition where nonaxisymmetric instability plays the major role of instability during the electrospinning process, the jet is likely to be stretched uniformly [38]. The solution jets fly to the collector under electrostatic force, being accompanied by solvent evaporation, and deposit on the collector as dry fibers at last.

1.2.2 Effects of electrospinning parameters

Electrospinning process, fiber morphology, fibrous structure, and fiber production rate are governed by a number of parameters, e.g. applied voltage, flow rate, nozzle diameter, collecting distance, solution properties (e.g. polymer molecular weight, concentration, electrical conductivity, surface tension, solvent properties), and ambient conditions (temperature, humidity) (figure 1.4). Although, many of these electrospinning parameters are interdependent and there are interactions between them, we can derive a general trend of influences of some parameters on the electrospinning process.

1.2.2.1 Applied voltage

Applied high voltage is an essential factor in electrospinning, without which jet initiation does not happen. When the applied voltage is low, the electrostatic force is insufficient to overcome surface tension of the solution droplet, and as a result, no jet is stretched out and dripping happens. With increasing applied voltage, the electrostatic force increases, and eventually leads to jet initiation, and the electrospinning process starts. Generally speaking, the fiber diameter decreases with increasing voltage, and attributable to the growing stretching force [40, 41]. Applied voltage plays a far more important role in needleless electrospinning than in nozzle electrospinning. It can decrease the fiber diameter and increase the fiber production rate because high voltage can increase the jet number on the free solution surface of a spinneret and escalate the jet flying speed [42, 43].

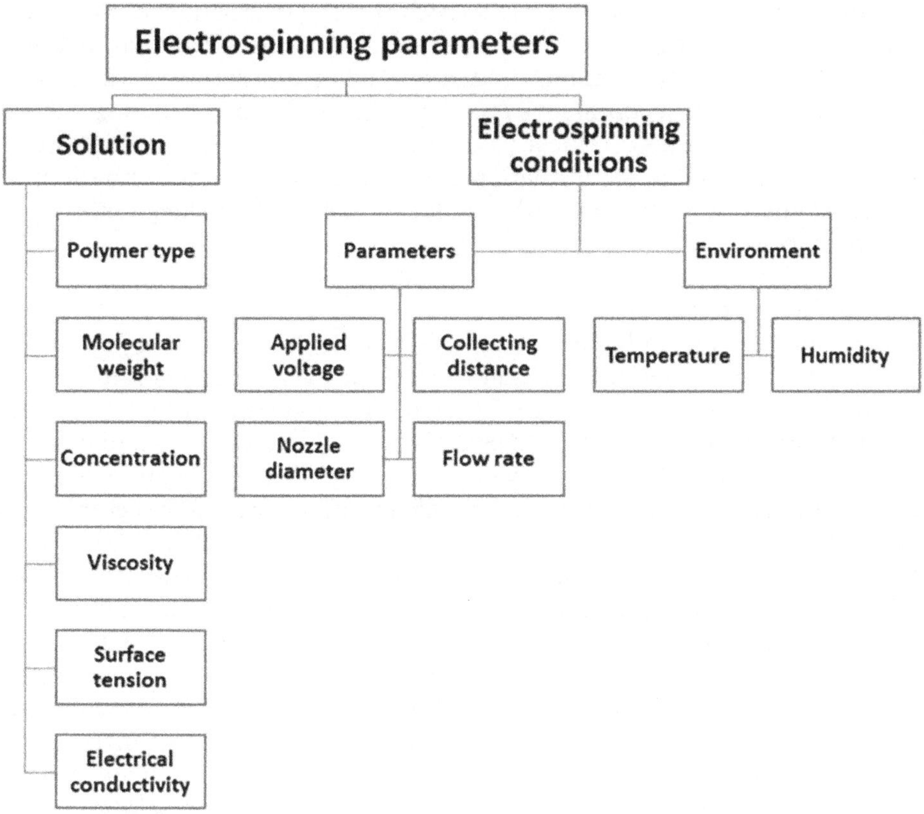

Figure 1.4. Important electrospinning parameters and conditions during the electrospinning process.

1.2.2.2 Collecting distance

In general, an optimum distance for electrospinning should be long enough for fiber stretching and solvent evaporation. Under the circumstance that applied voltage is constant, the change of distance between nozzle and collector will affect the electric field and fiber diameter accordingly. Another important influence of changing distance is on fiber morphology and structure. The solvent in the solution jet needs to have sufficient time to evaporate and turn the jets into dry nanofibers. When the collecting distance is very short, interconnected nanofibers are often collected [42–44]. This phenomenon happens more frequently in needleless electrospinning due to instantaneous generation of a large number of solution jets. It has also been found that interconnected nanofibers collected at a short distance could benefit the mechanical strength of nanofiber mats with improved and more durable performance in energy harvesting applications [44].

1.2.2.3 Flow rate

The flow rate of solution in nozzle electrospinning has a direct effect on fiber production rate: a large flow rate results in a high production rate and a small flow rate results in a low production rate. A large flow rate will generally produce coarser nanofibers [45] because more solution is drawn out at the same time. It has been proposed that four electrospinning conditions may be observed with the increasing flow rate: discontinuous, continuous (stable), intermittent, and dripping. When the solution flow rate exceeds a critical value, the congestion of excess solution at the nozzle tip can affect the jet formation process, leading to the formation of an unstable jet, beaded fibers, and other defects, such as branching, splitting, and flattened fibers [46].

1.2.2.4 Solution properties

Polymer molecular weight has a substantial effect on electrospinning performances. In general, increasing molecular weight enables the polymer solution to be electrospun into uniform fibers at relatively low concentrations because a high molecular weight induces a large degree of chain entanglement [47]. It was found that polyvinyl alcohol (PVA) solutions produced fibers with morphologies ranging from beaded fiber, uniform fiber, to coarse non-uniform fibers with increasing concentration [48]. In another study, PA6 solution produced droplets at 5 wt% concentration, merged fibers at 15 wt% concentration, and smooth fibers at 25 wt% concentration [49].

When the solution concentration is very low, electrospray happens instead of electrospinning. At low solution concentrations, electrospinning usually produces defective fibers (discontinuous, merged, or beaded) because the surface tension of the solution overcomes the viscoelastic forces and electrostatic drawing force. When the solution concentration is sufficiently high, the chain entanglement of macromolecules is enough to overcome surface tension, and the fiber diameter increases with the rising concentration as there is more solid content in the solution [47, 50]. However, a too-high concentration makes electrospinning difficult due to high viscoelasticity [51], especially for needleless electrospinning that stops jet production at high solution concentrations. In addition, the critical voltage for electrospinning may go up when increasing the solution concentration [52]. The minimum solution concentration to produce smooth nanofibers is dependent on the polymer type, polymer molecular weight, and solvent used.

The addition of surfactant and salt alters the surface tension and electrical conductivity of polymer solutions, and their influences on electrospinning may vary in different conditions. It has been reported that the addition of lithium chloride (LiCl), sodium nitrate ($NaNO_3$), sodium chloride (NaCl), and calcium chloride ($CaCl_2$) salts can increase solution conductivity, thus reducing electrospun PAN fiber diameter (the reduction is proportional to solution conductivity) [53]. On the contrary, the addition of salt in PA6 solution was found to increase fiber diameter; this is attributed to the increase in viscoelastic force within the solution jet [47]. The addition of surfactant can reduce solution surface tension, and as a result produce thinner nanofibers with better uniformity [54, 55]. Introducing surfactant (e.g. dodecylbenzene sulfonic acid, tetrabutylammonium chloride) in the solution could also produce nanofibers with special morphologies, e.g. tree-like [56] and nano-net morphologies [57].

1.2.2.5 Temperature and humidity of environment

Electrospun nanofibers normally have smaller fiber diameter with rising environment temperature during electrospinning because of the declined viscosity and surface tension of solution [58]. When using a low temperature (200–220 K) to prepare poly (lactic acid-*co*-glycolic acid) nanofibers, the porosity of electrospun nanofibers can be improved by four times because the ice crystals formed at low temperature serve as a removable void template to create additional pores [59]. Humidity can also affect fiber diameter and morphology [58, 60, 61]. It has been reported that polystyrene nanofibers prepared in a high-humidity environment exhibit porous morphology, because solvent evaporation from solution jets has a cooling effect and causes moisture condensation on the fiber surface, leading to breath-figure self-assembly [61].

In the following section of this chapter, state-of-the-art progresses of electrospinning techniques and apparatuses will be summarized (figure 1.5), covering nozzle electrospinning techniques (e.g. multicomponent electrospinning, multinozzle electrospinning, near-field electrospinning, melt electrospinning), needleless electrospinning techniques, and methods of controlled nanofiber deposition (e.g. aligned nanofibers, nanofiber yarns).

1.3 Nozzle electrospinning

In a nozzle spinneret-based electrospinning process, a spinning solution is transported to fiber electrospinning sites through a capillary channel or multiple channels.

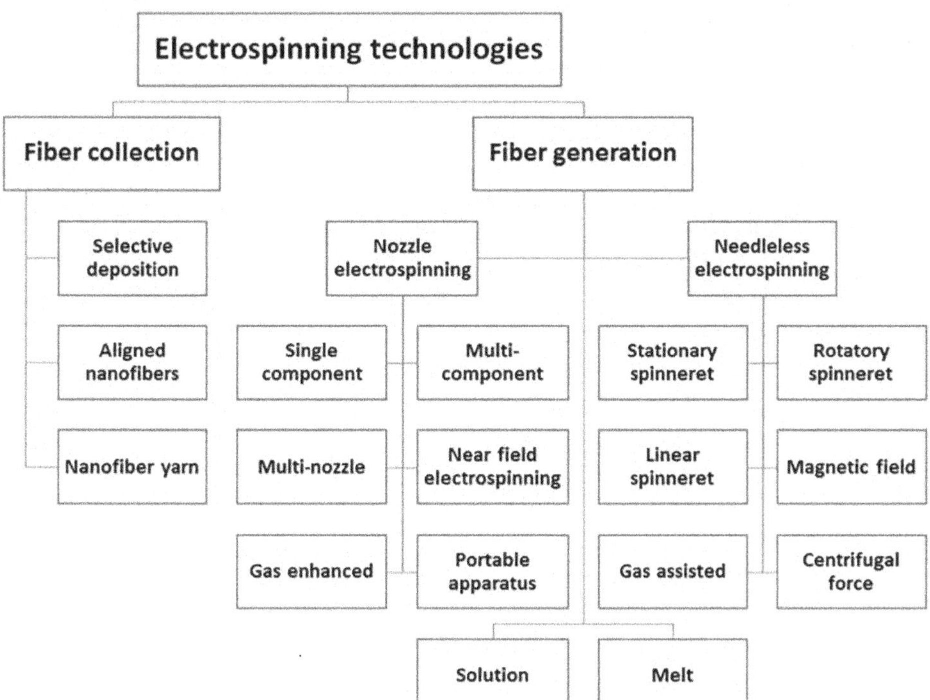

Figure 1.5. A summary of different electrospinning techniques.

This kind of electrospinning technique has many unique features, e.g. enabling incessant solution feeding and continuous electrospinning, the enclosed system avoiding unnecessary evaporation and maintaining the solution stability, tunable capillary diameter with easily controlled fiber diameter, and producing uniform nanofibers.

1.3.1 Single-component electrospinning

Single-component electrospinning refers to the circumstance that spinning solution is supplied through a single capillary nozzle, and jet initiation occurs at the nozzle tip when high voltage is applied. Although positive high voltage is generally used in electrospinning, negative high voltage has also been applied for making electrospun nanofibers. Tong *et al* investigated the effect of high voltage polarity on fiber quality of poly(hydroxybutyrate-*co*-hydroxyvalerate) (PHBV) nanofibers. They noticed that PHBV fiber diameter increased with increasing applied voltage for positive voltage electrospinning, but decreased for negative electrospinning; however, water contact angle, tensile strength, and stiffness of the nanofibers were barely affected by the polarity of applied voltage [54]. In another work of electrospinning zein/soy protein isolate (95/5), the diameter of positive electrospun nanofibers was smaller than that of negative electrospun nanofibers. With the increasing applied voltage, positive electrospinning produced thinner nanofibers while negative electrospinning did not show any clear effect on the fiber diameter [62].

In addition to the conventional setup where high voltage is connected to the nozzle spinneret, high voltage can also be connected to the collector to perform electrospinning. This works based on the inductive effect produced from the high potential applied by the collector. It has been found that when high voltage is connected to the collector and the nozzle is grounded, electrospinning proceeds successfully but requires higher critical voltage to start the spinning process, and results in enlarged fiber diameter and lower nanofiber production rate [63].

Direct current (DC) voltage is usually used for electrospinning when the high voltage power supply has either a positive high voltage output or a negative output. Alternating current (AC) voltage can not only implement electrospraying/electrospinning, but also has better controllability of fiber deposition. It has been found that AC electrospinning of poly(ethylene oxide) (PEO) significantly subsided the whipping phenomenon with a high degree of fiber alignment [64]. Unlike the conventional DC electrospinning in which a funnel shaped nanofiber mesh is formed, a visible thread can be observed emerging downstream from the needle in the AC electrospinning process. Furthermore, the fibrous thread is not attracted by the grounded electrode, and it can be easily deflected and collected [65].

1.3.2 Multicomponent electrospinning

In the textile industry, a single fiber with two components distributed in the radial direction, for example core–sheath and islands-in-the-sea structures, can find many unique applications. Although it is possible to obtain core–sheath-structured nanofibers induced by phase separation [66] or emulsion electrospinning [67, 68], normal single-component electrospinning usually produces an electrospun nanofiber without an obvious core–sheath structure. This is attributable to the fact that two solutions inside the nozzle (diameter < 1 mm) have small Reynolds numbers. They can be identified as laminar flows and there is no mixing between them; as a result, they maintain their initial injection states [69]. Because of this, it is possible to conduct bicomponent electrospinning and fabricate nanofibers with tunable composition in the radial direction [70].

In a core–sheath electrospinning spinneret, an inner nozzle of small diameter and an outer nozzle of large diameter are positioned axially with the inner nozzle slightly protruding out of the outer nozzle. During the electrospinning process, the sheath solution is fed into the outer nozzle and the core solution is fed into the inner nozzle. At the nozzle tip, the outer solution forms a thin sheath that enclose the inner solution. Under the action of electrostatic force, both solutions are pulled into a compound Taylor cone of core–sheath structure. Then the core–sheath solution jet is stretched into thin core–sheath fibers by the electric field and deposited on the collector [70–72]. When the inner solution is replaced with a liquid (usually octane), hollow nanofibers (figure 1.6(a) and (b)) can be obtained after removing the liquid from the resulting core–sheath electrospun nanofibers [71]. Thanks to core–sheath electrospinning, many unspinnable materials, e.g. polydimethylsiloxane (PDMS) [73] and medicine [74], can be successfully processed into ultrathin fibers.

When two components distribute inside the bicomponent electrospinning nozzle in a side-by-side way, the fabricated nanofibers have a side-by-side structure [75]. Interestingly, self-crimped polyacrylonitrile nanofibers (figure 1.7) were produced when the polyurethane component in polyacrylonitrile/polyurethane side-by-side nanofibers was removed; this could provide an efficient way to tailor the structure of electrospun nanofibers [76].

Multicomponent nanofibers (figure 1.8) can be obtained via simply increasing the number of components inside the spinneret [77]. However, further increasing the component number to produce an islands-in-the-sea structure may be difficult

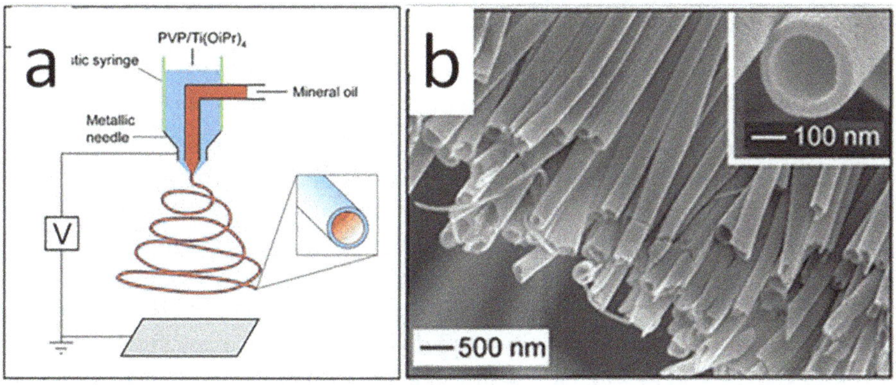

Figure 1.6. (a) Core–sheath electrospinning setup, and (b) SEM images of electrospun hollow nanofibers. Reproduced with permission from Royal Society of Chemistry [71].

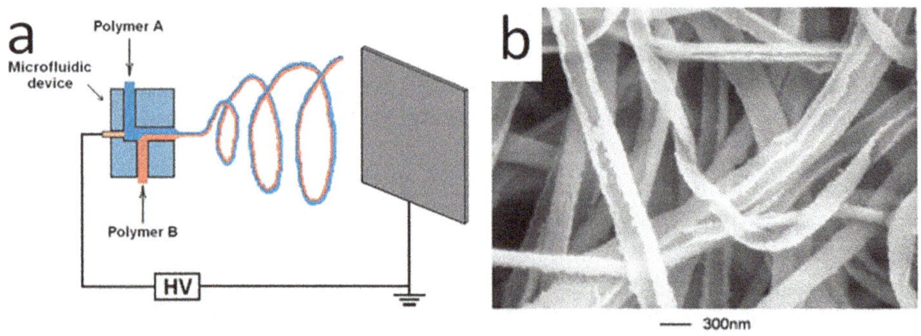

Figure 1.7. (a) Side-by-side electrospinning setup, and (b) SEM image of side-by-side bicomponent nanofibers with one component removed. Reproduced with permission from Wiley [76].

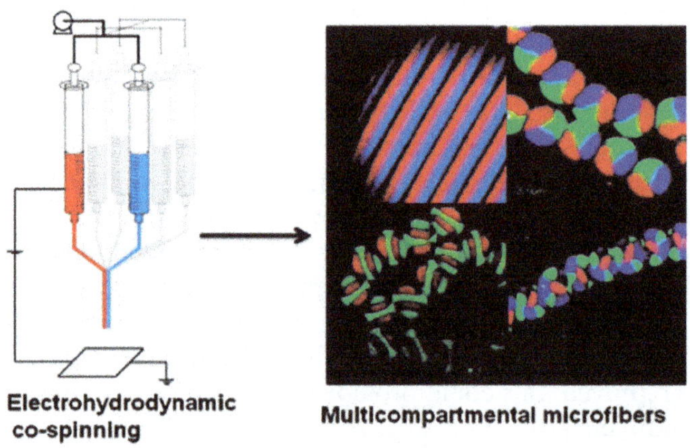

Figure 1.8. Schematic drawing of a multicompartmental electrospinning setup and prepared fibers. Reproduced with permission from American Chemical Society [77].

because the diameter of the nozzle is normally small (<1 mm); this makes it impractical to further prepare more complex spinneret, and demands large pressure to feed all the components evenly.

1.3.3 Multinozzle and porous spinneret

The single nozzle electrospinning generally has a low nanofiber productivity of $<0.3\ g^{-1}\ h^{-1}$ per nozzle, which is far below the requirement for industrial nanofiber production. A direct and practical way of improving the nanofiber production rate is to increase the number of nozzles for electrospinning, namely by employing multinozzle electrospinning [78–81]. However, there is strong electrostatic repulsion between the jets in multinozzle electrospinning, which can easily affect the electrospinning process, causing poor fiber quality and an uneven fibrous membrane (figure 1.9(a)). Different nozzle position arrangements have been developed to reduce electrostatic repulsion and improve multinozzle electrospinning performances [82]. An auxiliary electrode can interfere with the electric field in multinozzle

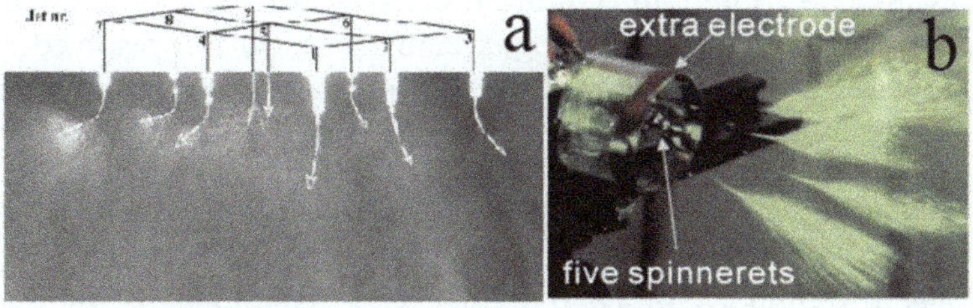

Figure 1.9. (a) A multinozzle electrospinning setup. Reproduced with permission from Elsevier [82]. (b) Cylindrical auxiliary electrode-assisted multinozzle electrospinning. Reproduced with permission from Elsevier [85].

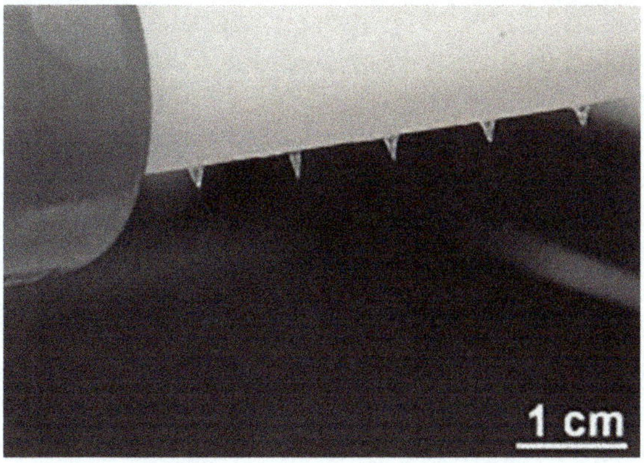

Figure 1.10. Horizontal tube electrospinning. Reproduced with permission from Elsevier [86].

electrospinning and improve its distribution (figure 1.9(b)); this improves electrospinning performances [83–85]. The presence of the external electrode could shrink the fiber deposition area and improve the fiber production rate.

Using a cylindrical tube with channels throughout the tube wall is an effective way to increase the solution channel number, thus increasing fiber production rate [86, 87]. It is still based on the capillary effect, which transports the spinning solution from inside of the tube to the outside, and the electrospinning happens on the tube surface. Tube electrospinning exhibits improved electrospinning productivity in comparison with single needle electrospinning, attributable to numerous nozzle number (figure 1.10). Although it is possible to improve the nanofiber production rate easily by increasing the tube length and number of channels, the gap between adjacent channels cannot be too small because strong jet interference can result in a nanofiber belt instead of a fiber web [86].

In electrospinning, ejected solution jet carries a large amount of charges, which drive the jet stretching and fiber deposition on the collector. The interference among solution jets in multijet electrospinning cannot be completely eliminated. In addition, solution and electrospinning conditions (e.g. solution type, solution concentration, applied voltage) influence the electrostatic repulsion between ejected jets, and it is difficult to find an optimized distance between solution channels that can be adaptable to all electrospinning circumstances. If not resolved successfully, the electrostatic repulsion problem will be a serious barrier to the industrialization of multijet electrospinning.

1.3.4 Near-field electrospinning

Near-field electrospinning can be regarded as a technique of integrating nanolithography [88, 89] and electrospinning, and is a good example of interdisciplinary technological convergence. In a normal electrospinning system, the collecting distance is usually over 5 cm and the whipping stability is evident. Near-field electrospinning working on a much lower applied voltage can precisely control electrospun fiber deposition [90, 91]. In near-field electrospinning, the jet whipping instability is eliminated or greatly restricted due to a very short collecting distance (<5 cm). When a collector is controlled by a computer program with precisely positioned movement, the collected fibers can form predesigned patterns. Near-field electrospinning also has a few demerits; for example, it produces fibers with a much larger diameter than those in conventional electrospinning due to the insufficient jet stretching, and it has a small nanofiber production rate. The differences between near-field electrospinning and conventional electrospinning are presented in table 1.1.

Figure 1.11(a) illustrates a near-field electrospinning process, and the SEM image in figure 1.11(b) shows a collected nanofiber. The nanofibers fabricated by near-field electrospinning has a narrower diameter distribution than those fabricated by the conventional electrospinning process. This technique greatly expands the application fields of the electrospun nanofiber with ordered structures, e.g. into the fields of nanogenerators, tissue engineering, wearable sensors, and microelectromechanical systems [93–97].

Because of the precise fiber deposition in near-field electrospinning, it is feasible to fabricate a patterned 3D fibrous structure [95]. At microscale (figure 1.12), these fibers prepared by near-field electrospinning show precise deposition and they formed a well-organized 3D fibrous structure.

Table 1.1. Comparison of conventional electrospinning and near-field electrospinning [92].

	Conventional electrospinning	Near-field electrospinning
Material	Solution, polymer melt	Solution, polymer melt
Applied voltage (kV)	10–30	0.2–12
Collecting distance (mm)	50–500	0.5–50
Fiber diameter (μm)	0.01–1	0.05–30

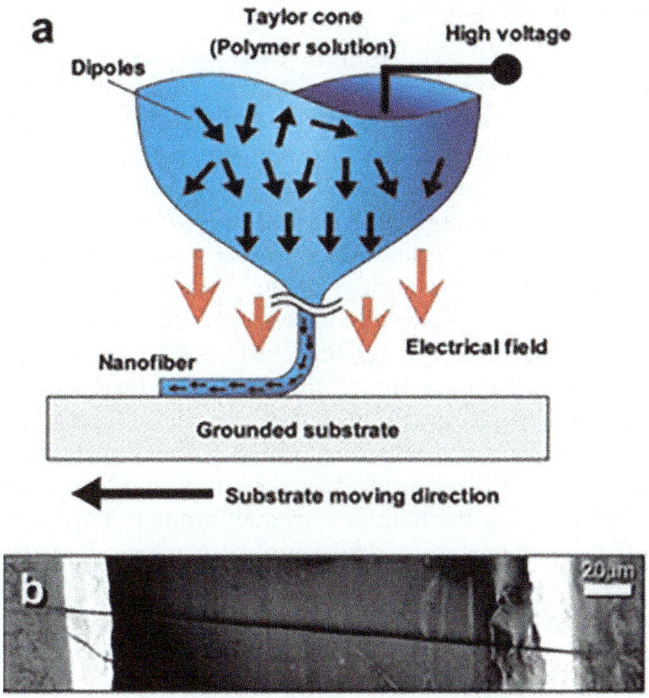

Figure 1.11. (a) Schematic diagram of near-field electrospinning process, and (b) SEM image of a single polyvinylidene fluoride nanofiber formed across two electrodes. Reproduced with permission from American Chemical Society [93].

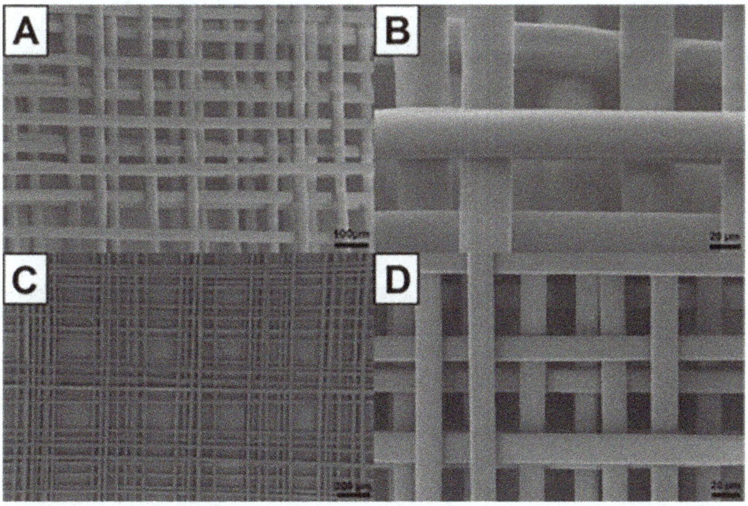

Figure 1.12. (a) SEM images of deposited poly(2-ethyl-2-oxazoline) fibrous structures. Reproduced with permission from Elsevier [98].

1.3.5 Gas-enhanced electrospinning

Applying auxiliary gas to an electrospinning process can diminish the influence of Coulombic repulsion force in multinozzle electrospinning [99]. As a result, gas-enhanced electrospinning can produce thinner nanofibers [100–102], improve nanofiber collection [103–105], and increase nanofiber production rate [106]. It has been found that the application of nitrogen gas containing solvent vapor can help to eliminate the whipping motion of the solution jet and lead to the collection of highly aligned nanofibers on the fast-rotating cylindrical collector [103]. The Zetta electrospinning system was developed in another work; it can process both polymer solution and polymer melt into nanofibers. Figure 1.13 shows that Zetta electrospinning uses airflow to enhance electrospinning with a much-improved nanofiber production rate [106].

When preparing thick nanofiber membranes by extending electrospinning time, it is very easy to build up electrostatic charges on the collector because polymer nanofibers are electrically nonconductive. These accumulated electrostatic charges can weaken the electric field, prevent fiber deposition, and produce coarser fibers. Solvent accumulation inside the nanofibrous membrane is another problem, which, if not removed rapidly, can cause jointed nanofibers or even film after a long period of electrospinning. Gas-enhanced electrospinning can efficiently solve these problems [107]. Airflow has also been applied to assist melt electrospinning (figure 1.14), resulting in 10% thinner polylactic acid fibers compared to fibers produced by unassisted melt electrospinning [100].

1.3.6 Melt electrospinning

Melt electrospinning technology was reported in as early as 1936 in a patent filed by Charles Norton from the Massachusetts Institute of Technology [108]. In spite of its early appearance, melt electrospinning had not drawn as much attention as solution electrospinning because it needs to maintain an elevated temperature during the electrospinning process and polymer melts have low conductivity and high viscosity. In recent developments, various heating methods have been applied in melt electrospinning, e.g. electrical heating [109, 110], air heating [111, 112], circulating fluid heating [113], and laser heating [114, 115]. Melt electrospinning does not involve any solvent and is an environmentally friendly process, which makes it

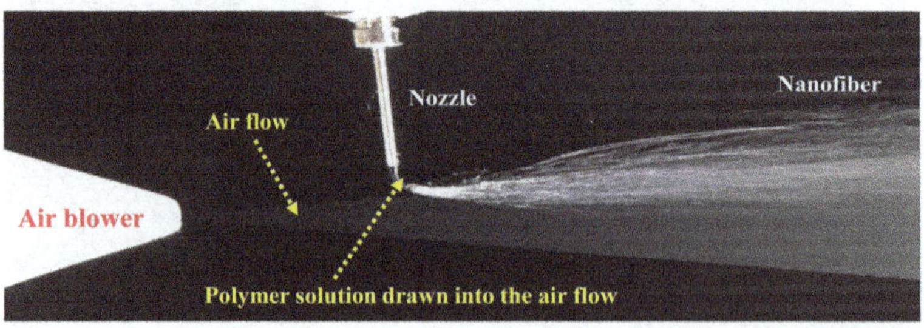

Figure 1.13. Zetta electrospinning process. Reproduced with permission from American Chemical Society [106].

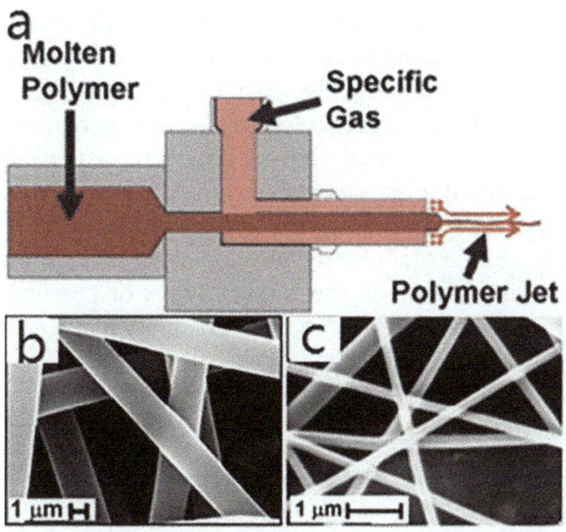

Figure 1.14. (a) Schematic diagram of gas-enhanced melt electrospinning; SEM images of fabricated polylactic acid fibers (b) without a gas-assisted system and (c) with a gas-assisted system. Reproduced with permission from Elsevier [100].

favorable for various applications, e.g. filtration [116], sensors [117], textiles [118], and especially biomedical applications [113, 119–121]. Conversely, solution electrospinning uses flammable, toxic solvents; its operation can cause environmental issues and hazard risks, which indicate that melt electrospinning could play a more important role in nanofiber production and applications.

1.4 Needleless electrospinning

In spite of great efforts to improve electrospinning productivity, conventional capillary electrospinning can only produce a very limited quantity of nanofibers, and is unable to meet the ever-increasing demand for industrial applications. Needleless electrospinning offers a solution to this problem. Needleless electrospinning (also referred to as free surface electrospinning) is a special type of electrospinning. Instead of forming a jet from a capillary tip, needleless electrospinning forms a large number of solution jets directly from an open liquid surface. Needleless concept appeared in as early as 2004 [122], when Yarin reported upward needleless electrospinning of nanofibers from a two-layer system. Since then, this technology has attracted tremendous interests attributable to its ability for large-scale nanofiber production.

During a needleless electrospinning process, numerous jets are generated instantaneously from the spinneret surface, exempt from influences of the capillary effect [123]. The electrospinning liquid (electrically conductive) on the spinneret surface self-organizes on a mesoscopic scale to form waves, and jet initiation happens from wave crests when the applied voltage exceeds a critical value. The electric field intensity profile around the spinneret and in the electrospinning zone play a far more important role in needleless electrospinning than in needle electrospinning, in terms of jet initiation, jet stretching, nanofiber productivity, and morphology.

Most of the works on needleless electrospinning have been focusing on understanding how spinneret design affects electrospinning performances and nanofiber production. It has been established that a large curvature can generate a high-intensity electric field, which increases fiber production rate. Based on their motility, needleless electrospinning spinnerets can be classified into the following categories: stationary needleless spinneret, linearly moving needleless spinneret, and rotatory needleless spinneret. In addition to these three types of needleless spinnerets, there are some other needleless electrospinning designs working on different jet initiation mechanisms.

1.4.1 Stationary needleless spinnerets

Stationary needleless spinnerets refer to those spinnerets that do not move during the electrospinning process while the spinning solution is fed into them or the spinnerets work in batch mode. This kind of electrospinning setup is relatively simple and the electrospinning processes can be implemented with minimum efforts. Spinnerets in this category include, wire, twisted wire, conical wire, bowl, sharp edge, cylinder, slit, curved slot, stepped pyramid, and cleft (shown in table 1.2).

1.4.2 Rotatory needleless spinnerets

During electrospinning using stationary spinnerets, there is no effective control over the solution distribution on the spinneret surface. As a result, electrospinning conditions and fibrous membrane uniformity may vary as electrospinning continues. When rotatory needleless spinnerets are used for electrospinning, the spinnerets can spread the solution evenly onto their surface through rotation and ensure continuous electrospinning. The solution layer thickness can be easily regulated by the rotating speed with improved electrospinning stability and nanofiber uniformity. Jet initiation in needleless electrospinning can be summarized in four stages. (1) A thin solution layer is formed on the spinneret surface because of spinneret rotation. (2) Rotation causes perturbations on the solution layer, inducing the formation of conical spikes. (3) When high voltage is applied, these spikes centralize the electric force, intensifying perturbation to form Taylor cones. (4) Solution jets are stretched out from Taylor cones, resulting in nanofiber formation. Table 1.3 lists the common rotatory spinnerets and fiber-generating sites, including cylinder (roller), disc, ball, coil, cone, and wire frame.

With the help of finite element analysis (FEA), the electric field intensity profile in the electrospinning area can be simulated and used for optimizing spinneret design [42, 136, 137]. It has been found that an auxiliary structure on the primary spinneret structure can centralize the electric field around with increased intensity. Therefore, auxiliary structures were introduced onto common needleless spinnerets to enhance electric field strength and improve electrospinning performances; these structures include the Von Koch curve fractal structure [142], needles on a disc or helix slice [143, 144], barbed roller [145], probed cylinder [146], and threaded rod [147].

Table 1.2. A summary of stationary needleless electrospinning spinnerets and their spinning process.

Spinneret	Fiber generation	Fiber generation area	Ref
Cleft	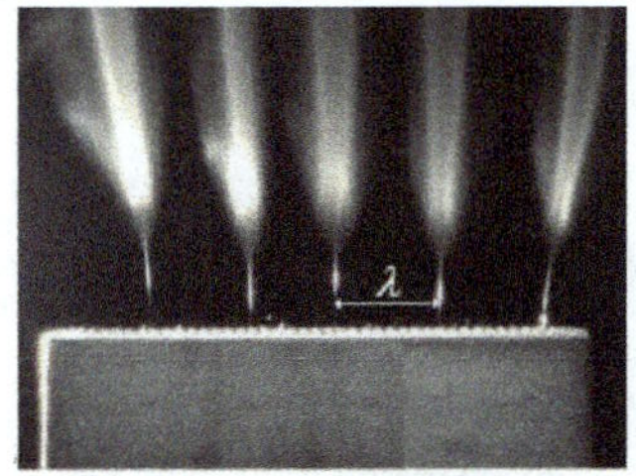 (Reproduced with permission from AIP Publishing)	Cleft surface	[123]
Conical coil	 (Reproduced with permission from Wiley)	Coil wire surface and gap	[43]
Plate	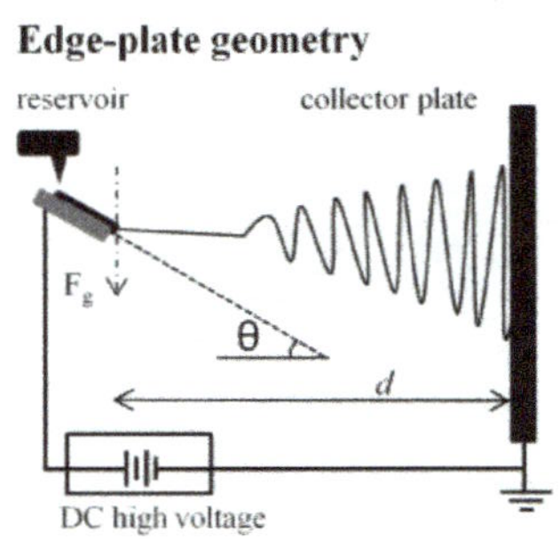 (Reproduced with permission from Elsevier)	Plate edge	[124]

Bowl	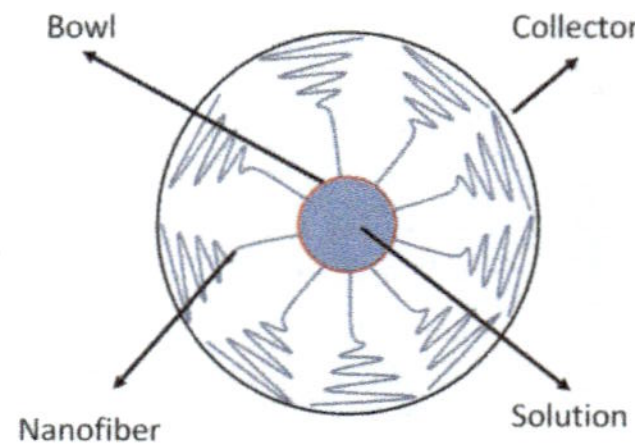	Bow edge	[125]
Slit	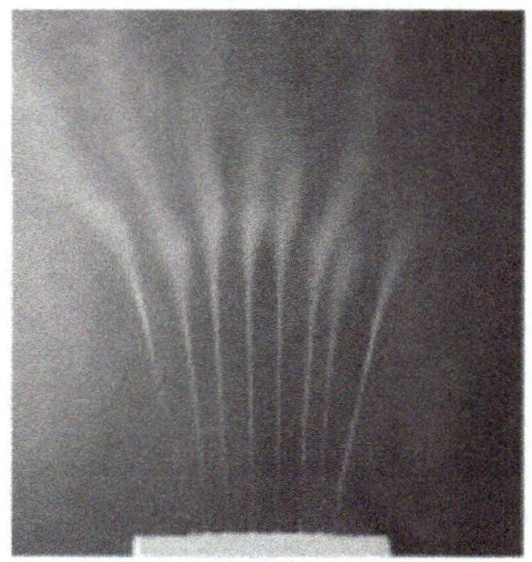	Slit	[126–129]
Stepped pyramid	 (Reproduced with permission from Elsevier)	Pyramid edge	[130, 131]

(Continued)

Table 1.2. (*Continued*)

Spinneret	Fiber generation	Fiber generation area	Ref
Twisted wire	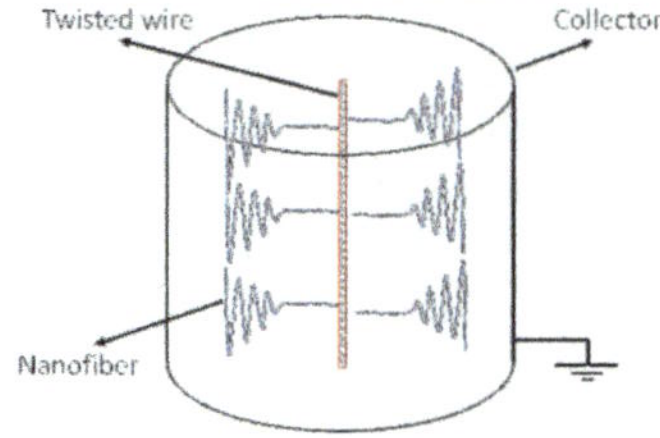	Wire surface	[132]
Curved slot	 (Reproduced with permission from Springer Nature)	Slot	[133]

Table 1.3. A summary of rotatory needleless electrospinning spinnerets and their spinning process.

Spinneret	Fiber generation	Fiber generation area	Ref
Cylinder roller	 (Reproduced with permission from Wiley)	Cylinder surface	[42, 134]
Disc	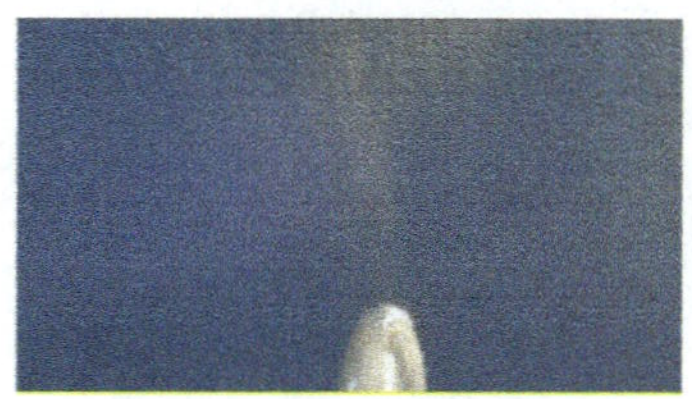 (Reproduced with permission from Wiley)	Disc rim	[42]
Ball	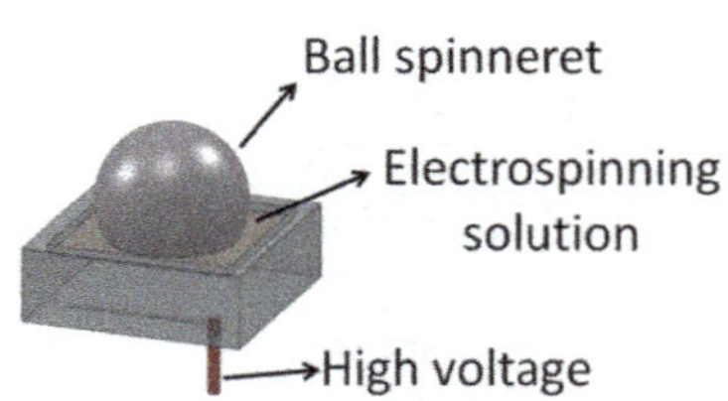	Ball surface	https://sncfibers.com [135]

(*Continued*)

Table 1.3. (*Continued*)

Spinneret	Fiber generation	Fiber generation area	Ref
Coil	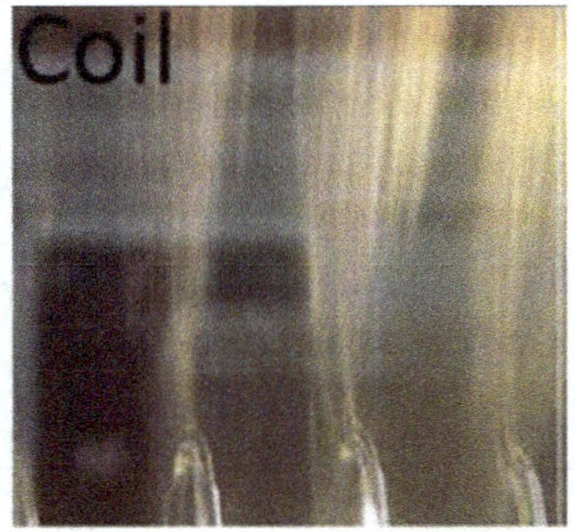	Coil wire surface	[136, 137]
Cone	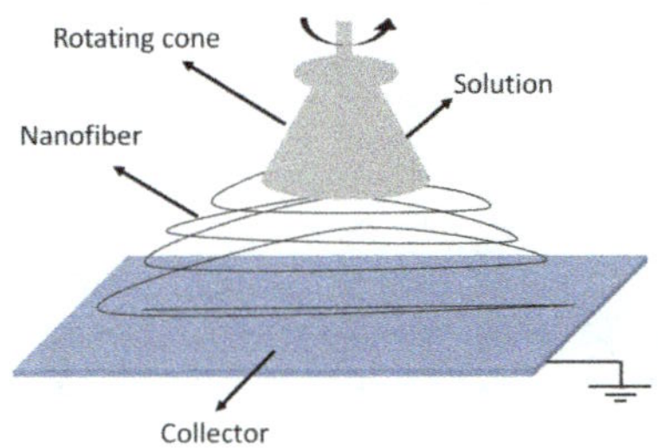	Cone edge	[138]
Wire frame	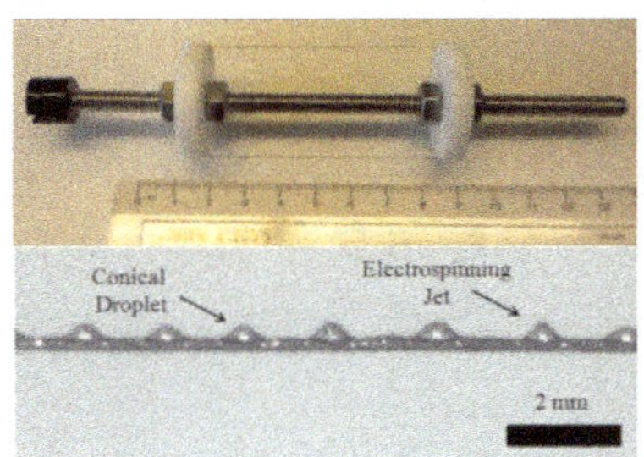 (Reproduced with permission from Elsevier)	Wire surface	[139–141]

1.4.3 Linear needleless spinneret

The jet initiation sites on linear needleless spinnerets are distributed in parallel to the linear direction of the spinnerets, and a typical design is wire electrospinning. Although the wire spinneret remains stationary, there is a solution-feeding device moving linearly along the wire to feed spinning solution evenly along the wire, thus ensuring continuous electrospinning. In another design, a horizontal bead chain was used to produce nanofibers, and the chain moved in parallel to the flat collector. The beads, as an auxiliary structure, can centralize the electric field and improve electrospinning performances [148].

The applied voltage needed to start jet initiation and electrospinning in needleless electrospinning is usually much higher than that in conventional needle electrospinning because it requires a much higher electrostatic force to stretch out a solution jet from the small curvature surface of the needleless spinneret [42]. In a recent work of electro-aerodynamic field aided needleless electrospinning, two auxiliary forces of additional electric field and airflow were used to assist the electrospinning process (figure 1.15). The high-intensity electric field was generated between the slot and the inductive electrode (5 cm distance). In this way, electrospinning ran at a voltage equivalent to that in needle electrospinning (e.g. 10–30 kV) [149]. The auxiliary airflow diverted the nanofibers away from the inductive electrodes and directly to the collector.

With the fast advances of electrospinning technology, commercialized needleless electrospinning technologies are already on the market (e.g. technologies produced by Elmarco, Fanavaran Nano-meghyas, Revolution Fibers, SPUR Company, Shanghai Yuntong Nanomaterials Technology Co., Ltd, Stellenbosch Nanofiber Company, INOVENSO). Such technologies produce nanofibers based on different mechanisms but all have a large-scale nanofiber production capacity.

1.4.4 Magnetic field-assisted needleless electrospinning

Apart from the above-mentioned needleless spinnerets that relying on specific geometric shapes to conduct electrospinning, there are also a number of needleless electrospinning technologies that utilize additional forces such as gas blowing, a magnetic field, or centrifugal force to implement electrospinning.

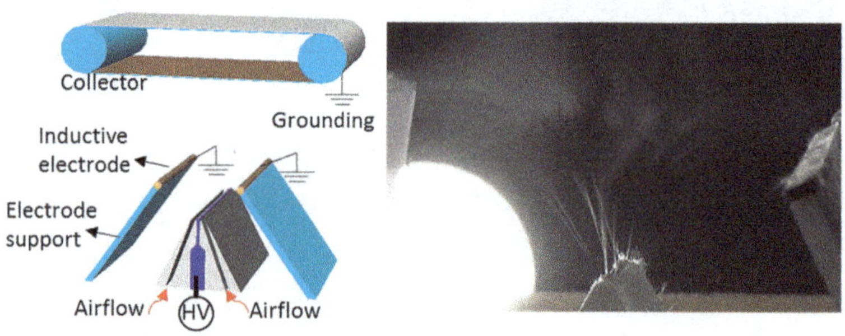

Figure 1.15. Schematic drawing of electro-aerodynamic needleless spinning setup and its electrospinning process (HV means high voltage). Reproduced with permission from IOP Publishing [149].

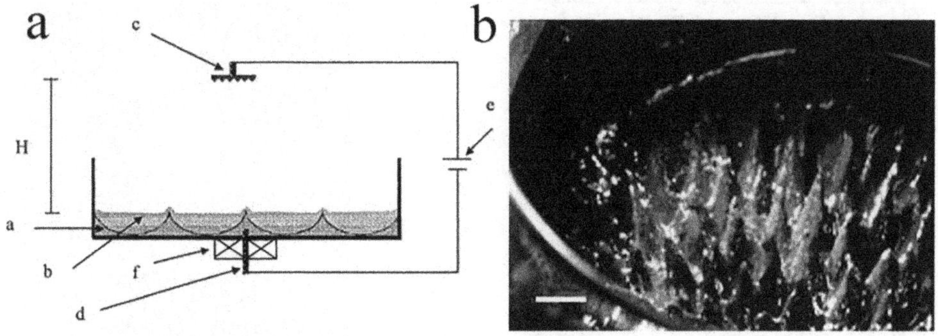

Figure 1.16. (a) Schematic drawing of magnetic field-assisted needleless electrospinning. (b) Spikes formed on the silicone oil-based magnetic fluid under the action of a permanent magnet. Reproduced with permission from Elsevier [122].

Yarin and Zussman [122] reported an electrospinning technique that used a magnetic field to initiate the jet formation (figure 1.16). The spinning liquid comprised two layers: a bottom ferromagnetic fluid layer, and a top polymer solution layer. When an external magnetic field was applied to the ferromagnetic fluid and an electric field applied to the polymer solution, the ferromagnetic fluid triggered the formation of steady vertical spikes, which perturbed at the interlayer interface and solution layer. When the applied voltage was high enough, solution jets were stretched out from the spikes.

1.4.5 Gas-assisted needleless electrospinning

Because of low cost and high safety, airflow has been used to assist jet initiation in electrospinning. A gas-jet electrospinning technique (also referred to as bubble electrospinning) was developed in 2007 [150]; it used gas to create bubbles on the liquid surface, which increased surface curvature and facilitated jet initiation. Since the introduction of this, gas-initiated electrospinning has attracted great interest [151, 152]. In nozzle electrospinning [153] and needleless electrospinning [154], high-speed gas can improve nanofiber production rate because airflow can enhance solution jet stretching, thus facilitating jet initiation. In addition to the gas-assisted solution electrospinning, melt electrospinning can also benefit from additional airflow [100, 155].

1.4.6 Centrifugal force-assisted needleless electrospinning

Centrifugal spinning has been developed over many years, and was originally extensively used for producing glass fibers [156]. Recently, centrifugal spinning has been applied to prepare nanofibers [157–159]. Many parameters in centrifugal electrospinning, e.g. voltage, spinneret rotation speed, solution feed rate, distance between spinning head and collector, and solution concentration, can affect the electrospinning process and nanofiber quality. The combination of mechanical rotation and electric field makes it very effective in fabricating aligned nanofibers

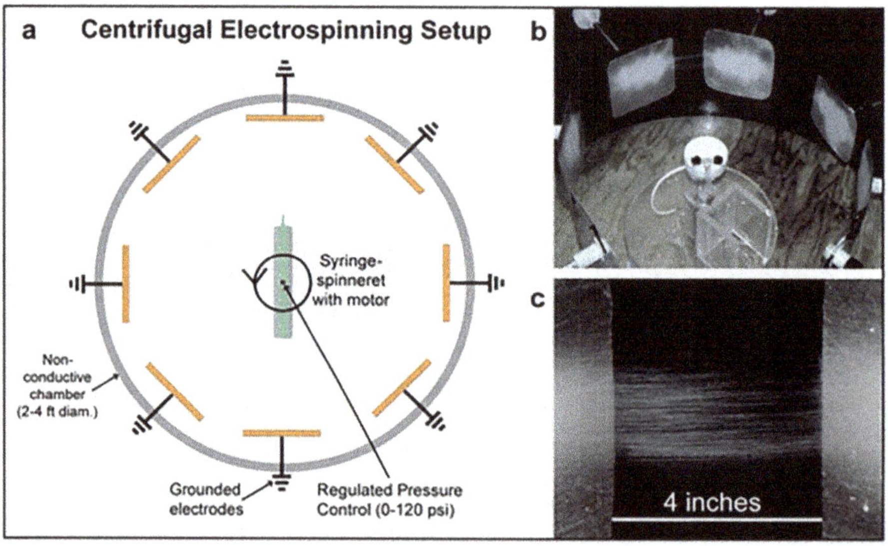

Figure 1.17. (a) Schematic drawing and (b) digital photo of centrifugal electrospinning process; (c) SEM image of centrifugal poly(vinylidene fluoride) (PVDF) fibers deposited between two grounded electrodes. Reproduced with permission from Royal Society of Chemistry [162].

(figure 1.17) [160–162]. The combination of centrifugal spinning and electrospinning has many advantages [162, 163]:

- For high-concentration solution or polymer melt, it is hard to achieve jet initiation in normal electrospinning because of the high viscosity; however, centrifugal force can easily transport these fluids.
- Higher nanofiber production rate.
- Finer nanofibers.
- Lower jet initiation voltage.

1.5 Nanofiber collection

Electrospun nanofibers are generally collected as randomly distributed nanofiber mats, the morphology of which is similar to a nonwoven fibrous mat. In lots of studies, such a fibrous structure was described as a fiber web, fiber sheet, nonwoven fiber, and membrane. To meet specific applications, nanofibers are collected into different structures from nonwoven fibers. Numerous setups have been developed to manipulate nanofiber deposition. As a result, aligned nanofibers, nanofiber yarn, and 3D nanofiber structures have been achieved.

1.5.1 Selective nanofiber deposition

Due to the large amount of electrostatic charges carried by nanofibers, there is significant Coulombic repulsion force between these fibers. As a result, the nanofiber deposition area on the collector is usually large and with low fiber collection efficiency. An auxiliary electric field could improve the control over nanofiber

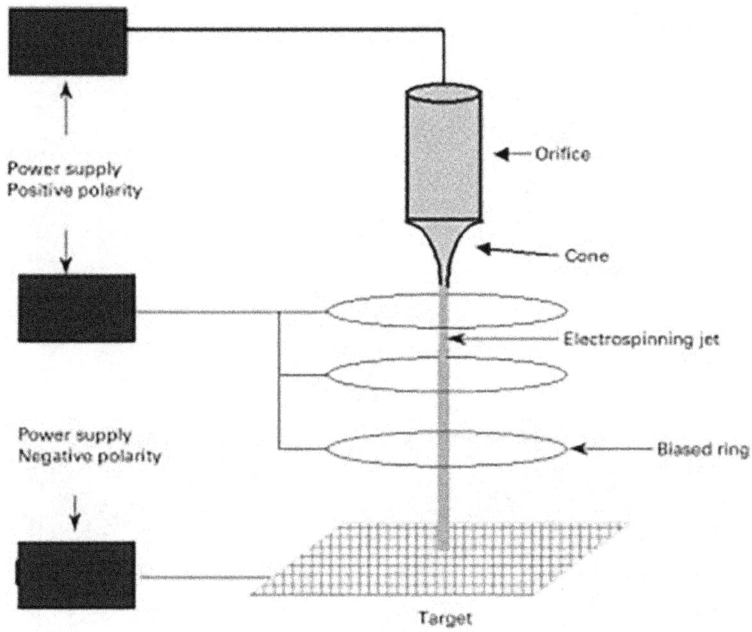

Figure 1.18. Auxiliary electric field-assisted fiber deposition. Reproduced with permission from Elsevier [164].

deposition [83, 84]. Charged rings have been used to restrain nanofiber disposition [164] to a small area; the schematic diagram of this electrospinning setup is shown in figure 1.18. When an auxiliary electrode was used in electrospinning to control fiber deposition, nanofibers in a 2D pattern or 3D structure could be fabricated [165, 166]. During the electrospinning process, the Taylor cone was stabilized and the jet whipping was converged sufficiently [165]. This method could provide a practical strategy for the fabrication of nanofibers with elaborate structures.

By electrospinning nanofibers onto a columnar-shaped collector, instead of flat collector as usual, tubular nanofibers can be obtained [167, 168]; these will have promising potentials in biomedical and industrial applications. Figure 1.19 shows complex nanofiber tubes collected on columnar collectors.

In addition, when a patterned substrate is used to collect nanofibers, the electric field does not distribute uniformly on the substrate, the collected nanofibers show predesigned arrangement on the substrate [169–171]. Figure 1.20(a) shows that the tips on the substrate can centralize the electric field, and as a result, more nanofibers are drawn to these tips and collected PEO nanofibers show a patterned structure (figure 1.20(b)).

1.5.2 Aligned nanofibers

When nanofibers are collected unidirectionally, the tensile strength of nanofiber mats in the fiber-aligned direction can be significantly improved. Collecting nanofibers using a high-speed rotating drum collector or disc is an efficient way to obtain aligned nanofibers [103, 172, 173]. Aligned nanofibers can also be collected using a

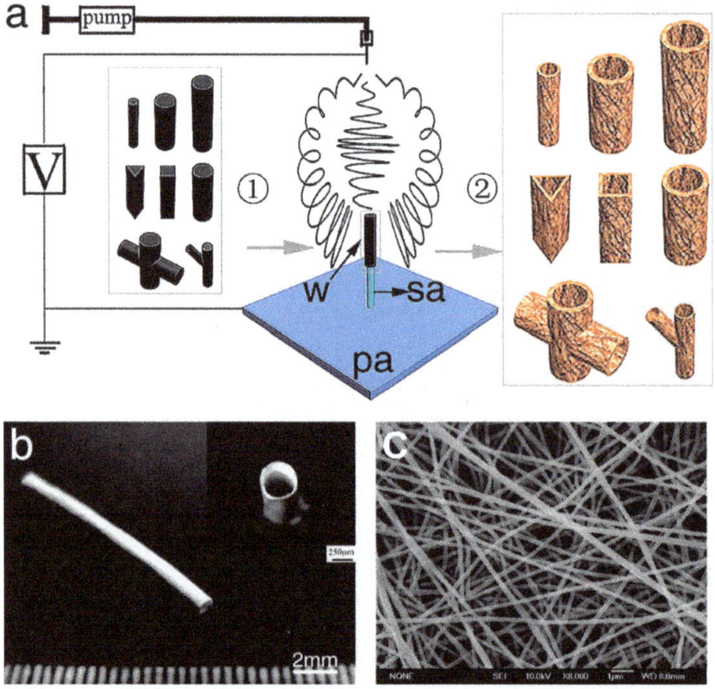

Figure 1.19. (a) Schematic illustration of columnar collectors for fabricating fibrous tubes. (①: columnar collectors and ②: fibrous tubes.) (b) Fabricated fibrous tube (diameter = 500 μm, inset is the cross-section image). (c) SEM image of nanofibers. Reproduced with permission from American Chemical Society [168].

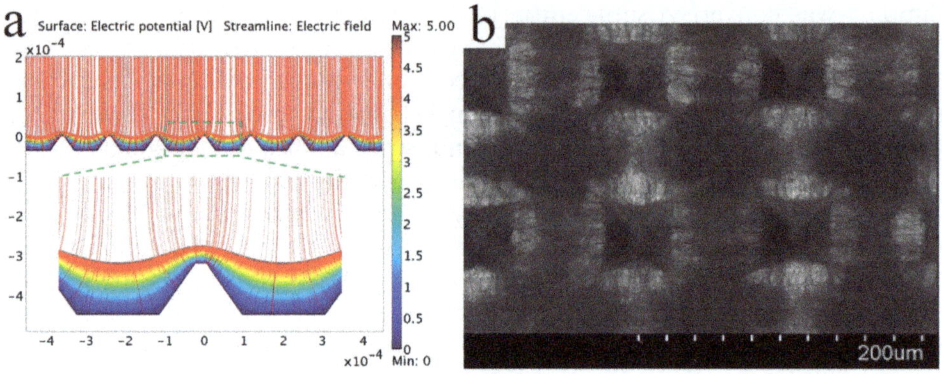

Figure 1.20. (a) Simulation of electric potential and electrical field distribution on the substrate, (b) SEM image of PEO fiber deposited on a patterned substrate for 5 min. Reproduced with permission from American Chemical Society [169].

stationary parallel electrodes collector [174–176] or tip collector [177]. Figure 1.21 shows aligned short nanofibers collected by two parallel Si substrates. When using two magnet bars as the collector, it is also possible to collect aligned wavy polymeric nanofibers [178].

In another work, an electrode with counter polarity voltage applied was used to govern fiber deposition [179]. Fiber placement and alignment on both microscale and nanoscale can be achieved through controlling the shape and magnitude of the electric field of the counter electrode. This technology demonstrates the ability to make nanofiber membranes with tailored porosity (figure 1.22).

AC voltage has the potential of minimizing jet whipping instability and enhancing fiber alignment. A method named biased AC electrospinning used amplified AC voltage to perform electrospinning with improved fiber alignment. When the electrospinning process is driven by DC voltage, it has a high level of whipping instability due to the action of repulsive Coulombic force. Meanwhile, AC power carries both positive and negative charges, and both attractive and repulsive Coulombic forces take effect during electrospinning and lead to a weakening of overall whipping instability [180].

Figure 1.21. SEM image of aligned nanofibers collected across two parallel Si substrates. Reproduced with permission from American Chemical Society [174].

Figure 1.22. A straight fiber being collected with auxiliary counter electrode, and SEM image of collected fibers. Reproduced with permission from American Chemical Society [179].

1.5.3 Nanofiber yarns

The first attempt in electrospinning nanofiber yarns goes back to the 1930s; the involved apparatus contained a fiber spinning wheel and a fiber yarn collecting device [181]. Recently, the production of uniaxial nanofiber bundles or twisted nanofiber yarns by electrospinning has drawn increasing interest because yarns can be woven or knitted into 2D fabrics or 3D structures with tailored structure, mechanical strength, and porosity. Nanofibers collected in a yarn form can find applications in the traditional textile industry, and creates possibilities for a greater number of new applications. Especially, the large-scale production of nanofiber yarns is becoming more and more imperative. The most popular setups for nanofiber yarn/bundle production are listed in table 1.4, and can be divided into the following categories:

- Collect aligned nanofibers in a short length first (e.g. two rotating discs, high-speed collector) and then twist them into twisted short nanofiber yarns [182–185].
- Collect short twisted yarns directly using two rotating tube collectors [186].
- Collect continuous nanofiber bundles with the help of airflow [187, 188].
- Use a water bath as the collector to obtain continuous nanofiber bundles [189–191].
- Use an auxiliary electrode to govern nanofiber collection for obtaining nanofiber bundles [192–194].
- Apply both positive and negative potentials in electrospinning to improve nanofiber alignment and collect continuous nanofiber bundles [195–197].
- Use a rotating funnel as the collector to obtain continuously twisted nanofiber yarns directly [198–203].
- Wrap electrospun nanofibers around conventional filaments or yarns to obtain composite nanofiber yarns [204, 205].
- Use AC potential to electrospin nanofibers instead of DC potential, and then collect continuous nanofiber bundles [206].

In addition to solution electrospinning, direct yarn production has also been realized in melt electrospinning. Polypropylene nanofiber yarn was continuously manufactured by a melt electrospinning method, wherein suction airflow was used to facilitate the formation of aligned nanofiber strand. The twisting of yarn strands was realized through a tailored rotating collector [188]. In another interesting work, AC electrospinning was used to generate nanofiber bundles directly in the absence of a collector [206]. Smoke-like nanofibers were generated from a rod electrode. Because of the existence of both positive changes and negative changes, they are easy to manipulate and can be effectively wound into nanofiber yarns (figure 1.23).

1.6 Summary and outlook

The science and engineering of nanofiber electrospinning technologies have advanced rapidly in recent years. Many successful attempts to improve electrospinning yield, nanofiber and nanofiber membrane quality, controlled nanofiber

Table 1.4. A summary of nanofiber yarn electrospinning setups, mechanisms, and features.

Electrospinning setup example	Mechanism	Continuous production (Y/N)	Collected form	Representative work
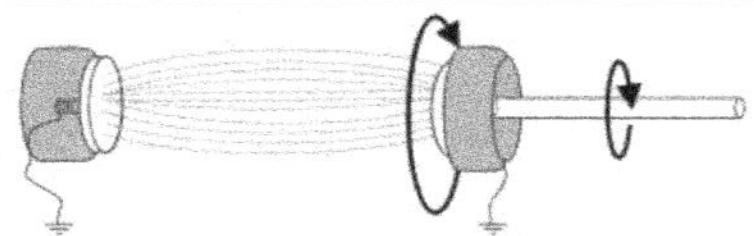 (Reproduced with permission from Elsevier)	Two separated rotating disc	N	Twisted yarn	[182]
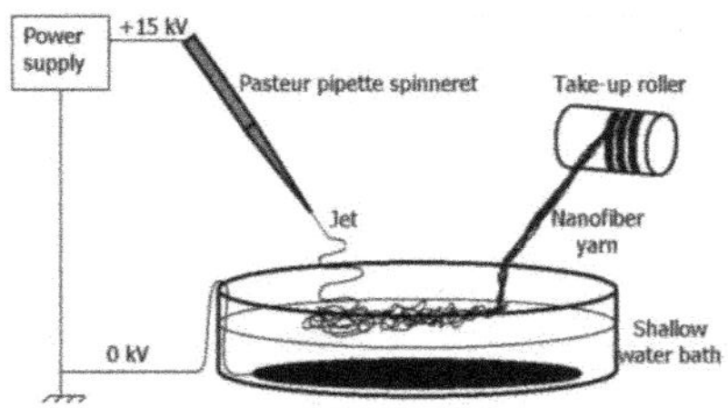 (Reproduced with permission from Elsevier)	Water bath collection	Y	Nanofiber bundle	[189, 190]
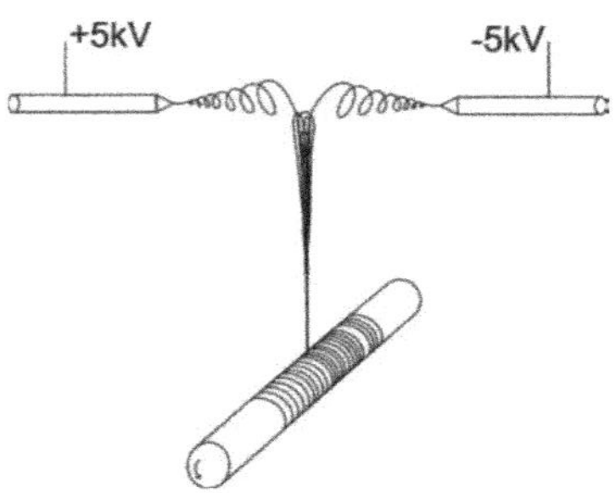 (Reproduced with permission from Elsevier)	Opposite polarity electrospinning	Y	Nanofiber bundle	[195–197]

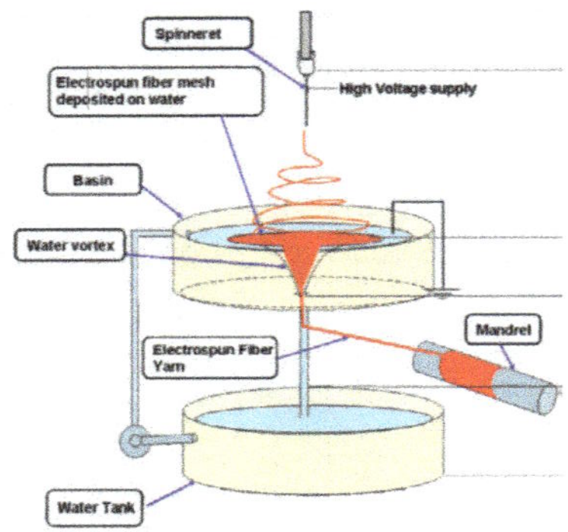 (Reproduced with permission from Elsevier)	Water bath collection	Y	Nanofiber bundle	[191]
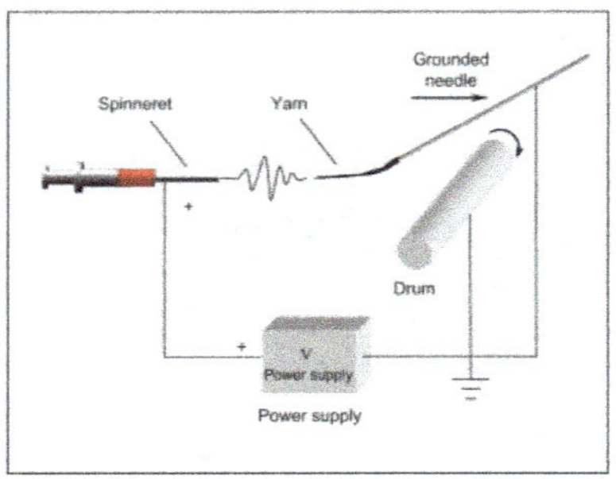 (Reproduced with permission from Elsevier)	Grounded bar-induced collection	Y	Nanofiber bundle	[192, 193]
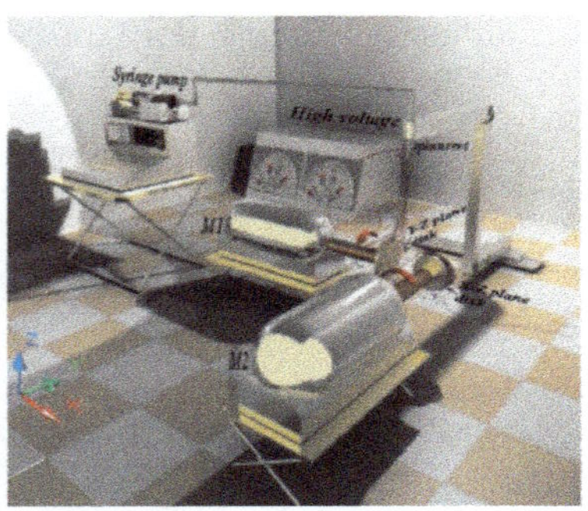 (Reproduced with permission from Elsevier)	Two rotating discs perpendicular to each other	Y	Nanofiber bundle	[194]

(Continued)

Table 1.4. (*Continued*)

Electrospinning setup example	Mechanism	Continuous production (Y/N)	Collected form	Representative work
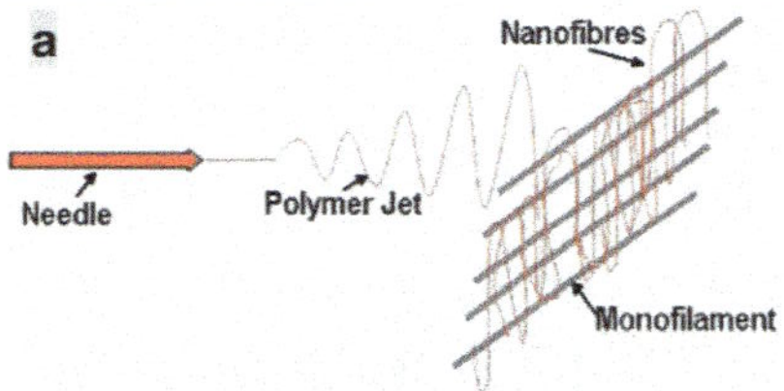 (Reproduced with permission from Elsevier)	Parallel filament collection	Y	Nanofiber-covered filaments	[204]
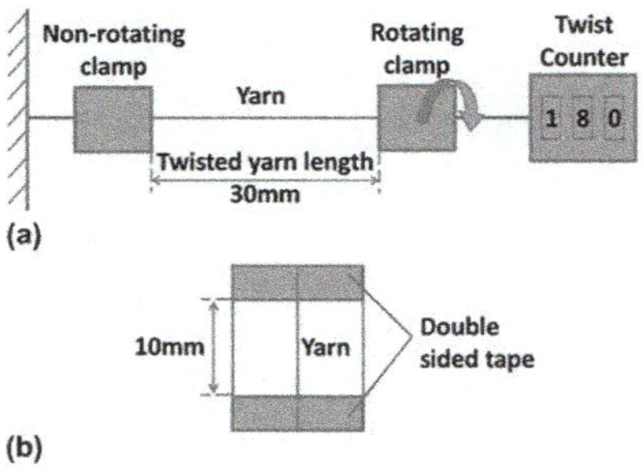 (Reproduced with permission from Cambridge University Press)	Short aligned nanofiber strip	N	Twisted yarn	[183–185]
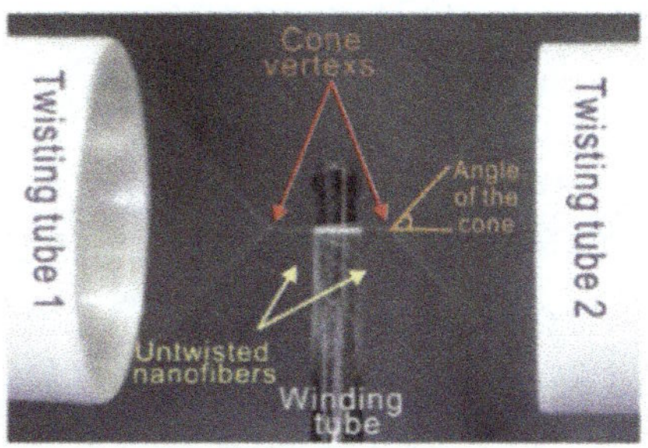 (Reproduced with permission from Elsevier)	Two rotating tubes and a winding tube	Y	Twisted yarn	[186]

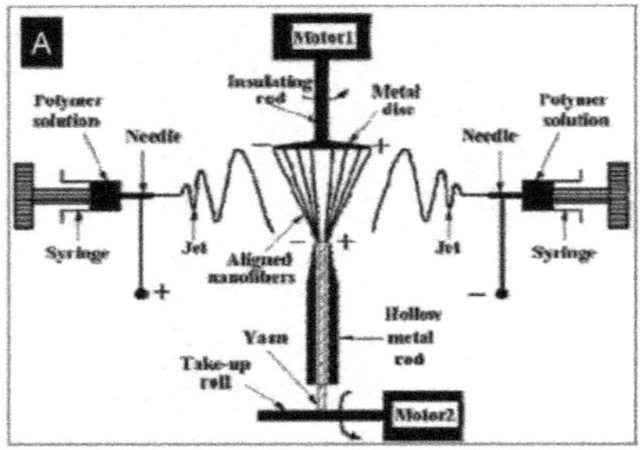 (Reproduced with permission from Elsevier)	Opposite polarity electrospinning, rotating funnel collection	Y	Twisted yarn	[198–201]
 (Reproduced with permission from Wiley)	Needleless electrospinning, rotating collection	Y	Nanofiber-covered yarns	[205]

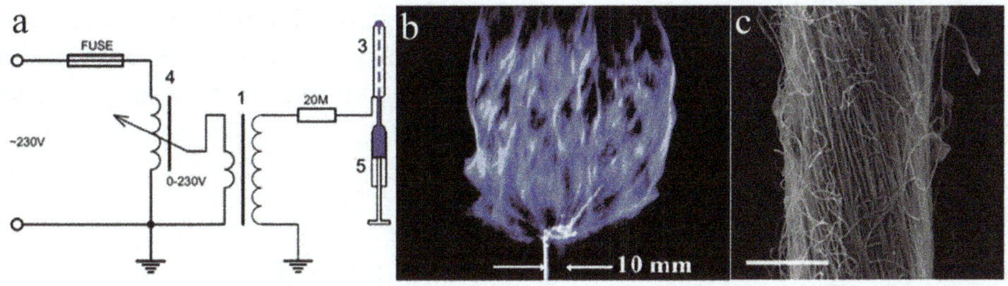

Figure 1.23. (a) Schematic diagram of the AC electrospinning setup, (b) generated polyvinyl butyral nanofiber plume, and (c) SEM image of nanofiber yarn (scale bar = 5 μm). Reproduced with permission from Royal Society of Chemistry [206].

deposition, and secondary nanofibrous structure have contributed to the exciting development of various advanced electrospinning technologies. Commercial production of electrospun nanofibers has also been realized.

In spite of the enormous amount of efforts made to further electrospinning technology, there are still some major challenges in this field, e.g. small production rate of nanofiber yarn, high electrical hazard risk associated with needleless electrospinning, and large fiber diameter in melt electrospinning. Future research on electrospinning should primarily focus on the large-scale production of high-quality nanofibers with improved safety and efficiency.

1. Near-field electrospinning has the ability of producing delicate fiber structures in 2D or 3D arrangements; however, the fiber diameter is much larger than that of conventional needle electrospinning and the fiber production rate is much lower. An important trend in near-field electrospinning is to become more productive, e.g. though multijet synchronous printing, and to have the ability of fabricating nanofiber-based real structures.
2. The current yarn electrospinning technology is either based on needle electrospinning that has a low nanofiber production rate or has a low yarn production efficiency in needleless electrospinning. The large-scale production of nanofiber yarn can greatly expand the application of electrospun nanofibers.
3. The needleless electrospinning technologies generally require high voltage to initiate jet ejection and ensure continuous electrospinning. The high voltage can lead to electrical discharge, and even fire/explosion hazards. The utilization of auxiliary fields, e.g. a magnetic field, airflow field, or centrifugal force, will effectively facilitate the electrospinning process. Results of this would include reducing the critical voltage, thinning the fiber diameter, and improving nanofiber production rate; therefore, this is a promising future direction for energy-efficient, safe production of nanofibers on a large scale.
4. Melt electrospinning is an environmentally friendly process, and melt electrospun fibers are of great importance in biomedical, filtration, and textile areas. Due to the deficiency of free charge in polymer melt, the fiber diameter of melt electrospinning is normally at the micrometer scale. The

utilization of airflow or centrifugal force can effectively improve melt jet stretching and produce thin melt electrospun fibers. The production of thinner fibers in melt electrospinning will be an important research direction in the future.

References

[1] Chen X *et al* 2012 Nanomaterials for renewable energy production and storage *Chem. Soc. Rev.* **41** 7909–37

[2] Burda C *et al* 2005 Chemistry and properties of nanocrystals of different shapes *Chem. Rev.* **105** 1025–102

[3] Alivisatos A P 1996 Perspectives on the physical chemistry of semiconductor nanocrystals *J. Phys. Chem.* **100** 13226–39

[4] Chinnappan A *et al* 2017 An overview of electrospun nanofibers and their application in energy storage, sensors and wearable/flexible electronics *J. Mater. Chem.* C **5** 12657–73

[5] Ramakrishna S *et al* 2006 Electrospun nanofibers: solving global issues *Mater. Today* **9** 40–50

[6] Suja P S *et al* 2017 Electrospun nanofibrous membranes for water purification *Polym. Rev.* **57** 467–504

[7] Zhu M *et al* 2016 Electrospun nanofibers membranes for effective air filtration *Macromol. Mater. Eng.* **302** 1600353

[8] Chen S *et al* 2017 Recent advances in electrospun nanofibers for wound healing *Nanomedicine* **12** 1335–52

[9] Sill T J and von Recum H A 2008 Electrospinning: applications in drug delivery and tissue engineering *Biomaterials* **29** 1989–2006

[10] Li D and Xia Y 2004 Electrospinning of nanofibers: reinventing the wheel? *Adv. Mater.* **16** 1151–70

[11] Gibson P, Schreuder-Gibson H and Rivin D 2001 Transport properties of porous membranes based on electrospun nanofibers *Colloids Surf.,* A **187–188** 469–81

[12] Ma P X and Zhang R 1999 Synthetic nano-scale fibrous extracellular matrix *J. Biomed. Mater. Res.* **46** 60–72

[13] Liu G, Qiao L and Guo A 1996 Diblock copolymer nanofibers *Macromolecules* **29** 5508–10

[14] Hartgerink J D, Beniash E and Stupp S I 2001 Self-assembly and mineralization of peptide-amphiphile nanofibers *Science* **294** 1684–8

[15] Martin C R 1996 Membrane-based synthesis of nanomaterials *Chem. Mater.* **8** 1739–46

[16] Ward G F 2001 Meltblown nanofibres for nonwoven filtration applications *Filtr. Separat.* **38** 42–3

[17] Weinberg M, Dee G and Harding T 2006 Flash spun web containing sub-micron filaments and process for forming same *US Patent* 20060135020

[18] Fedorova N and Pourdeyhimi B 2007 High strength nylon micro- and nanofiber based nonwovens via spunbonding *J. Appl. Polym. Sci.* **104** 3434–42

[19] Doshi J and Reneker D H 1995 Electrospinning process and applications of electrospun fibers *J. Electrostat.* **35** 151–60

[20] Srinivasan G and Reneker D H 1995 Structure and morphology of small diameter electrospun aramid fibers *Polym. Int.* **36** 195–201

[21] William G 1600 *de Magnet* (trans. by P E Mottelay), ed. B. 2

[22] Formhals A 1934 Process and apparatus for preparing artificial threads *US Patent* 1,975,504
[23] Formhals A 1944 Method and apparatus for spinning *US Patent* 2,349,950
[24] Taylor G 1964 Disintegration of water drops in an electric field *Proc. R. Soc. Lond. Ser. A Math. Phys. Sci.* **280** 383–97
[25] Taylor G 1969 Electrically driven jets *Proc. R. Soc. Lond., Ser.* A **313** 453–75
[26] Kowalewski T A, Hiller W J and Behnia M 1993 An experimental study of evaporating small diameter jets *Phys. Fluids* A **5** 1883–90
[27] Huang Z-M *et al* 2003 A review on polymer nanofibers by electrospinning and their applications in nanocomposites *Compos. Sci. Technol.* **63** 2223–53
[28] Sir Geoffrey Taylor F R S 1966 Studies in electrohydrodynamics. I. The circulation produced in a drop by an electric field *Proc. R. Soc. Lond., Ser.* A **291** 159
[29] Subbiah T *et al* 2005 Electrospinning of nanofibers *J. Appl. Polym. Sci.* **96** 557–69
[30] Rutledge G C and Fridrikh S V 2007 Formation of fibers by electrospinning *Adv. Drug Deliv. Rev.* **59** 1384–91
[31] Reneker D H and Yarin A L 2008 Electrospinning jets and polymer nanofibers *Polymer* **49** 2387–425
[32] Fong H and Reneker D H 1999 *Electrospinning and the Formation of Nanofibers* 225–46
[33] Reneker D H *et al* 2000 Bending instability of electrically charged liquid jets of polymer solutions in electrospinning *J. Appl. Phys.* **87** 4531–47
[34] Yarin A L, Koombhongse S and Reneker D H 2001 Bending instability in electrospinning of nanofibers *J. Appl. Phys.* **89** 3018–26
[35] Hohman M M *et al* 2001 Electrospinning and electrically forced jets. I. Stability theory *Phys. Fluids* **13** 2201–20
[36] Hohman M M *et al* 2001 Electrospinning and electrically forced jets. II. Applications *Phys. Fluids* **13** 2221–36
[37] Yarin A L, Koombhongse S and Reneker D H 2001 Taylor cone and jetting from liquid droplets in electrospinning of nanofibers *J. Appl. Phys.* **90** 4836–46
[38] Lin T *et al* 2004 The charge effect of cationic surfactants on the elimination of fibre beads in the electrospinning of polystyrene *Nanotechnology* **15** 1375–81
[39] Han T, Reneker D H and Yarin A L 2007 Buckling of jets in electrospinning *Polymer* **48** 6064–76
[40] Ribeiro C *et al* 2010 Influence of processing conditions on polymorphism and nanofiber morphology of electroactive poly(vinylidene fluoride) electrospun membranes *Soft Mater.* **8** 274–87
[41] Beachley V and Wen X 2009 Effect of electrospinning parameters on the nanofiber diameter and length *Mater. Sci. Eng.,* C **29** 663–8
[42] Haitao N, Tong L and Xungai W 2009 Needleless electrospinning. I. A comparison of cylinder and disk nozzles *J. Appl. Polym. Sci.* **114** 3524–30
[43] Wang X *et al* 2009 Needleless electrospinning of nanofibers with a conical wire coil *Polym. Eng. Sci.* **49** 1582–6
[44] Shao H *et al* 2015 Robust mechanical-to-electrical energy conversion from short-distance electrospun poly(vinylidene fluoride) fiber webs *ACS Appl. Mater. Interfaces* **7** 22551–7
[45] Xiao-Peng T *et al* 2014 Effect of flow rate on diameter of electrospun nanoporous fibers *Therm. Sci.* **18** 1447–9

[46] Barua B and Saha M C 2015 Investigation on jet stability, fiber diameter, and tensile properties of electrospun polyacrylonitrile nanofibrous yarns *J. Appl. Polym. Sci.* **132** 41918
[47] Mit-uppatham C, Nithitanakul M and Supaphol P 2004 Ultrafine electrospun polyamide-6 fibers: effect of solution conditions on morphology and average fiber diameter *Macromol. Chem. Phys.* **205** 2327–38
[48] Koski A, Yim K and Shivkumar S 2004 Effect of molecular weight on fibrous PVA produced by electrospinning *Mater. Lett.* **58** 493–7
[49] Hekmati A H *et al* 2013 Effect of needle length, electrospinning distance, and solution concentration on morphological properties of polyamide-6 electrospun nanowebs *Text. Res. J.* **83** 1452–66
[50] Biber E *et al* 2010 Effects of electrospinning process parameters on nanofibers obtained from Nylon 6 and poly (ethylene-n-butyl acrylate-maleic anhydride) elastomer blends using Johnson SB statistical distribution function *Appl. Phys.* A **99** 477–87
[51] Demir M M *et al* 2002 Electrospinning of polyurethane fibers *Polymer* **43** 3303–9
[52] Ding B *et al* 2002 Preparation and characterization of nanoscaled poly(vinyl alcohol) fibers via electrospinning *Fibers Polym.* **3** 73–9
[53] Qin X-H *et al* 2006 Effect of different salts on electrospinning of polyacrylonitrile (PAN) polymer solution *J. Appl. Polym. Sci.* **103** 3865–70
[54] Ho-Wang T and Min W 2010 Electrospinning of fibrous polymer scaffolds using positive voltage or negative voltage: a comparative study *Biomed. Mater.* **5** 054110
[55] Tong L *et al* 2004 The charge effect of cationic surfactants on the elimination of fibre beads in the electrospinning of polystyrene *Nanotechnology* **15** 1375
[56] Li Z *et al* 2016 Fabrication of a polyvinylidene fluoride tree-like nanofiber web for ultra high performance air filtration *RSC Adv.* **6** 91243–9
[57] Yang S *et al* 2011 Controllable fabrication of soap-bubble-like structured polyacrylic acid nano-nets via electro-netting *Nanoscale* **3** 564–8
[58] Hardick O, Stevens B and Bracewell D G 2011 Nanofibre fabrication in a temperature and humidity controlled environment for improved fibre consistency *J. Mater. Sci.* **46** 3890–8
[59] Simonet M *et al* 2007 Ultraporous 3D polymer meshes by low-temperature electrospinning: use of ice crystals as a removable void template *Polym. Eng. Sci.* **47** 2020–6
[60] De Vrieze S *et al* 2008 The effect of temperature and humidity on electrospinning *J. Mater. Sci.* **44** 1357
[61] Fashandi H and Karimi M 2012 Pore formation in polystyrene fiber by superimposing temperature and relative humidity of electrospinning atmosphere *Polymer* **53** 5832–49
[62] Phiriyawirut M *et al* 2008 Morphology of electrospun mats of soy protein isolate and its blend *Adv. Mater. Res.* **55** 733–6
[63] Ali U, Wang X and Lin T 2012 Effect of nozzle polarity and connection on electrospinning of polyacrylonitrile nanofibers *J. Text. Inst.* **103** 1160–8
[64] Kessick R, Fenn J and Teppe G 2004 The use of AC potentials in electrospraying and electrospinning processes *Polymer* **45** 2981–4
[65] Maheshwari S and Chang H 2009 Assembly of multi-stranded nanofiber threads through AC electrospinning *Adv. Mater.* **21** 349–54
[66] Chen G *et al* 2015 Core–shell structure PEO/CS nanofibers based on electric field induced phase separation via electrospinning and its application *J. Polym. Sci., Part A: Polym. Chem.* **53** 2298–311

[67] Yang Y *et al* 2008 Release pattern and structural integrity of lysozyme encapsulated in core–sheath structured poly(dl-lactide) ultrafine fibers prepared by emulsion electrospinning *Eur. J. Pharm. Biopharm.* **69** 106–16
[68] Yang Y *et al* 2007 Structural stability and release profiles of proteins from core-shell poly (DL-lactide) ultrafine fibers prepared by emulsion electrospinning *J. Biomed. Mater. Res. Part A* **86A** 374–85
[69] George M C and Braun P V 2009 Multicompartmental materials by electrohydrodynamic cojetting *Angew. Chem. Int. Ed.* **48** 8606–9
[70] Sun Z *et al* 2003 Compound core–shell polymer nanofibers by co-electrospinning *Adv. Mater.* **15** 1929–32
[71] McCann JT L D and Xia Y N 2005 Electrospinning of nanofibers with core-sheath, hollow, or porous structures *J. Mater. Chem.* **15** 735–8
[72] Zussman E *et al* 2006 Electrospun polyaniline/poly(methyl methacrylate)-derived turbostratic carbon micro-/nanotubes *Adv. Mater.* **18** 348–53
[73] Niu H *et al* 2014 Ultrafine PDMS fibers: preparation from *in situ* curing-electrospinning and mechanical characterization *RSC Adv.* **4** 11782–7
[74] Yang C *et al* 2016 Electrospun pH-sensitive core–shell polymer nanocomposites fabricated using a tri-axial process *Acta Biomater.* **35** 77–86
[75] Gupta P and Wilkes G L 2003 Some investigations on the fiber formation by utilizing a side-by-side bicomponent electrospinning approach *Polymer* **44** 6353–9
[76] Lin T, Wang H and Wang X 2005 Self-crimping bicomponent nanofibers electrospun from polyacrylonitrile and elastomeric polyurethane *Adv. Mater.* **17** 2699–703
[77] Bhaskar S and Lahann J 2009 Microstructured materials based on multicompartmental fibers *J. Am. Chem. Soc.* **131** 6650–1
[78] Varesano A, Carletto R A and Mazzuchetti G 2009 Experimental investigations on the multi-jet electrospinning process *J. Mater. Process. Technol.* **209** 5178–85
[79] Yang E, Shi J and Xue Y 2010 Influence of electric field interference on double nozzles electrospinning *J. Appl. Polym. Sci.* **116** 3688–92
[80] Ding B *et al* 2004 Fabrication of blend biodegradable nanofibrous nonwoven mats via multi-jet electrospinning *Polymer* **45** 1895–902
[81] Ying Y *et al* 2008 Electrospun uniform fibres with a special regular hexagon distributed multi-needles system *J. Phys. Conf. Ser.* **142** 012027
[82] Theron S A *et al* 2005 Multiple jets in electrospinning: experiment and modeling *Polymer* **46** 2889–99
[83] Yang Y *et al* 2010 A shield ring enhanced equilateral hexagon distributed multi-needle electrospinning spinneret *IEEE Trans. Dielectr. Electr. Insul.* **17** 1592–601
[84] Zheng Y and Zeng Y 2014 Electric field analysis of spinneret design for multihole electrospinning system *J. Mater. Sci.* **49** 1964–72
[85] Kim G, Cho Y-S and Kim W D 2006 Stability analysis for multi-jets electrospinning process modified with a cylindrical electrode *Eur. Polym. J.* **42** 2031–8
[86] Varabhas J S, Chase G G and Reneker D H 2008 Electrospun nanofibers from a porous hollow tube *Polymer* **49** 4226–9
[87] Dosunmu O O *et al* 2006 Electrospinning of polymer nanofibres from multiple jets on a porous tubular surface *Nanotechnology* **17** 1123
[88] Piner R D *et al* 1999 'Dip-Pen' nanolithography *Science* **283** 661
[89] Huo F *et al* 2008 Polymer pen lithography *Science* **321** 1658

[90] Sun D *et al* 2006 Near-field electrospinning *Nano Lett.* **6** 839–42

[91] Hellmann C *et al* 2009 High precision deposition electrospinning of nanofibers and nanofiber nonwovens *Polymer* **50** 1197–205

[92] He X-X *et al* 2017 Near-field electrospinning: progress and applications *J. Phys. Chem.* C **121** 8663–78

[93] Chang C *et al* 2010 Direct-write piezoelectric polymeric nanogenerator with high energy conversion efficiency *Nano Lett.* **10** 726–31

[94] Fuh Y K, Chen S Z and He Z Y 2013 Direct-write, highly aligned chitosan-poly(ethylene oxide) nanofiber patterns for cell morphology and spreading control *Nanoscale Res. Lett.* **8** 97

[95] Brown T D, Dalton P D and Hutmacher D W 2011 Direct writing by way of melt electrospinning *Adv. Mater.* **23** 5651–7

[96] Liu Z H *et al* 2015 Crystallization and mechanical behavior of the ferroelectric polymer nonwoven fiber fabrics for highly durable wearable sensor applications *Appl. Surf. Sci.* **346** 291–301

[97] SangHoon L *et al* 2007 Chip-to-chip fluidic connectors via near-field electrospinning *2007 IEEE 20th Int. Conf. on Micro Electro Mechanical Systems (MEMS)*

[98] Hochleitner G *et al* 2014 High definition fibrous poly(2-ethyl-2-oxazoline) scaffolds through melt electrospinning writing *Polymer* **55** 5017–23

[99] Wojasiński M, Goławski J and Ciach T 2017 Blow-assisted multi-jet electrospinning of poly-L-lactic acid nanofibers *J. Polym. Res.* **24** 76

[100] Zhmayev E, Cho D and Joo Y L 2010 Nanofibers from gas-assisted polymer melt electrospinning *Polymer* **51** 4140–4

[101] Hsiao H-Y *et al* 2011 Effect of air blowing on the morphology and nanofiber properties of blowing-assisted electrospun polycarbonates *J. Appl. Polym. Sci.* **124** 4904–14

[102] An S *et al* 2014 Supersonically blown ultrathin thorny devil nanofibers for efficient air cooling *ACS Appl. Mater. Interfaces* **6** 13657–66

[103] Kiselev P and Rosell-Llompart J 2012 Highly aligned electrospun nanofibers by elimination of the whipping motion *J. Appl. Polym. Sci.* **125** 2433–41

[104] Varesano A, Montarsolo A and Tonin C 2007 Crimped polymer nanofibres by air-driven electrospinning *Eur. Polym. J.* **43** 2792–8

[105] Chen C *et al* 2016 Use of electrospinning and dynamic air focusing to create three-dimensional cell culture scaffolds in microfluidic devices *Analyst* **141** 5311–20

[106] Tanioka A and Takahashi M 2016 Highly productive systems of nanofibers for novel applications *Ind. Eng. Chem. Res.* **55** 3759–64

[107] Kim G H and Yoon H 2008 A direct-electrospinning process by combined electric field and air-blowing system for nanofibrous wound-dressings *Appl. Phys.* A **90** 389–94

[108] Norton C L 1936 Method of and apparatus for producing fibrous or filamentary material *Google Patents*

[109] Lyons J, Li C and Ko F 2004 Melt-electrospinning part I: processing parameters and geometric properties *Polymer* **45** 7597–603

[110] Hunley M T *et al* 2008 Taking advantage of tailored electrostatics and complementary hydrogen bonding in the design of nanostructures for biomedical applications *Macromol. Symp.* **270** 1–7

[111] Dalton P D *et al* 2006 Direct *in vitro* electrospinning with polymer melts *Biomacromolecules* **7** 686–90

[112] Qin C-C *et al* 2015 Melt electrospinning of poly(lactic acid) and polycaprolactone microfibers by using a hand-operated Wimshurst generator *Nanoscale* **7** 16611–5
[113] Kim S J *et al* 2010 Fabrication and characterization of 3-dimensional PLGA nanofiber/microfiber composite scaffolds *Polymer* **51** 1320–7
[114] Li X *et al* 2012 Preparation and characterization of PLLA/nHA nonwoven mats via laser melt electrospinning *Mater. Lett.* **73** 103–6
[115] Ogata N *et al* 2007 Poly(lactide) nanofibers produced by a melt-electrospinning system with a laser melting device *J. Appl. Polym. Sci.* **104** 1640–5
[116] Li X *et al* 2014 Effect of oriented fiber membrane fabricated via needleless melt electrospinning on water filtration efficiency *Desalination* **344** 266–73
[117] Kim J *et al* 2014 Preparation and gas-sensing properties of pitch-based carbon fiber prepared using a melt-electrospinning method *Res. Chem. Intermed.* **40** 2571–81
[118] Lee S and Obendorf S K 2006 Developing protective textile materials as barriers to liquid penetration using melt-electrospinning *J. Appl. Polym. Sci.* **102** 3430–7
[119] Pham Q P, Sharma U and Mikos A G 2006 Electrospinning of polymeric nanofibers for tissue engineering applications: a review *Tissue Eng.* **12** 1197–211
[120] Brown T D *et al* 2012 Design and fabrication of tubular scaffolds via direct writing in a melt electrospinning mode *Biointerphases* **7** 13
[121] Hacker C *et al* 2013 Functionally modified, melt-electrospun thermoplastic polyurethane mats for wound-dressing applications *J. Appl. Polym. Sci.* **131** 40132
[122] Yarin A L and Zussman E 2004 Upward needleless electrospinning of multiple nanofibers *Polymer* **45** 2977–80
[123] Lukas D, Sarkar A and Pokorny P 2008 Self-organization of jets in electrospinning from free liquid surface: a generalized approach *J. Appl. Phys.* **103** 084309
[124] Thoppey N M *et al* 2010 Unconfined fluid electrospun into high quality nanofibers from a plate edge *Polymer* **51** 4928–36
[125] Thoppey N M *et al* 2011 Edge electrospinning for high throughput production of quality nanofibers *Nanotechnology* **22** 345301
[126] Sharma U *et al* 2012 Electrospinning process for manufacture of multi-layered structures *US Patent* 2012/0193836 Al
[127] Yan X *et al* 2015 Slit-surface electrospinning: a novel process developed for high-throughput fabrication of core-sheath fibers *PLoS One* **10** e0125407
[128] Yan X *et al* 2012 High-throughput needleless electrospinning of core-sheath fibers *The Fiber Society Conf. Proc.*
[129] Ucar N, Mehmet U and Kizildag N 2013 Design of a novel nozzle prototype for increased productivity and improved coating quality during electrospinning *Tekst. Konfeksiyon* **23** 199
[130] Jiang G, Zhang S and Qin X 2013 High throughput of quality nanofibers via one stepped pyramid-shaped spinneret *Mater. Lett.* **106** 56–8
[131] Jiang G and Qin X 2014 An improved free surface electrospinning for high throughput manufacturing of core-shell nanofibers *Mater. Lett.* **128** 259–62
[132] Holopainen J *et al* 2015 Needleless electrospinning with twisted wire spinneret *Nanotechnology* **26** 025301
[133] Yan G *et al* 2017 Curved convex slot: an effective needleless electrospinning spinneret with low solvent evaporation loss *J. Mater. Sci.* **52** 11749–58
[134] Jirsak O *et al* 2009 Method of nanofibres production from a polymer solution using electrostatic spinning and a device for carrying out the method *US Patent* US7585437B2

[135] Niu H and Lin T 2012 Fiber generators in needleless electrospinning *J. Nanomater.* **2012** 725950
[136] Niu H, Wang X and Lin T 2012 Upward needleless electrospinning of nanofibers *J. Eng. Fibers Fabrics* **7** 17–22
[137] Wang X *et al* 2012 Needleless electrospinning of uniform nanofibers using spiral coil spinnerets *J. Nanomater.* **2012** 785920
[138] Lu B *et al* 2010 Superhigh-throughput needleless electrospinning using a rotary cone as spinneret *Small* **6** 1612–6
[139] Bhattacharyya I *et al* 2016 Free surface electrospinning of aqueous polymer solutions from a wire electrode *Chem. Eng. J.* **289** 203–11
[140] Forward K M and Rutledge G C 2012 Free surface electrospinning from a wire electrode *Chem. Eng. J.* **183** 492–503
[141] Forward K M, Flores A and Rutledge G C 2013 Production of core/shell fibers by electrospinning from a free surface *Chem. Eng. Sci.* **104** 250–9
[142] Yang W *et al* 2016 Optimal spinneret layout in Von Koch curves of fractal theory based needleless electrospinning process *AIP Adv.* **6** 065223
[143] Liu Z, Chen R and He J 2016 Active generation of multiple jets for producing nanofibres with high quality and high throughput *Mater. Des.* **94** 496–501
[144] Liu Z, Ang K K J and He J 2017 Needle-disk electrospinning inspired by natural point discharge *J. Mater. Sci.* **52** 1823–30
[145] Cengiz-Çallıoğlu F 2014 Dextran nanofiber production by needleless electrospinning processe *Polymers* **14** 5–13
[146] Moon S, Gil M and Lee K J 2017 Syringeless electrospinning toward versatile fabrication of nanofiber web *Sci. Rep.* **7** 41424
[147] Zheng G *et al* 2018 Self-cleaning threaded rod spinneret for high-efficiency needleless electrospinning *Appl. Phys.* A **124** 473
[148] Liu S L *et al* 2014 Needleless electrospinning for large scale production of ultrathin polymer fibres *Mater. Res. Innov.* **18** S4-833–7
[149] Guilong Y *et al* 2018 Electro-aerodynamic field aided needleless electrospinning *Nanotechnology* **29** 235302
[150] Liu Y and He J H 2007 Bubble electrospinning for mass production of nanofibers *Int. J. Nonlinear Sci. Numer. Simul.* **8** 393
[151] Pu C *et al* 2013 Double-nozzle air-jet electrospinning for nanofiber fabrication *J. Appl. Polym. Sci.* **131** 40040
[152] Chen R-X, Li Y and He J-H 2014 Mini-review on Bubbfil spinning process for mass-production of nanofibers *Matéria* **19** 325–43
[153] Um I C *et al* 2004 Electro-spinning and electro-blowing of hyaluronic acid *Biomacromolecules* **5** 1428–36
[154] Wang X, Lin T and Wang X 2014 Use of airflow to improve the nanofibrous structure and quality of nanofibers from needleless electrospinning *J. Ind. Text.* **45** 310–20
[155] Zhmayev E, Cho D and Joo Y L 2010 Modeling of melt electrospinning for semi-crystalline polymers *Polymer* **51** 274–90
[156] Jones F R and Huff N T 2009 The structure and properties of glass fibres *Handbook of Textile Fibre Structure* ed S J Eichhorn *et al* (Cambridge: Woodhead Publishing), ch 9, pp 307–52
[157] Badrossamay M R *et al* 2010 Nanofiber assembly by rotary jet-spinning *Nano Lett.* **10** 2257–61

[158] Sarkar K *et al* 2010 Electrospinning to Forcespinning™ *Mater. Today* **13** 12–4
[159] Weitz R T *et al* 2008 Polymer nanofibers via nozzle-free centrifugal spinning *Nano Lett.* **8** 1187–91
[160] Liu S L, Long Y Z, Zhang Z H, Zhang H D, Sun B and Zhang J C 2013 Assembly of oriented ultrafine polymer fibers by centrifugal electrospinning *J. Nanomater.* **2013** 713275
[161] Dabirian F *et al* 2011 A comparative study of jet formation and nanofiber alignment in electrospinning and electrocentrifugal spinning systems *J. Electrostat.* **69** 540–6
[162] Edmondson D *et al* 2012 Centrifugal electrospinning of highly aligned polymer nanofibers over a large area *J. Mater. Chem.* **22** 18646–52
[163] Dabirian F, Ravandi S A H and Pishevar A R 2010 Investigation of parameters affecting PAN nanofiber production using electrical and centrifugal forces as a novel method *Curr. Nanosci.* **6** 545–52
[164] Deitzel J M *et al* 2001 Controlled deposition of electrospun poly(ethylene oxide) fibers *Polymer* **42** 8163–70
[165] Kim G H 2006 Electrospinning process using field-controllable electrodes *J. Polym. Sci., Part B: Polym. Phys.* **44** 1426–33
[166] Kim H-Y *et al* 2010 Nanopottery: coiling of electrospun polymer nanofibers *Nano Lett.* **10** 2138–40
[167] Thomas V, Zhang X and Vohra Y K 2009 A biomimetic tubular scaffold with spatially designed nanofibers of protein/PDS® bio-blends *Biotechnol. Bioeng.* **104** 1025–33
[168] Zhang D M and Chang J 2008 Electrospinning of three-dimensional nanofibrous tubes with controllable architectures *Nano Lett.* **8** 3283–7
[169] Ding Z, Salim A and Ziaie B 2009 Selective nanofiber deposition through field-enhanced electrospinning *Langmuir* **25** 9648–52
[170] Salim A, Son C and Ziaie B 2008 Selective nanofiber deposition via electrodynamic focusing *Nanotechnology* **19** 375303
[171] Park S M *et al* 2018 Direct fabrication of spatially patterned or aligned electrospun nanofiber mats on dielectric polymer surfaces *Chem. Eng. J.* **335** 712–9
[172] Katta P, Alessandro M, Ramsier R D and Chase G G 2004 Continuous electrospinning of aligned polymer nanofibers onto a wire drum collector *Nano Lett.* **4** 2215–8
[173] Theron A, Zussman E and Yarin A L 2001 Electrostatic field-assisted alignment of electrospun nanofibres *Nanotechnology* **12** 384–90
[174] Li D, Wang Y and Xia Y 2003 Electrospinning of polymeric and ceramic nanofibers as uniaxially aligned arrays *Nano Lett.* **3** 1167–71
[175] Li D, Wang Y and Xia Y 2004 Electrospinning nanofibers as uniaxially aligned arrays and layer-by-layer stacked films *Adv. Mater.* **16** 361–6
[176] Yang D *et al* 2007 Fabrication of aligned fibrous arrays by magnetic electrospinning *Adv. Mater.* **19** 3702–6
[177] Rafique J *et al* 2007 Electrospinning highly aligned long polymer nanofibers on large scale by using a tip collector *Appl. Phys. Lett.* **91** 063126
[178] Liu Y *et al* 2010 Magnetic field-assisted electrospinning of aligned straight and wavy polymeric nanofibers *Adv. Mater.* **22** 2454–7
[179] Carnell L S *et al* 2008 Aligned mats from electrospun single fibers *Macromolecules* **41** 5345–9
[180] Sarkar S, Deevi S and Tepper G 2007 Biased AC electrospinning of aligned polymer nanofibers *Macromol. Rapid Commun.* **28** 1034–9

[181] Anton F 1934 Process and apparatus for preparing artificial threads *US Patent* US50028330A

[182] Dalton P D, Klee D and Möller M 2005 Electrospinning with dual collection rings *Polymer* **46** 611–4

[183] Nakashima R *et al* 2011 Mechanical properties of poly(vinylidene fluoride) nanofiber filaments prepared by electrospinning and twisting *Adv. Polym. Tech.* **32** E44–52

[184] Zhou Y *et al* 2012 Strip twisted electrospun nanofiber yarns: structural effects on tensile properties *J. Mater. Res.* **27** 537–44

[185] Chawla S, Naraghi M and Davoudi A 2013 Effect of twist and porosity on the electrical conductivity of carbon nanofiber yarns *Nanotechnology* **24** 255708

[186] Yan H, Liu L and Zhang Z 2011 Continually fabricating staple yarns with aligned electrospun polyacrylonitrile nanofibers *Mater. Lett.* **65** 2419–21

[187] Ko F *et al* 2003 Electrospinning of continuous carbon nanotube-filled nanofiber yarns *Adv. Mater.* **15** 1161–5

[188] Ma X *et al* 2017 Continuous manufacturing of nanofiber yarn with the assistance of suction wind and rotating collection via needleless melt electrospinning *J. Appl. Polym. Sci.* **134**

[189] Smit E, Bűttner U and Sanderson R D 2005 Continuous yarns from electrospun fibers *Polymer* **46** 2419–23

[190] Tian L, Yan T and Pan Z 2015 Fabrication of continuous electrospun nanofiber yarns with direct 3D processability by plying and twisting *J. Mater. Sci.* **50** 7137–48

[191] Teo W-E *et al* 2007 A dynamic liquid support system for continuous electrospun yarn fabrication *Polymer* **48** 3400–5

[192] Dabirian F, Hosseini Y and Ravandi S A H 2007 Manipulation of the electric field of electrospinning system to produce polyacrylonitrile nanofiber yarn *J. Text. Inst.* **98** 237–41

[193] Wang X *et al* 2008 Continuous polymer nanofiber yarns prepared by self-bundling electrospinning method *Polymer* **49** 2755–61

[194] Bazbouz M B and Stylios G K 2008 Novel mechanism for spinning continuous twisted composite nanofiber yarns *Eur. Polym. J.* **44** 1–12

[195] Huan Pan L L, Hu L and Cui X 2006 Continuous aligned polymer fibers produced by a modified electrospinning method *Polymer* **47** 4901–4

[196] Hajiani F, Jeddi A A and Gharehaghaji A A 2012 An investigation on the effects of twist on geometry of the electrospinning triangle and polyamide 66 nanofiber yarn strength *Fibers Polym.* **13** 244–52

[197] Maleki H *et al* 2013 Influence of the solvent type on the morphology and mechanical properties of electrospun PLLA yarns *Biofabrication* **5** 035014

[198] Ali U *et al* 2012 Direct electrospinning of highly twisted, continuous nanofiber yarns *J. Text. Inst.* **103** 80–8

[199] Wu S-H and Qin X-H 2013 Uniaxially aligned polyacrylonitrile nanofiber yarns prepared by a novel modified electrospinning method *Mater. Lett.* **106** 204–7

[200] He J *et al* 2013 Continuous twisted nanofiber yarns fabricated by double conjugate electrospinning *Fibers Polym.* **14** 1857–63

[201] He J-X *et al* 2013 Fabrication of continuous nanofiber yarn using novel multi-nozzle bubble electrospinning *Polym. Int.* **63** 1288–94

[202] Afifi A M *et al* 2010 Electrospinning of continuous aligning yarns with a 'funnel' target *Macromol. Mater. Eng.* **295** 660–5

[203] Shuakat M N and Lin T 2016 Direct electrospinning of nanofibre yarns using a rotating ring collector *J. Text. Inst.* **107** 791–9

[204] Zhou F-L, Gong R-H and Porat I 2010 Nano-coated hybrid yarns using electrospinning *Surf. Coat. Technol.* **204** 3459–63

[205] Niu H *et al* 2014 Composite yarns fabricated from continuous needleless electrospun nanofibers *Polym. Eng. Sci.* **54** 1495–502

[206] Pokorny P *et al* 2014 Effective AC needleless and collectorless electrospinning for yarn production *Phys. Chem. Chem. Phys.* **16** 26816–22

Chapter 2

Introduction to piezoelectricity and electrospun piezoelectric materials and devices

T Rodrigues-Marinho, A C Lima, P Martins, P Costa and S Lanceros-Mendez

Piezoelectric materials are increasingly being used in a wide range of applications. Single crystals, ceramics, and polymers are all being used as sensors, actuators, and for energy harvesting applications. The most-used piezoelectric inorganic materials are lead zirconate titanate ($Pb(Zr_{1-x}, Ti_x)O_3$, also called PZT) and barium titanate ($BaTiO_3$) for ceramics and poly(vinylidene fluoride) (PVDF) for polymers. Polymer-based composites including highly dielectric and piezoelectric ceramic nano- or microparticles are also largely investigated to take advantage of both polymers and ceramics. Among the different processing methods, electrospinning allows the development of piezoelectric nano- and microfibers, thus opening a distinct range of applications in electronics, energy generation and storage, and membranes for water and air remediation.

2.1 Introduction

From the beginning of the new millennium, and more particularly during the last decade, strong advances have taken place in science and technology resulting from an interplay between different innovation areas and disciplines [1, 2].

This approach has been particularly successful in the development, optimization, and application of smart and multifunctional materials, taking advantage of knowledge in biology, physics, chemistry, nanotechnology, materials science, as well as mechanical and electrical engineering. This leads not just to an increasing number of high-performance smart and multifunctional materials, but also to the emergence of disruptive applications based on them [2, 3].

Further, smart materials associated with the Internet of Things (IoT) concept represent the innovative vehicle to reach wireless, sustainable, and interconnected autonomous smart systems and cities [3].

doi:10.1088/978-0-7503-2005-4ch2

There are different types of smart materials, including shape memory, piezoelectric, piezoresistive, magnetorheological, magnetostrictive, magnetoelectric, electrorheological, and chromic [4–9]. Each of them exhibits a specific response to external stimulus, which can be used for a variety of applications such as in sensors, actuators, energy harvesting, and the biomedical field [2].

The mechanoelectrical transduction capability of piezoelectric materials, which converts mechanical energy into electrical energy or vice versa, allows us to accurately detect physical parameters, such as pressure, mass, or acceleration [10], being therefore applied in a large variety of devices; such devices include force, touch, and strain sensors, self-powered devices, acoustic and ultrasonic tools, precision positioning instruments, fluid injection systems, and damping controllers [9]. All of these applications are based on the piezoelectric effect, which was discovered in 1880 by the Curie brothers [11].

Materials showing the piezoelectric effect are typically clarified as a function of the piezoelectric coefficients d_{ij}, quantifying the mechanical to electrical conversion. This is exemplified in figure 2.1.

$BaTiO_3$, PZT, and related materials are the most used piezoelectrics due to their highest piezoelectric response (figure 2.1). Nevertheless, some disadvantages, such as mechanical brittleness or high rigidity, hinder their application in devices that require high flexibility [12, 13]. To solve such issues, polymer-based piezoelectric composites have emerged. Within all possible composites, the ones with 0–3 connectivity, i.e. randomly dispersed piezoceramic particles in a nonpiezoelectric polymer matrix, exhibit the lowest piezoelectric response (20–30 times less) when compared with polymer–ceramic composites with 1–3 connectivity, i.e. ceramic fibers in a nonpiezoelectric polymer matrix, or 2–2 connectivity, i.e. ceramic laminates in a nonpiezoelectric polymer matrix [12].

More recently, piezoelectric polymers such as PVDF and its copolymers, such as poly(vinylidene fluoride-*co*-trifluoroethylene) (PVDF–TrFE), are being increasingly

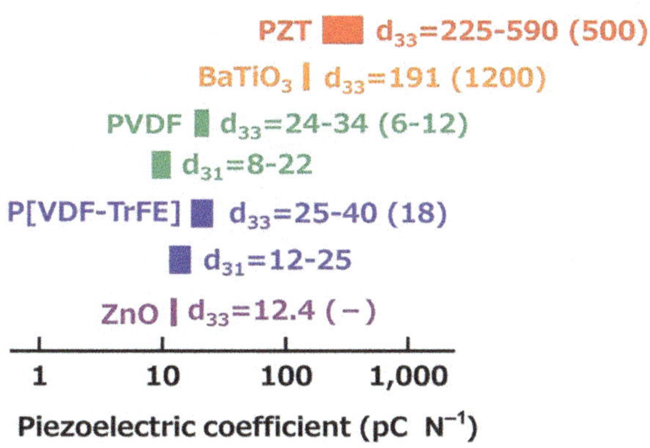

Figure 2.1. Piezoelectric coefficient ranges of various piezoelectric materials; d_{33} and d_{31} are the moduli of the longitudinal and transverse piezoelectric coefficients. The values in parentheses correspond to the relative permittivity. Numerical figures are represented as absolute values. Reproduced with permission from Narita *et al* [12].

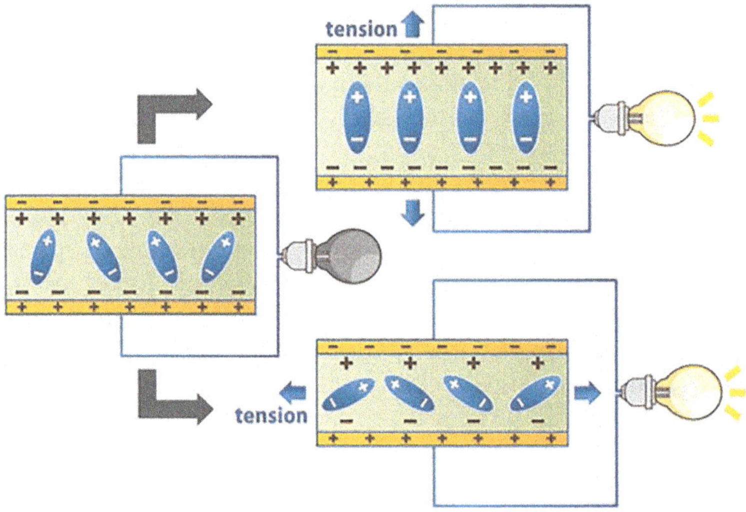

Figure 2.2. Piezoelectric effect for energy harvesting. Image taken with permission from Narita *et al* [12].

used due to their flexibility, high elastic compliance, strong electrical dipole moment of the PVDF monomer unit (5–8 × 10^{-30} C m), high chemical and radiation resistance, light weight, additive manufacturing compatibility, and low cost [2, 14].

Nowadays, piezoelectric materials have become increasingly explored to effectively harvest energy from different mechanical sources such as human motion, low-frequency seismic vibrations, and acoustic noise (figure 2.2). The harvested energy is being used to power sensors, biosensors, and actuators [15], and is particularly suitable for powering wearable electronic devices [16].

2.2 Piezoelectricity: fundamentals and energy generation

The word piezoelectricity comes from the Greek *piezein*, which means 'to press', and therefore can be understood as the capability of specific materials to develop an electrical voltage under the application of a certain mechanical stress.

The piezoelectric effect is strongly related to the crystalline structure of materials. From the 32 crystal classes, 21 are noncentrosymmetric, 20 of them showing piezoelectricity. Of the 20 piezoelectric classes, half of them are pyroelectric, developing a voltage as a response to temperature variation, and some of them can also be ferroelectric (the dipolar moment can be reversed in an applied electric field) [17].

The piezoelectricity is thus established in nonsymmetric crystals with polar structures, and can be described as the capability of these materials to develop an electrical voltage under the application of a mechanical stimulus; or, the opposite: changing their geometry under an electric voltage field. The piezoelectric effect can be described as direct or inverse. The direct effect takes place when the material is subjected to external mechanical stimulus and the dipoles are oriented to cause the development of positive and negative charges on each surface, resulting in an electric field within the material. The inverse piezoelectric effect describes the geometric changes of the crystal when it is subjected to an external electric field. The reorientation of the dipolar moments under electric field causes the deformation of the crystal, which is linearly proportional to the electric field applied [18].

The direct and inverse piezoelectric effects were discovered in 1880 (by Pierre and Jacques Curie) and 1881 (by Lippmann), respectively, and can be described by the following constitutive equations [18]:

$$D_i = d_{ikl}T_{kl} + \varepsilon_{ik}^{T}E_k, \tag{2.1}$$

$$s_{ij} = S_{ijkl}^{E}T_{kl} + d_{kij}E_k, \tag{2.2}$$

where the subscripts i, j, k, and l take values of 1, 2, and 3. S and T are the stress and strain tensors, respectively. The electric displacement and field vectors are D and E, respectively. Furthermore, the elastic compliance matrix at a constant electric field is s, d is the piezoelectric strain coefficient, and ε represents the dielectric permittivity at constant stress. In both equations, d represents the charge variation induced by a given force in the absence of an electric field (short-circuit electrical condition), or the deflection caused by an applied voltage in the absence of an applied force (stress-free mechanical condition) [18], as illustrated in figure 2.3.

2.2.1 Piezoelectric coefficients

The most used piezoelectric coefficients are d_{31} and d_{33}, i.e. the transversal and longitudinal coefficients, respectively (figure 2.4). They are typically represented in units of meters per volt (m V^{-1}) or coulombs per newton (C N^{-1}). Each of the coefficients describe the electric polarization generated in one direction due to the application of a transverse or parallel applied stress.

A piezoelectric material converts mechanical force into electrical energy (or vice versa) and the conversion efficiency is quantified by the electromechanical coupling

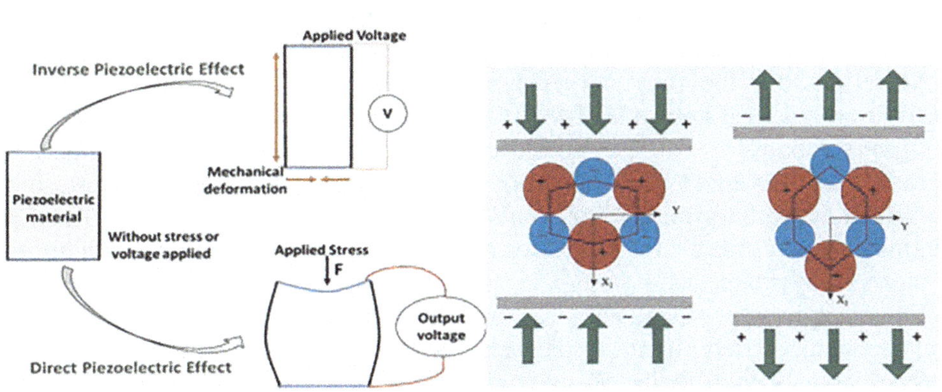

Figure 2.3. (a) Schematic representation of the direct and inverse piezoelectric effects. (b) Illustration of charges induced by the direct piezoelectric effect under mechanical compressive and tensile stress. Adapted with permission from [19].

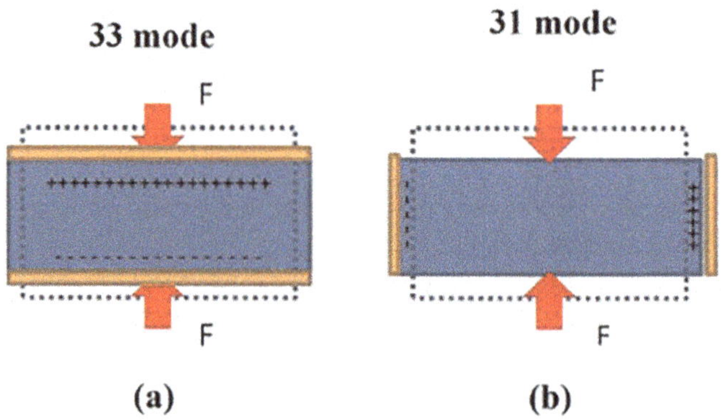

Figure 2.4. The piezoelectric transduction modes. Adapted with permission from [20].

coefficient, k, being closely related to the bandwidth of resonant devices. This coefficient is expressed as [21]

$$k_{ij} = \frac{d_{ij}}{\sqrt{\varepsilon_{ii}^{\sigma} S_{jj}^{E}}}. \tag{2.3}$$

2.2.2 Energy harvesting

Energy harvesting focuses on the recovery of environmental energy that would be otherwise wasted from sources such as force/pressure, heat, and vibration/movement. Typically, that energy can be used for powering low-power devices, in particular for applications in remote or inaccessible zones, where battery replacement is complicated or even impossible [22].

Piezoelectric, thermoelectric, and electromagnetic systems are the most used systems to harvest energy from wasted sources, and the harvested power typically ranges from nano- to microwatts per cm^2.

Vibrations are the most targeted source for energy harvesting using microelectromechanical systems (MEMS). The most used mechanical to electrical transduction mechanisms are piezoelectric, electrostatic, and electromagnetic [23]. Ceramics are the most used materials for this application, independently of the transduction mechanism, leading to μW of output harvested power [23]. However, ceramics are not suitable for large area applications due to high cost and limitations in mechanical properties.

Thus, polymers are being explored, and, in particular, PVDF and its copolymers are becoming widely used in piezoelectric applications. Further, PVDF-based composites with high dielectric materials (ceramics or single crystals) allow us to combine the mechanical properties of polymers and the high dielectric constant of brittle ceramic materials. Polymer-based piezoelectric devices can produce output voltages and currents near some hundreds of volts and tens of micro amperes [22–25].

2.3 Introduction to electrospinning technique

The development of nanotechnology platforms allowing us to produce nanofiber materials has a strong influence in different areas of science and engineering, contributing to society with applications in electronics, energy storage, renewable energy, water and air filtration, oil and gas industries, the food industry, the pharmaceutical industry, healthcare and consumer goods, among others [26, 27].

There are different methods for the fabrication of nanofibers, such as electrospraying [28], self-assembly [29], sol–gel [30], rotary jet spinning [31], electrospinning (allowing the production of random, aligned, core–shell, and vertical nanofibers) [32–34], phase separation [35] and melt-blown protocols [36], among others.

Electrospinning, due to its high efficiency, has emerged as one of the preferred techniques for nanofiber production. The electrospinning process is continuous and scalable. A large variety of materials can be processed by this technique with outstanding control over nanofiber dimensions and orientation. Additionally, the synthesis of nanoparticles by electrospraying overcomes limitations associated with other fabrication processes, such as emulsion/evaporation, coacervation, spray-drying, nanoprecipitation, and microfluidic methods. Therefore, synthesis of nanomaterials by electrospinning or electrospraying is currently an essential component of nanomaterial research [37].

Electrospinning and electrospraying are technologies where a polymeric solution can be spun or sprayed by the application of a high electric field to obtain fibers or spheres, respectively [38, 39]. These electrodynamic processes (figures 2.5(a) and (b)) use almost the same setup and equipment, which consists of four main components [38, 40]:

I. a high-voltage source (typically from 1–30 kV) mostly operated in DC mode, though an AC mode is also possible;
II. a blunt-ended stainless-steel needle or capillary;
III. a syringe pump;
IV. a grounded collector, either a flat plate or rotating drum.

In polymer electrospinning, a strong electric field is applied to a droplet formed by a polymer solution or polymer melt at the tip of a die, which acts as one of the

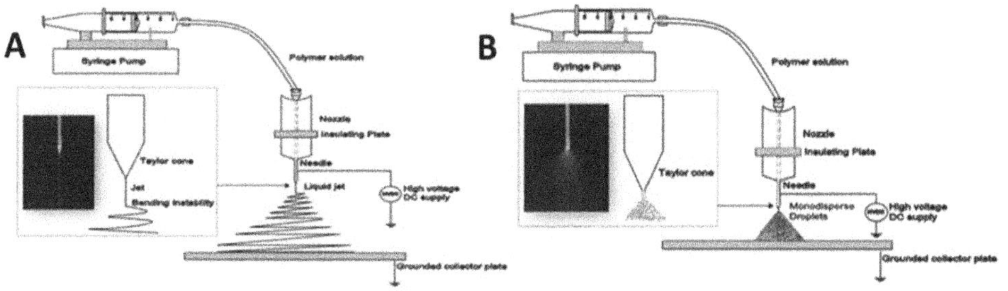

Figure 2.5. A typical laboratory scale, (a) electrospinning setup, and (b) electrospraying setup. Images taken with permission from [39].

electrodes. The charging of the fluid leads to a conical deformation of the droplet (Taylor cone) and finally to the ejection of a jet from the tip of the cone [41].

The charged jet is accelerated toward the counter electrode and thins rapidly, due to the elongation and evaporation of the solvent, until solid fibers are deposited into the substrate located on top of the counter electrode [42]. This process is characterized by a rapid and strong elongational deformation of the spinning jet due to a bending instability, which occurs during the course of the fiber formation. The deformation of the jet normally gives rise to nanofibers that display a strong orientation of the chain molecules and crystals. Such orientations cause significant increases in the mechanical stiffness and strength of the fibers [41].

Electrospraying, as illustrated in figure 2.5(b), is a variation of the electrospinning method, and was initially applied in medicine with the aim to produce materials able to encapsulate drugs [43]. Vonnegut and Neubauer [44] were pioneers in applying voltage to produce micrometer-sized spherical particles for application as sprays. Today, this technology is expanding due to its simplicity and low cost [43].

The major distinction between electrospinning and electrospraying is the polymer content in the polymeric solution. When the solution concentration is high, the jet from the Taylor cone is stabilized, and elongation occurs due to whipping instability [45]. If the solution concentration is low, the jet is destabilized and fine droplets are formed. These highly charged droplets are self-dispersed in space, preventing droplet agglomeration and coagulation.

Furthermore, evaporation of the solvent leads to contraction and solidification of droplets, resulting in solid polymeric particles deposited on the grounded collector [39].

Figure 2.6 shows some of the advantages of electrospun and electrosprayed products.

Nanofibers show interesting electric, thermal, optical, magnetic, and mechanical properties due to effects related to their small size, increased interface, and surface area. Furthermore, nanofibers prepared by electrospinning shown high surface

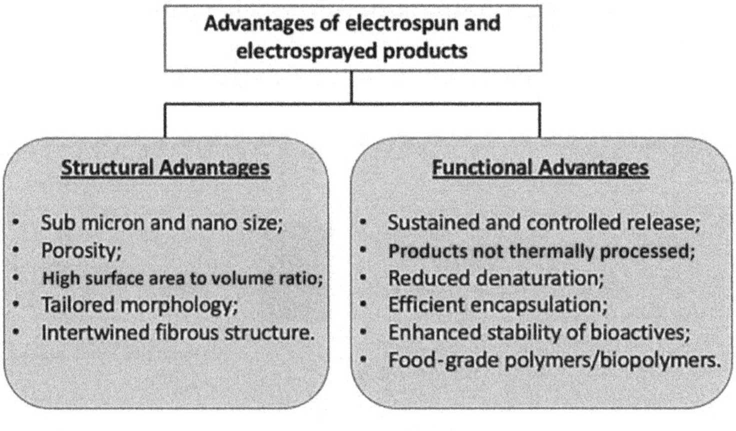

Figure 2.6. Main advantages of electrospun and electrosprayed materials.

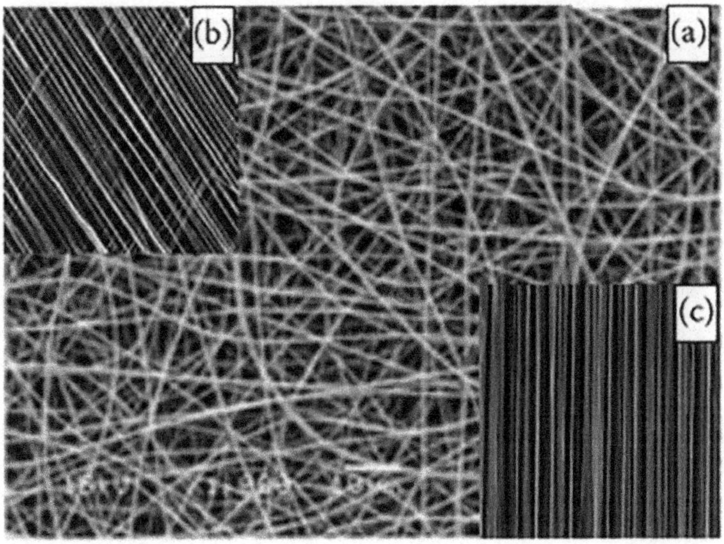

Figure 2.7. Scanning electron micrographs of electrospun (a) random nanofiber, (b) aligned fibers at a specific angle, and (c) aligned fibers. Image taken with permission from [47].

area-to-volume ratio, high overall porosity of the fiber mats, and a unique web structure that are essential, for example, for energy conversion and storage applications [46].

Electrospinning also allows the production of nanofibers with different orientations and hierarchical structures. The orientation of nanofibers includes randomly oriented and aligned structures as shown in figure 2.7 [47].

Particularly, well-aligned nanofibers are often required for energy harvesting devices or scaffolds for some tissue engineering applications [48, 49]. The fiber collector plays an essential role in obtaining nanofiber orientation. Thus, several methods have been developed to enhance alignment of electrospun nanofibers by using special designed collectors, such as rotating drums [50], metal frames [51], or two conductive substrates separated by a gap [52]. In addition, a liquid system combined with a rotating mandrel has been developed to produce a continuous yarn of electrospun fibers [53].

Nowadays, a large variety of polymers have been successfully electrospun into ultrafine fibers with diameters ranging from <3 nm to over 1 μm [54].

2.3.1 Electrospinning of piezoelectric materials

Polymers are typically dissolved into solvents before electrospinning, and the same polymer can be dissolved in different solvents to tune electrospun processing conditions and therefore final fiber mat properties. A summary of some relevant polymers that have been successfully electrospun into fibers is presented in table 2.1 [37] together with the used solvents, polymer concentrations, and proposed or perspective applications of the corresponding fibers.

Table 2.1. Representative polymer that has been electrospun from solution.

Polymer	Solvent	Concentration	Application	Ref.
PLA	Dichloromethane	5 wt%	Sensor, filter	[56]
PANI/PS	Camphor sulfonic acid	2 wt%	Conductive fiber	[57]
Silk-like polymer with fibronectin functionality	Formic acid	0.8–16.2 wt%	Implantable device	[58]
PVDF	Dimethylformamide:	20 wt%	Flat ribbons	[59]
	dimethylacetamide	20% (w/w)	Biomedical	[60]
CA	Acetone, acetic acid, dimethylacetamide	12.5%–20%	Membrane	[61]
PLGA	Tetrahydrofuran: dimethylformamide	1 g/20 ml	Tissue engineering	[61]
Nylon-4,6, PA-4,6	Formic acid	10 wt%,	Transparent composite	[62]
EVA	Isopropanol/water: 70/30 (%v/v)	2.5%–20% (w/v)	Biomedical	[63]
PVP	Tetrahydrofuran	20, 60% (w/v)	Antimicrobial agent	[64]
PAA–PM	Dimethylformamide	26 wt%	Optical sensor	[65]
PS	Tetrahydrofuran	15 wt%	Catalyst, filter	[66]
PEO	Distilled water: ethanol	4 wt%	Microelectronic wiring, interconnects	[67]

Abbreviation: Polylactic acid, PLA; polyaniline, PANI; polystyrene, PS; cellulose acetate, CA; polylactic-*co*-glycolic acid, PLGA; poly(ethylene-*co*-vinyl alcohol), EVA; polyvinyl phenol, PVP; poly(acrylic acid)–poly (pyrene methanol), PAA–PM.

Among polymers, piezoelectric polymers stand out as an interesting material for a new generation of applications including wearable smart materials, energy harvester devices, sensors, and actuators [55].

From all the piezoelectric polymers, PVDF and its copolymers are the ones with larger piezo, pyro, and ferroelectric responses. PVDF has five crystalline phases (figure 2.8), and among these five crystalline phases, the α phase is nonpolar [68] and the β phase is polar, providing the highest electroactive response [41, 69]. The polar β phase can be obtained through mechanical stretching [70, 71], low-temperature solvent evaporation [72], high electric field poling [73], the addition of specific fillers including ionic liquids [74], ferrites [75] or other conductive additives [76, 77], among others [2, 78, 79].

The formation of continuous fibers from PVDF solutions or melts under strong electrostatic fields allows the production of ultrafine piezoelectric PVDF nanofibers. Thus, randomly oriented electrospun PVDF nanofibrous membranes lead to a piezoelectric output of 140 mV under 5 N loading [80]. Aligned PVDF nanofibers on a polyethylene terephthalate (PET) substrate under an applied loading frequency of 7 Hz generate an average peak voltage output value of 76 mV [81]. Additionally, a piezoelectric unit based on two outermost polydimethylsiloxane (PDMS) layers,

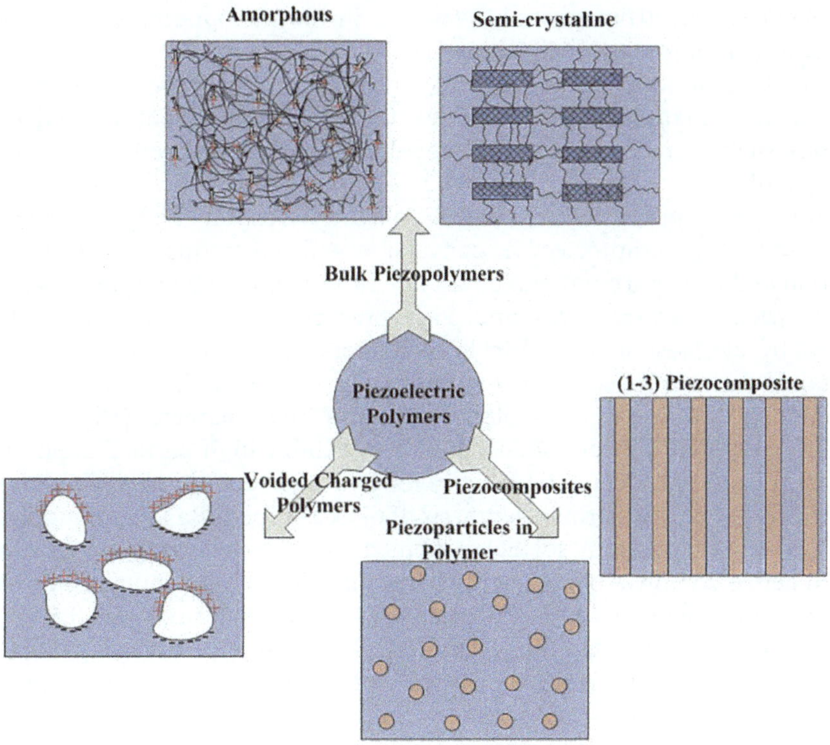

Figure 2.8. Schematic representation of piezoelectric polymer types: bulk piezopolymers that can be divided into amorphous dipolar polymers and semicrystalline polymers; polymer composites, with examples of fillers in the form of particles or columns within the polymer (known as 0–3 and 1–3 connectivity, respectively); and voided charged polymers, with surface-charged air voids within the polymer [21].

two aluminum electrodes, and a PVDF fibrous mat in the center reached a piezoelectric sensitivity of 178 mV kPa^{-1} [82].

Regarding inorganic piezoelectric materials, PZT is among the most used in sensors and actuators, nonvolatile ferroelectric memory devices, MEMS, and nanogenerators [83]. It shows, in general, higher performance than organic piezoelectric [83]. Nevertheless, synthetized bulk PZT ceramics are brittle materials and there is a growing demand for flexible PZT nanofiber materials [84]. Electrospinning is one of the most used methods to fabricate long fibers with uniform diameter and varied composition [85].

2.3.2 Energy harvesting based on electrospun nanofibers

With the limited supply of fossil fuels and the increasing demand of sustainable energy for economic development, it is essential to develop renewable energy technologies to sustain economic evolution. Capturing residual and typically wasted energy available in the environment produces an interesting possibility to power specific devices. The so-called energy harvesting involves collecting energy from vibrations, heat, electromagnetic waves, etc. in the surrounding environment, and converting them into electrical energy [12].

Due to their unique mechanical, electrical, and active properties, electrospun nanostructured materials are attractive for the development of active energy harvesters such as piezoelectrics, photovoltaics, hydrogen energy generators, and fuel cells [86].

As for their applications in solar cells, electrospun TiO_2 nanofibers have demonstrated high photoelectric conversion efficiency due to effective charge separation and transport and maximum light absorption, which is mainly attributed to high specific surface areas and high porosity [87]. Furthermore, electrodes prepared by electrospun Gd_2MSbO_7 nanofibers have demonstrated high specific capacitance and improved cycling stability due to their unique fiber morphology, including large surface area-to-volume ratio and small diameter [88].

Additionally, TiO_2 electrospun nanofibers exhibit high surface area, enhanced crystallinity, and photocatalytic activity for hydrogen evolution when compared to those made by hydrothermal synthesis [86]. Ce-doped $BaTiO_3$ nanofiber-based harvesters were prepared by sol–gel combined with the electrospinning method. The reported delivered power of the energy harvesting device was optimized to 14.37 μW with a load resistance of 100 MΩ when the Ce/Ba atomic ratio was 0.6%.

Regarding polymer-based fibers for piezoelectric energy harvesting [89] (PVDF–TrFE), nanofibers have been developed with diameters ranging from 1000 to sub-100 nm and 26 nW power output.

Aiming at improving the biocompatibility and biodegradability of the developed energy harvesters, poly(L-lactic acid) (PLLA) piezoelectric nanofibers were developed, demonstrating that an open-circuit voltage and short-circuit current with a strain deformation angle of 28.9° could reach 0.55 V and 230 pA [90]. The optimized prototype harvester worked well for harvesting human joint motion energy with a maximum peak power of 19.5 nW.

Regarding composites, $BaTiO_3$ nanoparticles embedded in polyurethane and PVDF–TrFE nanofibers were developed. The resulting nanofibers showed a high stretchability (40%) and high mechanical durability (9000 stretching cycles at 30% strain), producing a peak open-circuit voltage (V_{oc}) and short-circuit current (I_{sc}) of 9.3 V and 189 nA, respectively. When placed over the knee cap of a subject when walking, the harvester generated a maximum V_{oc} of 10.1 V.

Different contents of PZN–PZT nanoparticles of ≈54 nm were embedded into a matrix of PVDF–TrFE by electrospinning [91], reporting a 3.4 V output voltage and 240 nA output current for 20 vol% nanoparticles incorporated into the polymer matrix.

Layers of MoS_2 were grown on cellulose paper by the hydrothermal method followed by the deposition of *in situ* poled PVDF nanofibers that generated a peak voltage of 50 V, short-circuit current of 30 nA, and average power of 0.18 mW cm^{-2} [92].

There are more than 200 piezoelectric materials that have been reported to be able to convert mechanical energy into energy, including quartz, $BaTiO_3$, lead titanate (PT), cadmium sulfide, PZT, lead lanthanum zirconate titanate (PLZT), lead magnesium niobate ($MgNb_2O_9Pb_3$; PMN), PVDF, polyvinyl fluoride, among others [93]. They allow for a large variety of functional responses and tailored processability, including, for some of them, in the form of electrospun fibers.

2.4 Piezoelectric materials

Synthetic materials are chiefly used in commercial applications due their higher figures of merit in terms of functional response. Materials with piezoelectric properties can be divided into different categories: single crystals, ceramics, polymers, and polymer composites (usually ceramics embedded into polymer matrices). Most energy harvesting applications use piezoelectric single crystals and ceramics, or composite materials, since polymers show lower piezoelectric properties. However, polymers can sustain higher strain due to their intrinsic flexibility and can be processed in a larger variety of forms and shapes by low-temperature methods; conversely, ceramics and single crystals are too rigid and brittle.

2.4.1 Piezoelectricity in polymers

There are different piezoelectric polymer categories: bulk, composites, and charged polymers (figure 2.9) [21]. The first category of piezoelectric polymers are the solid films with intrinsic piezoelectric properties [21]; the second category, polymer composites, are based on polymers reinforced with piezoelectric micro- and nanomaterials (mostly ceramic materials) [21]. These materials make use of the mechanical properties of polymers (flexibility or stretchability, low density, etc) and the high dielectric and piezoelectric properties of the fillers. Finally, the third category is related to voided charged polymers, in which the electroactive response

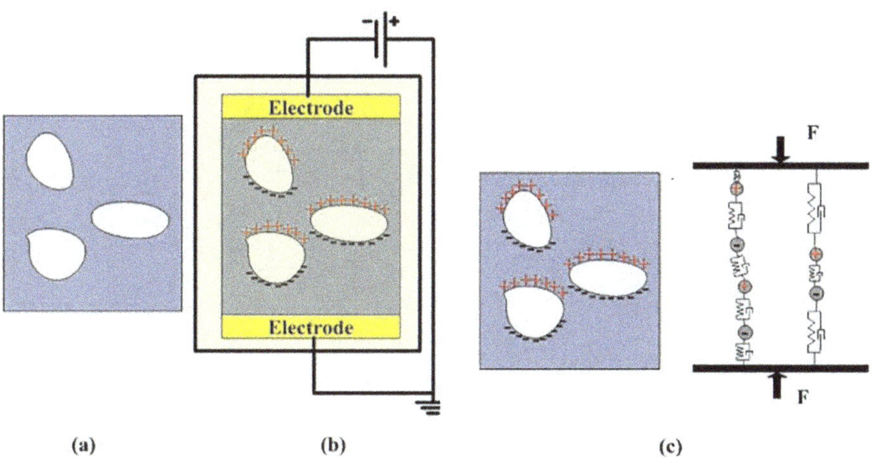

Figure 2.9. Illustration of piezoelectricity in voided charged polymers: voided polymer before charging (left), poling process to form the trapped dipoles (center), and charged polymer (right). Adapted from [21].

has a different origin: in these polymers, gas voids develop in the polymer and their surfaces are electrically charged to obtain internal dipoles.

Bulk piezoelectric polymers mainly consist of amorphous or semicrystalline polymers. These polymers must present molecular dipoles and these dipoles must be reoriented, by poling, within the bulk material and kept in their preferred orientation state [21].

Piezoelectric composites are developed using a polymer reinforced with piezoelectric inorganic particles. The polymer matrix can also be piezoelectric, increasing the coupling factor and dielectric properties of the composite [21]. The manufacturing of the polymer composites can lead to randomly oriented or specifically oriented fillers (figure 2.9).

Voided charged polymers are characterized by larger dielectric responses and a d_{33} larger than 20 000 pC N^{-1}, even larger than that of ceramic materials [21]. This structure was first presented in the early 1960s (figure 2.10) [94], based on the theoretical concept of embedding air voids within the polymer and poling the material to create internal charging in torn of the polymeric voids. By applying a large electric field in the polymer, the gas molecules within the voids get ionized and opposite charges are implanted on each side of the voids [94], leading to artificial dipoles that respond like piezoelectric materials to an external electric field or mechanical stimulus [21].

2.4.1.1 Main properties of piezoelectric polymers

Pioneer works in the 60s by Kawai in piezoelectric polymers led to strong research activity in PVDF and its respective copolymers [95]. These electroactive polymers are the most interesting ones used in sensors, actuators, energy harvesting, and biomedical applications [95].

Polymers present as their main advantage their mechanical properties, and as their main disadvantage low piezoelectric coefficients (usually, $d_{33} \leqslant 30$ pC N^{-1}). However, their piezoelectric stress constant is higher when compared to other

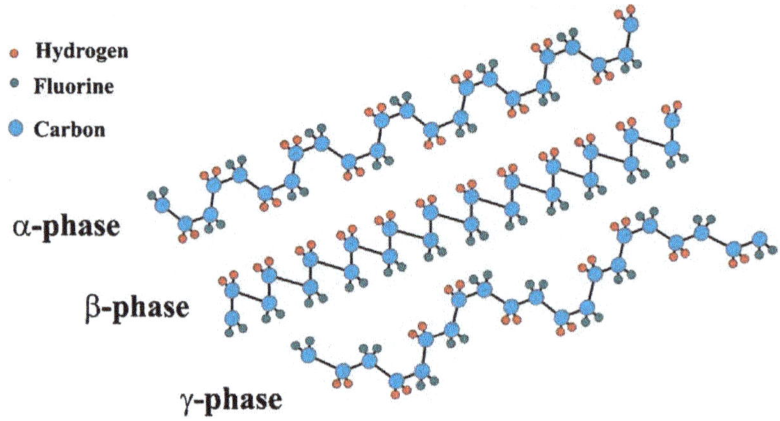

Figure 2.10. Schematic representation of the chain conformation for the α, β, and γ phases of PVDF [14].

piezoelectric materials, and thus are suitable for sensor devices. Other relevant advantages of polymers are their relatively low-cost processing, light weight (low density), low dielectric constant, and high dielectric breakdown.

2.4.1.2 Amorphous piezoelectric polymers

Amorphous polymers show lower piezoelectric properties when compared to semicrystalline materials. The piezoelectric effect in amorphous polymers is based on the fact that the polarization is rather in a quasistable state due to the freezing-in of molecular dipoles, and, therefore, it is not in a state of thermal equilibrium [18]. Thus, the application of an external electric field is generally required in order to reorient the molecular dipoles of the amorphous polymer. This polarization procedure must be applied above the glass temperature transition, where the dipoles have larger mobility [96].

Piezoelectric amorphous polymers include poly(acrylonitrile) (PAN), poly(vinylidene cyanide–vinyl acetate) (PVDCN/VAc), poly(phenyl ether nitrile), and poly(1-bici-clobutanecarbonitrile). The most studied amorphous polymers are vinylidene cyanide and vinyl acetate copolymers, which present suitable piezoelectric responses and large dielectric relaxation strengths [18].

The remaining amorphous polymers with piezoelectric properties present lower properties than semicrystalline polymers [96]. PVDCN/VAc and PAN present a similar structure and the largest remnant polarization and piezoelectric stress when compared to other polymers [96].

2.4.1.3 Semicrystalline polymers

Semicrystalline polymers present interesting piezoelectric properties, and several of them are already used in commercial applications.

2.4.1.3.1 PVDF and PVDF copolymers

The most studied and used semicrystalline polymer are PVDF and its copolymers since they are still the electroactive polymers with the best overall performance [14, 21].

PVDF is synthesized by the polymerization of CH_2–CF_2, the monomer of PVDF, showing a high electrical dipole moment of 8×10^{-30} C m. It can present different crystalline phases, the main ones being the nonpolar α phase (TGTG′—trans-gauche–trans-gauche—conformation) and the polar β phase (TTT—all trans—planar zigzag conformation) [14, 97] (figure 2.10). The highest piezoelectric response, ranging from −24 to −34 pC N^{-1} [14, 21] is obtained in the β phase. Other crystalline phases of PVDF include the δ (TGTG′—trans-gauche–trans-gauche), ε, and γ (T_3GT_3G') phases. The degree of crystallinity of PVDF ranges typically between 50% and 60%, and the glass transition temperature is around −35 °C [98].

PVDF in the polar β phase can be obtained by mechanical deformation of α-PVDF in order to obtain orientation of the molecular chains [18], by electrospinning [60], or by developing composites with $BaTiO_3$ [99], organic clays [100], ionic liquids [101], magnetic fillers [102], or other materials.

Copolymers and terpolymers of PVDF have been developed to tailor the electroactive response of the polymer [103]. Poly(vinylidene fluoride trifluoroethylene) (PVDF–TrFE) is synthesized through free radical polymerization of the two monomers [18]. It shows lower ferroelectric to paraelectric transition temperature than PVDF, and the transition temperature decreases with increasing TrFE contents in the copolymer [103]. The main advantage of copolymer PVDF–TrFE is the TrFE contents. It crystallizes directly in the electroactive phase, either by processing from solvent or from the melt. Furthermore, PVDF–TrFE shows larger electroactive responses, d_{33} and k_{33}, than PVDF [21].

Other copolymers include poly(vinylidene fluoride-*co*-hexafluoropropene) (PVDF–HFP) [104], poly(vinylidene fluoride–chloride trifluoride ethylene) [105], and poly(vinylidene fluoride–bromotrifluoroethylene [106, 107].

2.4.1.3.2 Other semicrystalline piezoelectric polymers

Thermoset polymers such as polyurea are constituted by hard and soft domains with microphase separation between them. Polyurea presents excellent overall properties, including stretchability, impact mitigation, chemical resistance, and a wide range of operating temperatures. Polyurea films are usually produced by vapor deposition polymerization in vacuum, exhibiting a piezoelectric constant of $d_{31} \approx 10$ pC N^{-1}, which is stable up to 200 °C [108].

Other semicrystalline polymers with excellent piezoelectric response in a wide temperature range are polyamides (known as odd-numbered nylons, including nylon-5, nylon-7, and nylon-11). The odd nylons present a monomer unit consisting of even numbers of methylene groups and one amide group with a dipole moment [108]. The amide dipoles align synergistically for the odd-numbered monomer, resulting in a net dipole moment of 4.9 debye and a piezoelectric constant above the glass transition temperature of $d_{31} \approx 17$ pC N^{-1}.

In addition to these polymers, electroactive biopolymers are a class of materials with applications in the biomedical area. Biopolymers show piezoelectric properties lower than those of the other polymer materials (less than ≈ 8 pC N^{-1}), with the piezoelectric effect mostly attributed to internal rotation of polar atomic groups linked to asymmetrical carbon atoms.

Biopolymers can be natural or synthetic. Natural ones have poor mechanical properties and are difficult to process. Natural biopolymers include DNA-oriented films, collagen, polypeptides like poly(γ-methylglutamate) and poly(γ-benzyl-L-glutamate), the poly(α-hydroxy acid)s family, and chitin [18]. Synthetic biopolymers include the poly(α-hydroxy acid)s family and its copolymers [18, 109].

2.4.1.4 Polymer-based composites

Piezoelectric composites are typically fabricated based on a polymer matrix due to their mechanical properties, and are reinforced by ceramic fillers with high dielectric response. Furthermore, the polymer can also be piezoelectric, allowing us to improve the overall piezoelectric properties of the composite.

One of the key parameters in a composite is its dimensional connectivity; figure 2.11 shows the 0–3 and 1–3 connectivity, which are the most used in device

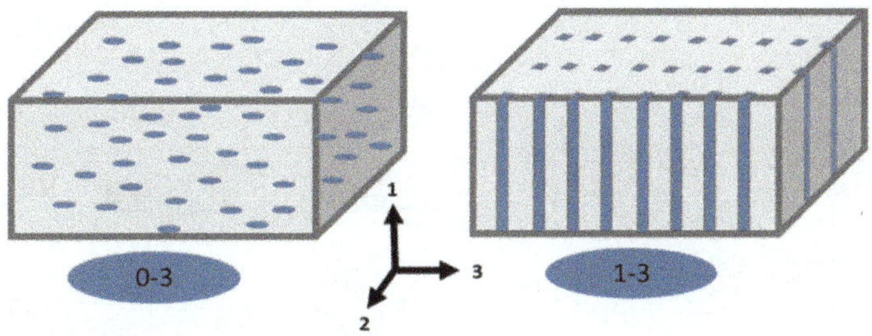

Figure 2.11. Piezoelectric composite material structures with 0–3 and 1–3 connectivity with polymer as matrix (gray) and dielectric fillers (blue spheres and bars).

applications. A composite with *m–n* connectivity has its piezoelectric phase connected in *n* dimensions for the polymer phase and in *m* dimensions for the filler [18].

The 0–3 connectivity is the most commonly used, and the challenge is obtaining larger filler contents in a polymeric matrix without losing mechanical properties. Homogeneous filler dispersion is the key in this type of composite. The two most common methods to produce 1–3 piezoelectric materials are the dice-and-fill and laminate-and-cut procedures [110].

Using two (or more) materials for the development of composites has advantages and disadvantages. On the one hand, there is the low cost, light weight, and flexibility of the polymers combined with the higher dielectric properties of the ceramics or single crystal materials; on the other hand, the thermal and chemical stability (for device applications) can be a limitation of composites.

Typical examples of ceramics used in polymer-based piezoelectric composites are $BaTiO_3$, PZT, and lead titanate doping lead magnesium niobite (PMN–PT) [18, 111]. These composites have been largely applied as ultrasonic transducers [18].

Regarding the polymers, elastomers and thermoplastics are widely used in piezoelectric composites due to their mechanical and dielectric properties [112].

2.4.1.5 Piezoelectric ceramics and single crystals

Piezoelectric materials with larger coefficients are ceramics, and single crystals and can be natural or man-made. One of the most abundant minerals on earth and a common natural piezoelectric material is quartz (SiO_2). Natural piezoelectric materials can be divided into two groups: minerals and organic materials. Quartz and Rochelle salt are the representatives of the first group, whereas bone and wood are the most known of the organic group.

In recent years there have been increasing studies on novel nanostructures, including nanoparticles, nanorods, nanotubes, nanowires, and nanoplates, which are expected to improve the piezoelectric properties of materials.

2.4.1.5.1 Piezoelectric ceramics

Piezoelectric ceramics are the most-used materials for practical applications. Ferroelectric oxides for the development of piezoelectric ceramics are classified into perovskite-type, tungsten bronze-type, and bismuth layer-structured compounds.

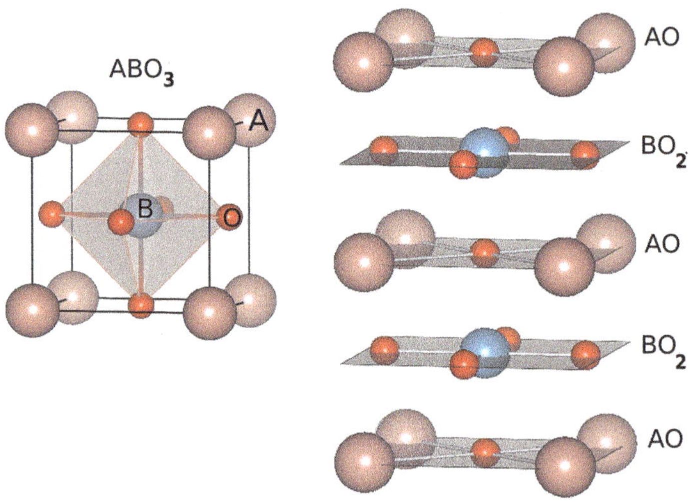

Figure 2.12. Left panel: The perovskite structure of general formula ABO_3. Right panel: The layered stacking of AO and BO_2 planes in a perovskite structure. Adapted from [113].

$BaTiO_3$ and PZT present high piezo- and dielectric properties, and they can be further improved by using specific dopants. Moreover, PZT and $BaTiO_3$ show high electromechanical coupling factors and display large bandwidth and sensitivity [18].

The structure of both ceramics is based on an ABO_3 perovskite structure, as shown in figure 2.12. Perovskite ABO_3 oxides structurally consist of (1 0 0) layers of AO and BO_2 with corner-sharing oxygen octahedra linked together in a cubic array with particular cations (Ti, Zr, Sn, Nb) occupying the central octahedral B-site, and larger cations (Pb, Ba, Sr, Ca, Na) filling the interstices between octahedra in the larger A-site.

The perovskite structures can be divided into three classes: $A^{I}B^{V}O_3$, $A^{II}B^{IV}O_3$, and $A^{III}B^{III}O_3$, where the Roman nomenclature represents the chemical valence of the cations.

PZT ceramic is a representative piezoelectric material used in microelectronic devices due to its large piezoelectric coefficients ($d_{33} \leqslant 1000$ pC N^{-1}) [114]. Zirconate (Zr) is essential in the crystalline symmetry of PZT that presents a tetragonal ferroelectric phase with a perovskite structure. The ratio between Zr/Ti near 52/48 leads to a structural change from a tetragonal to a rhombohedral phase [115]. The morphologic phase boundary (MPB) delimits both structural phases and, in the limit, both tetragonal and rhombohedral phases coexist (figure 2.14) [115]. The MPB state allows the maximum domain reorientation during poling, with larger remnant polarization in the materials [18]. By varying Zr and titanate content in the PZT ceramic, $Pb(Zr_{1-x}Ti_x)O_3$ can be antiferroelectric (at 230 °C) ($PbZrO_3$) and ferroelectric ($PbTiO_3$). As mentioned above, the Zr/Ti ratio of 52/48 leads to larger dielectric properties (figure 2.13), and it is thermally stable, with a Curie temperature (T_C) near 390 °C [116]. PZT ceramic can be doped with different ions (donor, acceptor, or isovalent). Donor- or acceptor-doped materials are known as 'soft' or

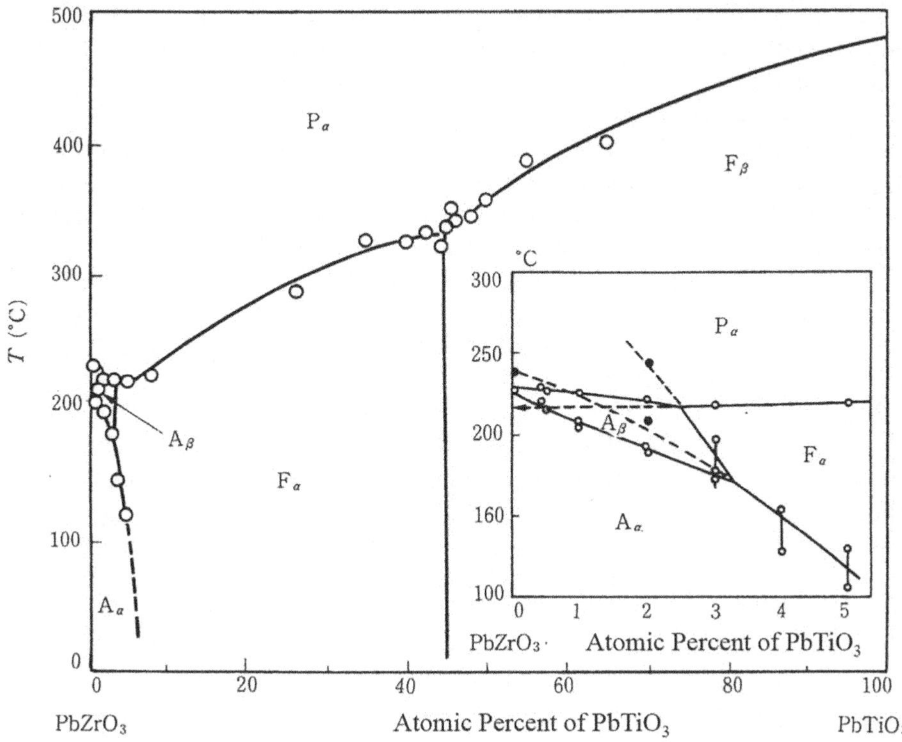

Figure 2.13. Phase diagram for the Pb(Zr,Ti)O_3 solid solution system [117].

'hard' PZT. An odd number in PZT, such as in PZT-5, creates Pb vacancies that allow domain motion and are used in high-sensitivity applications such as hydrophones or loudspeakers [116]. An even number in PZT, such as in PZT-8, are due to oxygen vacancies restricting the motion of the domain walls, and are used in high-power devices such as sonar or ultrasonic transducers [116]. Isovalent doping changes the ions of the same valence and may result in a decrease in the Curie temperature and an increase in the dielectric constant [116].

Lanthanum was found to improve the piezo- and dielectric properties of the PZT ceramics in ternary material; such a compound is called PLZT. PLZT ternary ceramics have been investigated for different lanthanum contents and Zr/Ti ratios, which exhibit a variety of ferric phases and cover all aspects of piezo-, pyro-, and ferroelectric ceramic properties [118]. Compared to undoped PZT, PLZT presents larger piezo- and dielectric values; however, the thermal properties (such as the Curie temperature) decrease with increasing lanthanum content in the ternary ceramic [119].

Another critical property of PZT (or other ceramics so-derived) is the toxicity of lead, and a lead-free line of investigation continues to discover environmentally friendly ceramics [18]. Among the lead-free materials, some based on $BaTiO_3$ perovskite-structured piezoelectrics, are included $(Bi_{1/2}Na_{1/2})TiO_3$, $(Bi_{1/2}K_{1/2})TiO_3$, potassium niobate ($KNbO_3$ or KN), or (K,Na) NbO_3, and their solid solutions have been actively studied as candidates for new lead-free piezoelectric ceramics [18]. Although environmentally friendly and showing suitable piezoelectric properties, to date there are no lead-free ceramic materials with better piezoelectric properties than PZT-based ceramics [18].

$BaTiO_3$, similar to PZT, shows interesting overall properties, with a large number of applications in commercial devices. The most relevant applications include capacitors, embedded capacitance in printed circuit boards, underwater transducers (sonars), thermistors with a positive temperature coefficient of resistivity, and electroluminescent panels [119]. The main properties of $BaTiO_3$ are the higher electromechanical coupling factor, $k_{33} \approx 0.50$, and a piezoelectric strain constant $d_{33} \approx 190$ pC N^{-1}. However, the Curie temperature is 130 °C, which is low for applications. Dopants such as lead and calcium have been used in $BaTiO_3$ ceramics to stabilize the tetragonal phase over a wider temperature range.

Other materials used as dopants in $BaTiO_3$ are sodium and potassium, and the ternary ceramic shows improved piezoelectric properties but low Curie temperature. In addition to PZT and $BaTiO_3$, bismuth sodium titanate ($(Bi_{1/2}Na_{1/2})TiO_3$; BNT) is a ferroelectric perovskite with high Curie temperature (≈290 °C) and exhibiting large remnant polarization and high piezoelectric performance ($d_{33} \approx$ 170–215 pC N^{-1}) [120] when compared to other lead-free piezoelectric ceramics [121]. BNT doped with niobium sodium oxide [121] improves the overall properties and processability of the ceramic, with the piezoelectric constant reaching $d_{33} > 100$ pC N^{-1} [122].

Replacing sodium with potassium titanate results in bismuth potassium titanate ($(Bi_{1/2}K_{1/2})TiO_3$; BKT), a lead-free ceramic with a perovskite structure of tetragonal symmetry and a high Curie temperature of approximately 380 °C, and a structure similar to that of BNT [123, 124]. Ternary ceramic materials combining BKT, BNT, and $BaTiO_3$ present interesting properties and continue to be investigated to optimize the overall properties of the three ceramics in one material. In this context, $(1 - x)$ $BaTiO_3$–xBKT shows larger piezoelectric properties, with piezoelectric constants $d_{31} \approx 350$ pC N^{-1} and $d_{33} \approx 450$ pC N^{-1} [124].

With a similar perovskite structure (ABO_3), an alternative lead-free ceramic material is $KNbO_3$. Doped with sodium, potassium sodium niobate (KNN) ceramics can show a high Curie temperature near $T_C = 430$ °C, and the piezoelectric properties of the family of the KNN lead-free ceramics can reach values up to $d_{33} \approx$ 500 pC N^{-1} [115, 122, 124]. KNN is considered one of the most promising alternatives to lead-containing ceramics, showing a high Curie temperature and electromechanical coupling factor, strong ferroelectricity, and a dielectric constant of $\varepsilon' \approx 500$. [18, 122]. KNN materials can be manufactured as nanocrystals, thus improving their properties [115]. Some examples of piezoelectric ceramics based on KNN is KNN–potassium copper titanate–manganese ($K_{0.5}Na_{0.5}NbO_3$–$K_{5.4}Cu_{1.3}Ta_{10}O_{29}$–$MnO_2$ or KNN–KCT–Mn), which shows good piezoelectric properties ($d_{33} \approx 90$ and $\varepsilon' \approx 300$ [18]) for commercial applications [115]. Among the different materials used as dopants for KNN, copper oxide improves the piezoelectric properties due to improved poling efficiency of the ceramic.

With interesting properties and being a lead-free ceramic, zinc oxide (ZnO) presents excellent mechanical properties, piezoelectric response ($d_{33} \approx 12.4$ and $\varepsilon' \approx 10.9$ [18]), and optical properties [125, 126]. Another characteristic of ZnO is its simple manufacturing in the form of nanoscale fibers, wires, and particles [126].

2.4.1.5.2 Piezoelectric single crystals

Single crystals were the first-discovered piezoelectric material. Quartz and lithium niobite ($LiNbO_3$) are the most commercially available single crystals. Quartz is SiO_2 crystalized in six-sided prisms, with piezoelectric properties when in the α phase and with a Curie temperature of 573 °C. Lithium niobite, with a chemical formula of ABO_3 and a Curie temperature of 1200 °C, is used in high-frequency devices and for high-temperature applications.

Rochelle salt ($NaKC_4H_4O_6$–$4H_2O$; sodium potassium tartrate tetrahydrate) shows a large piezoelectric behavior ($\varepsilon' > 100$ [127]) under very specific temperature (−18 and 24 °C) and humidity conditions [128], limiting its applicability.

Lead magnesium niobite ($Pb(Mg_{1/3}Nb_{2/3})O_3$; PMN) exhibits a high dielectric constant that can reach near 30 000 [18]. Lead zinc niobate ($Pb(Zn_{1/3}Nb_{2/3})O_3$; PZN) ferroelectric single crystals present excellent dielectric properties that can be used in sensor and actuator applications. Near the MPB at room conditions, they show large piezo- and dielectric constants as well as a good electromechanical coupling coefficient [129].

PMN–PT is an interesting single crystal family, with good dielectric and piezoelectric properties because of enhanced polarizability arising from the coupling near the MPB of the crystal. This family of crystals have dielectric properties similar to those of PZT ceramic but the Curie temperature is lower, near 130 °C [98]. A ternary system based on PMN–PT has been developed to overcome their intrinsic limitations, improving their mechanical and thermal properties [98].

Aluminum nitrite (AlN) and lithium tantalite ($LiTaO_3$) present good overall properties for applications. AlN thin films show a dielectric constant near $\varepsilon' \approx 10$ [130] and $d_{33} \approx 5$ [131] and can be used in surface acoustic wave applications. Compared to AlN, $LiTaO_3$ shows a higher dielectric and piezoelectric constant, $\varepsilon' \approx 44$–47 and $d_{33} \approx 14.6$, respectively [132], and has been applied in pyroelectric devices [132].

A summary of the main properties of the single crystal and ceramic materials presented so far is found in table 2.2.

2.5 Electrospun piezoelectric energy harvesting systems

The capability of harvesting or capturing wasted surrounding energy involves vehicles, industrial machines, or human activity, and converting it, not exclusively, into electrical energy describes the so-called energy harvesting systems. These systems use electroactive materials as well as piezoelectric, electromagnetic, or magnetostrictive methods to convert mechanical into electrical energy to enable self-powered microdevices or to supply energy to wireless sensor devices [143, 144]. Sensors with difficult access locations and in remote areas associated with the growing field of low-power electronics technology make energy harvesting devices good candidates for replacing batteries as power sources, since the incorporation of a self-sufficient power system is an advantage in terms of low-cost maintenance.

Table 2.2. Piezoelectric materials used in energy harvesting or related applications.

Materials	PE charge constant d_{33} (pC N^{-1})	PE charge constant d_{31} (pC N^{-1})	EM coupling factor k_{33}	Dielectric constant ε'	T_C (°C)	Ref.
Single crystals						
Quartz	2			4.5	—	[133]
$LiNbO_3$	40	—	0.6	47	1140	[134]
Rochelle salt	—	275	0.65	—	<24	[127, 128]
PMN	<2800	—	0.91–0.95	>5400	<150	[18]
PZN	—	—	—	9500	—	[129]
PMN–PT	1620	−760	0.93	7000	<155	[98, 135]
PIN–PMN–PT	2742	−1337	0.95	7244	200	[136]
PMN–PZT	1530	−718	0.92–0.93	>4850	≈ 216	[137]
PZN–PT	200	−1075	0.94	5100	—	[138]
AlN	5.1	−2	0.38	≈10	>2000	[130, 131]
$LiTaO_3$	14.6	728	0.32	44–47	620	[132]
Ceramics						
$PbTiO_3$	—	—	—	>300	≈490	[139]
PZT	472	—	0.71	714	340	[116, 140]
PZT-5A	375	—	0.71	1700	365	[116, 140]
PZT-5H	590	—	0.75	3400	190	[116, 140]
PZT-8	225	—	0.64	1000	300	[116, 140]
PLZT	—	—	0.45	1800	—	[119]
$BaTiO_3$	190	—	0.5	1700	115	[119]
ZnO	12.4	5.0	0.48	10.9	—	[18]
KNN	480	—	<0.6	<500	421	[122]
KNN–KCT–Mn	90	−33	0.4	300	404	[115]
BNT	64	—	—	302	310	[120]
BKT	74	—	—	—	380	[123, 124]
BNKT	185	—	—	868	292	[141]
BNLT	≈98	—	—	≈550	<370	[142]
BNT–NN	<80	—	—	<801	—	[110]
BNT–BZT	<147	—	—	<881	<251	[142]

Abbreviations: PIN–PMN–PT, lead indium niobate; PMN–PT, lead magnesium niobate–lead titanate; PZN–PT, lead zinc niobate–lead titanate.

Regarding environmental energy sources, dielectric elastomers demonstrated increased efficiency for electromechanical conversion from ocean waves [145].

Energy harvesting systems based on piezoelectricity are the most widely investigated ones due to high voltage output, high power density, easy materials processing, and production at different geometric scales [143]. Nowadays, there is increasing interest in flexible, deformable, and curved materials with a wide range of applicability. Therefore, the piezoelectric polymers that present all these characteristics along with low density, high flexibility, and high resistance to mechanical impacts are good choices for energy harvester systems. However, this class of materials presents lower piezoelectric constants than ceramic and single crystal counterparts. The most common piezoelectric polymer used is PVDF and its copolymers due to its easy processability, high chemical resistance, and flexibility.

In particular, strong advances in energy harvesting systems have been achieved based on electrospun materials.

Thus, concentric PVDF nanofibers were electrospun on a Cu foil in a polyvinyl chloride substrate and embedded in PDMS. These concentric nanofiber configurations allow the electrical energy generation from mechanical energy regardless of the direction of the applied force. The device achieved a 5 V output voltage, 400 nA current, and output power of 200 nW at a load resistance of 2 MΩ [146]. An energy harvester system based on electrospun PVDF nanofibers deposited on a PVDF film with an inner electrode of silver-deposited nylon filament (80 μm in diameter, and 25 $\Omega\ cm^{-1}$) reached a voltage, current, and power density of 0.52 V, 18.76 nA, and 5.54 $\mu W\ cm^{-3}$, respectively. The output signals were obtained under a cyclic compression of 0.02 MPa at 1.85 Hz, showing stability after 50 000 cycles. The improved interfacial properties between the PVDF fibers and film enhanced the performance of the device [147]. Higher performance of the energy harvester system is obtained by aligned PVDF nanofibers (average diameter of 350 ± 23 nm, figure 2.14(a)) due to the higher content of the β phase. Kang *et al* [148] deposited a PVDF electrospun nanofiber onto a PET substrate with Ag electrodes, as shown in figure 2.14(b). The device achieved an output voltage of 1.6 V at a frequency of 2 Hz.

Aligned PVDF fibers (average diameter ≈1.6 μm) were deposited by a hollow cylindrical near-field electrospinning process. The oriented fibers were transferred onto a PET substrate with top and bottom Cu foil electrodes. An output voltage of 76 mV and current of 39 nA were measured with a strain of 0.05% at 7 Hz. Schematic illustrations of the measurement systems are presented in figure 2.15 [149].

Pure PVDF–TrFE nanofibers were electrospun in a flexible plastic to develop a nanogenerator with an output voltage of 0.4 V under a cantilever pressure of 8 mN and at a frequency of 3 Hz [150]. An acoustic to electric energy conversion from a randomly oriented electrospun PVDF–TrFE nanofiber (diameter of 240 ± 40 nm and 20 μm thick) sandwiched between two gold-coated PET films (110 μm of thick) generated an output voltage, current, and power density of 14.5 V, 28.5 μA, and 306.5 $\mu W\ cm^{-3}$, respectively. The maximum values were obtained under a sound pressure of 115 dB at 210 Hz [151].

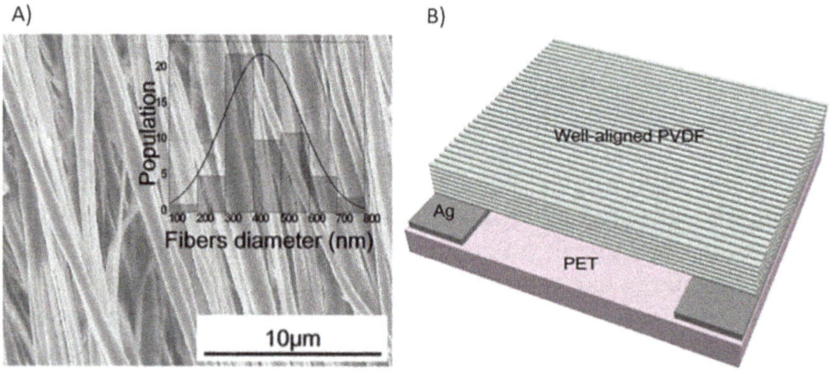

Figure 2.14. (a) SEM image of aligned PVDF nanofibers with the diameter distribution fitted with a Gaussian function; (b) illustration of the energy harvester system. Adapted from [148].

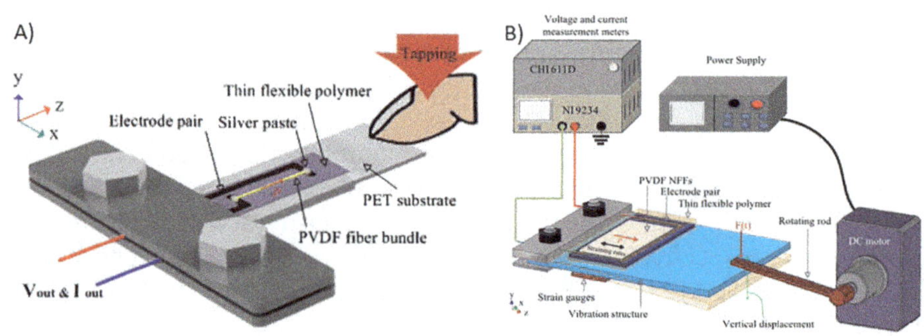

Figure 2.15. (a) Schematic illustration of the tapping measurement system of the energy harvesting system; (b) the vibration measurement system. Adapted from [149].

A flexible and hybrid energy harvesting system based on electrospun poled PVDF-TrFE fibers was sandwiched between two fiber-based electrospun electrodes based of polyurethane coated with silver nanowires and carbon nanotubes (10–30 μm in length) and packed with PDMS. The vertical structure combining the tribo- and piezoelectric method increases the performance of the device. When an external force of 4 Hz was applied to the triboelectric generator, an output voltage, current, and power density of 183 V, 1.9 μA, and 630 μW were obtained, respectively. In the case of the piezoelectric device, the values observed were 57.1 V, 2.95 μA, and 0.85 μW, respectively, at a load resistance of 600 MΩ. Due to the flexibility and sensitivity of the piezoelectric material, this device shows applicability in wearable healthcare monitoring systems for real-time monitoring of human physiological signals [152]. A polymer composite device composed of PVDF–TrFE and $BaTiO_3$ ceramic fillers was electrospun on interdigitated electrodes to study the influence of fillers in an electrospinning process for the generation of energy. The average diameter obtained was 450 ± 100 nm. A higher power output was obtained for the pure PVDF fibers, with 0.02 μW and 25 μW for low and high mechanical deformations, respectively.

The performance of the device was evaluated by periodic bending tests with a maximum vertical displacement of 7 mm at a 1 MΩ load resistance. The incorporation of ceramic fillers increased the mechanical stiffness and consequently reduced the power output [153]. An energy harvesting system with cylindrical shape based on PVDF electrospun fibers with Au-coated nanowire electrodes (figure 2.16) was obtained with increased performance due to the great flexibility, stretchability, high transmittance, and low sheet resistance of the electrodes. The device with a 0.34 cm diameter generated an output voltage and current of 4.1 V and 295 nA, respectively, at a strain of 0.5% and frequency of 5 Hz. In comparison, the same device in the same conditions but without rolling generated an output voltage and current of 1.8 V and 140 nA, respectively. The rolling step led to a higher performance of the device [154].

The addition of multiwalled carbon nanotubes (MWCNT) in electrospun PVDF fibers increase the surface conductivity and enhance the performance of the energy harvester system. Increases of 200% and 44.8% in output voltage and charging power are observed when compared with pure PVDF fibers. The best results were obtained for a 5 wt% MWCNT composition with a voltage of 6 V and capacitor charging power of 81.8 nW [155]. An energy harvesting system based on aligned PVDF fibers with $BaTiO_3$ nanoparticles (diameter $\approx$200 nm) incorporated and embedded with PDMS obtained an output voltage of 0.48 V at a displacement of 6 mm and frequency of 0.007 Hz for a composition of 16 wt% $BaTiO_3$ nanoparticles [156]. Flexible nanofiber composites of $BaTiO_3$ dispersed in PVDF–TrFE were electrospun and embedded in a spin-coating elastomer (PDMS) to generate and

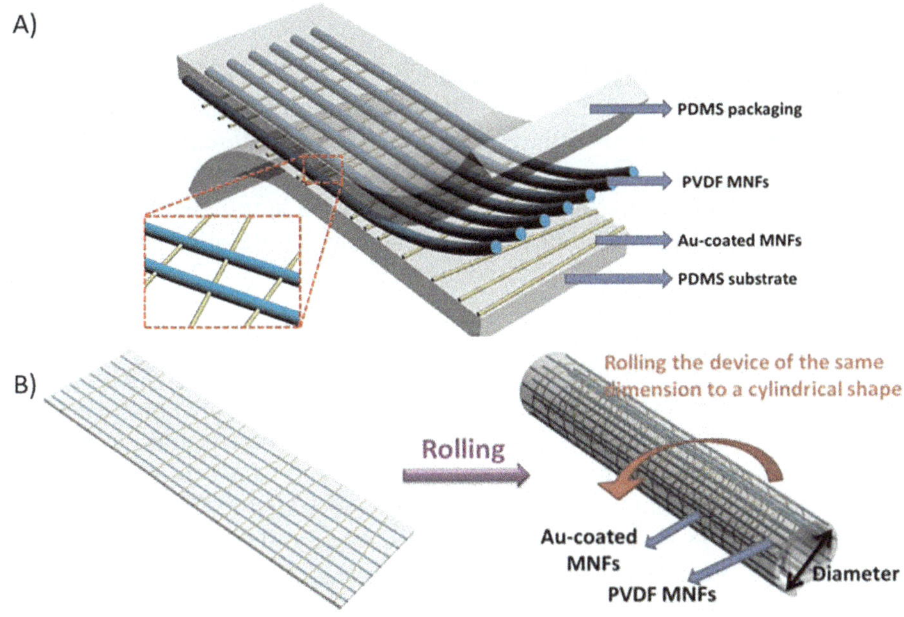

Figure 2.16. (a) Illustration of the piezoelectric energy harvester system; (b) rolling step to obtain a cylindrical energy harvester system. Adapted from [154].

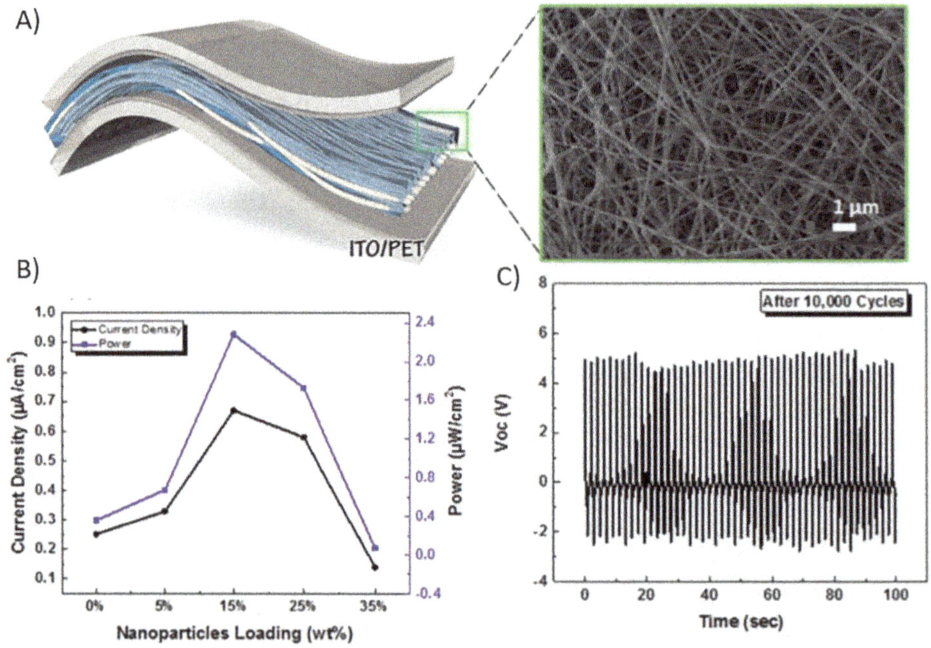

Figure 2.17. (a) Schematic of the energy harvester system and SEM image of 35 wt% $BaTiO_3$ in PVDF–TrFE nanofiber; (b) output current and power density for the energy harvesting system as a function of $BaTiO_3$ content at a force of 20 N; (c) open-circuit voltage output of a cycling tapping test of 15 wt% $BaTiO_3$ at 20 N after 10 000 cycles. Adapted from [157].

store energy during walking. This is presented in figure 2.17. The composite fibers had an average diameter of 250 nm. The elastomer provides protection for the nanofibers from mechanical damage, giving the device greater durability. The performance was maintained over 10 000 cycles and 1000 human steps, as shown in figure 2.17(c). The energy harvesting device has an output voltage of 25 V at a walking frequency of 0.6 Hz and a high mechanical durability under high loads (600 N) with 15 wt% $BaTiO_3$. This voltage can charge a 4.7 µF capacitor after approximately 72 steps. The self-poled fibers lead to a high energy harvesting performance, enhancing the voltage and current outputs up to 200%. This device has a low cost and simple fabrication process, small form factor, flexibility, durability, and transparency, which leads to it being a good candidate for application in self-powered wearable electronics [157]. The same polymeric matrix was embedded with $(Na_{0.5}K_{0.5})NbO_3$ (NKN) nanoparticles with a diameter between ≈30–150 nm, and electrospun fibers were sandwiched between an ITO-coated PET substrate. The best performance with 0.98 V and 78 nA for output voltage and current, respectively, was obtained by a 10 vol% NKN nanoparticles composition [158].

A composition of 20 vol% of $0.4Pb(Zn_{1/3}Nb_{2/3})O_3$–$0.6Pb(Zr_{0.5}Ti_{0.5})O_3$ (PZN–PZT) nanoparticles (≈54 nm) embedded in a PVDF–TrFE electrospun nanofiber sandwiched on a ITO-coated PET substrate exhibited a output voltage of ≈3.4 V and a current of ≈240 nA [159]. A small amount of graphene nanoplatelets (0.1 wt%) led to a significant increase in the β phase, the most piezoelectric phase of PVDF, resulting in a higher energy harvesting performance in a PVDF/graphene electrospun composite with a mean diameter of 107 ± 8 nm. For a pressure of 0.2 MPa and a frequency of 1 Hz, a composite with 0.1 wt% graphene increased the open-circuit voltage and short-circuit current from 3.8 V to 7.9 V and 2 μA to 4.5 μA, respectively. As a proof-of-concept, this sample was able to charge a 33 μF capacitor to 6.5 V and then turn on an LED light for more than 30 s. The incorporation of a higher amount of graphene leads to a decrease in the electrical output of the device [160].

A tri-composed energy harvesting device using electrospun $BaTiO_3$ nanoparticles and graphene nanosheets dispersed in a polymeric matrix (PVDF) was studied. The fibers were sandwiched between aluminum foils, copper strips, and PET substrates with 50 μm and 150 μm of thickness for top and bottom electrodes, respectively. The fibers present a thickness of 18–20 μm. The best performance was obtained with the composition of 0.15 wt% graphene nanosheets and 15 wt% $BaTiO_3$ nanoparticles; moreover, an open-circuit voltage of 11 V and an electric power of 4.1 μW was obtained under a load frequency and strain of 2 Hz and 4 mm, respectively. The device showed good stability of the open-circuit voltage after 1800 cycles. The improved output is due to the synergistic contribution of both fillers to the piezoelectric enhancement of the fibers. This was applied in a finger pressing–releasing process, and an output voltage of 112 V was obtained [161].

The incorporation of the platinum (Pt) nanoparticles (1.5 wt%) in PVDF electrospun nanofibers (diameter near 14 μm) sandwiched between a copper–nickel plate and polypropylene proved to be a good material for small-power consumer electronics. The energy harvester system reaches an output voltage, current density, and power density of 30 V, 6 mA cm^{-2} and 22 μW cm^{-2}, respectively under a load resistance of 1 MΩ. A good stability was also observed up to 90 000 cycles [162]. An electrospun nanofiber of the copolymer PVDF–HFP embedded with silver (Ag) nanoparticles (83 nm in diameter and layer thickness ≈150 ± 20 μm) sandwiched between two flexible electrodes achieved an output voltage of 3 V and a current density of 0.9 μA cm^{-2} in a periodic pressure of 15 kPa and frequency of 2 Hz [163]. The incorporation of Eu^{3+} to create a composite electrospun nanofiber based in Eu^{3+} doped PVDF–HFP/graphene led to a higher β phase content and an improvement in fiber crystallinity without any post poling process. The fibers were sandwiched between two conducting Ni–Cu–Ni-plated polyester fabric substrates. An output voltage and power of 9 V and 5.8 μW, respectively, were observed under finger-touch motion with 5.6 kPa stress (figure 2.19). The device was also tested under wind flows, showing an open-circuit output voltage of 4.5 V and a power of 2.4 μW for a wind velocity of 8.5 m s^{-1} and incident angle of 45° (figure 2.18). An acoustic sensitivity of 11 V Pa^{-1} for external pressure as low as ≈23 Pa of the device show the applicability of this device for sensors, including those used in personal identification, national security, and medical diagnosis [164].

A flexible energy harvester system based on oriented zinc sulfide (ZnS) embedded in electrospun PVA nanofibers (40–200 nm in diameter) showed an improved

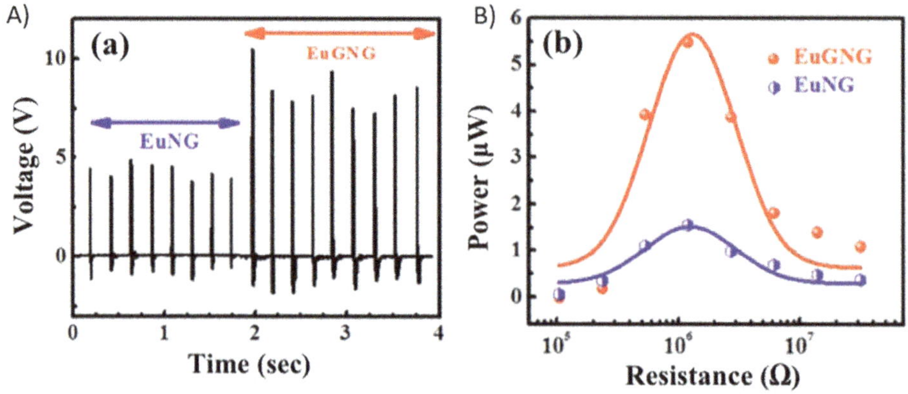

Figure 2.18. (a) Output voltage of Eu^{3+}-doped PVDF–HFP/graphene (EuGNG) and P(VDF–HFP)/graphene (EuNG) under repeated finger-touch motion; (b) instantaneous output power as a function of load resistance. Adapted from [164].

acoustic sensitivity of 2 V Pa^{-1}, leading to harvesting energy from random acoustic vibrations. The aluminum foil from the collector during electrospinning was used as the bottom electrode while a conductive adhesive carbon tape served as the top electrode. A maximum power density of 1.7 nW mm^{-3} was obtained at 40 MΩ of resistance. In a case of harvesting energy from sound vibration, when a sound pressure of 100 dB was applied, an output voltage of 4 V was measured [165].

Another polymeric matrix (PLLA) was used to produce electrospun nanofibers (100 nm to 2 μm in diameter) for an energy harvesting device on polyimide substrate with a gold comb electrode. This device showed an open-circuit voltage of 250 mV and short-circuit current of 96 nA for a displacement of 1 mm. To show the applicability on flexible energy harvesting systems, a voltage of 0.55 V and current of 230 nA was observed for a strain deformation of 28.9°. After 2800 cycles, the system showed stability of the electrical signals generated, and it was applied to a human joint bending; a power generation of 9.5 nW was then observed [166]. Aligned $BaTiO_3$ nanofibers with a diameter ≈100–200 nm was electrospun and transferred to a plastic substrate coated with an interdigital electrode and embedded in PDMS. An output voltage of ≈0.45 V was generated under a load resistance of 1 MΩ and a frequency of 45 Hz, corresponding to an output power of 60 nW. During the measurements, a power output of 6 nW was observed [167]. An energy harvesting system based on electrospun fibers of $Bi_{0.5}Na_{0.5}TiO_3$–$Bi_{0.5}K_{0.5}TiO_3$ (BNT–BKT) was applied for harvesting energy from human finger movement. The nanofibers (150 nm in diameter) were sandwiched between the bottom electrode of an ITO-coated PET film 100 μm thick and a 25 μm thick PDMS layer, as shown in figure 2.19(a). The top and bottom electrodes were identical. The output voltage, current, and power obtained at a 60 MΩ load resistance knocked by a finger was 3.5 V, 280 nA, and 37.2 nW, respectively, as shown in figure 2.19(b). The device was attached to a human finger, and a voltage and current output of 0.49 V and 13.5 nA were observed; a good mechanical stability after 100 cycles was also observed [168].

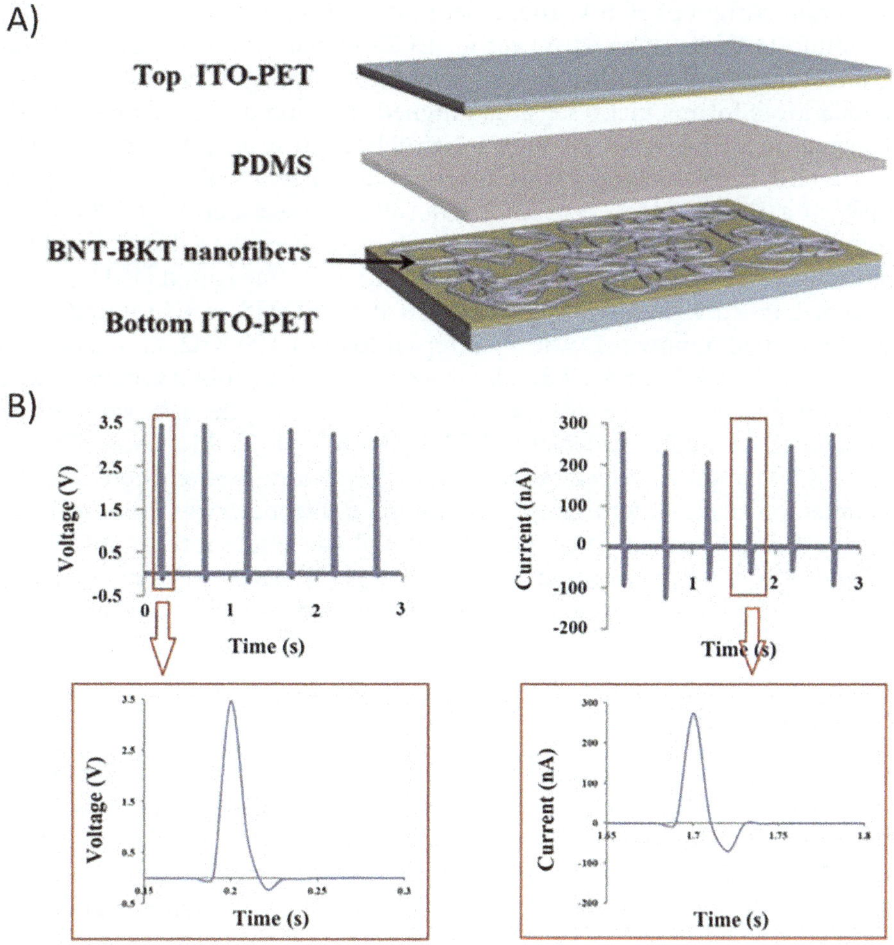

Figure 2.19. (a) Illustration of the BNT–BKT structure; (b) the open-circuit voltage (left) and short-circuit current (right) output under repeated finger-touch motion. Adapted from [168].

An energy harvesting system based on PZT nanofibers (60 nm in diameter and 500 μm in length) were deposited on interdigitated electrodes of platinum fine wire (diameter of 50 μm) arrays in a silicon substrate. To prevent damage and increase the lifetime of the nanogenerator, it was embedded in a PDMS matrix. The device at a load resistance of 6 MΩ and frequency of 40 Hz generated an output voltage and power of 1.63 V and 0.03 μW, respectively [169].

A different approach was introduced by Cui *et al* [170] based on electrospun PZT nanowires on a PDMS-coated magnetite (Fe_3O_4) substrate embedded on PDMS with top and bottom silver electrodes. Two types of noncontact magnetic force-driven nanogenerators were developed, one using the piezoelectric potential drop along the radial direction and the other using the potential drop along the axial direction of the nanowire; therefore, electricity could be generated without contact with the mechanical movement source. The device generated an output voltage of

3.2 V, output current of 50 nA, and a maximum power density of 70 μW cm^{-3}. The applicability of this system to power an LCD screen was demonstrated, and the noncontact magnetic force-driven nanogenerator opens the possibility of applying nanogenerators for magnetic sensing. Aligned PZT nanowires (diameter 370 nm) electrospun on a PET substrate with top and bottom silver electrodes generated a 6 V open-circuit voltage and 45 nA short-circuit current with an output power of 0.12 μW at a 100 MΩ load resistance. An output power density of 200 μW cm^{-3} was obtained. The piezoelectric layer was removed from the substrate and then an output voltage of 0.24 V and current of 2.5 nA were measured [171]. In another approach, aligned PZT nanowires embedded in PDMS were rotated to create vertically aligned nanowires. For an external load of 100 MΩ, an output voltage and current of 198 V and 17.8 μA, respectively, were observed. For a larger applied impact, a voltage of 209 V and current of 53 μA were observed, corresponding to a current density of 23.5 μA cm^{-2} [172]. Mn-doped $(Na_{0.5}K_{0.5})NbO_3$ (NKN) nanofibers electrospun on a PES substrate and embedded in PDMS, and contacted using interdigitated platinum electrodes, was produced for an energy harvesting system. An open-circuit voltage and short-circuit current of 0.3 V and 50 nA were measured, respectively [173]. Table 2.3 summarizes some energy harvesting system based on electrospun piezoelectric polymers and polymer composite fibers.

2.6 Final remarks

Electrospun piezoelectric nanofibers have attracted increasing attention due to their specific microstructure, excellent mechanical properties, and, in some cases, biocompatibility. The main concepts, materials, and device applications related to electrospun piezoelectric materials have been presented, mainly focusing on energy harvesting systems based on electrospun piezoelectric polymers and polymer composites.

Piezoelectric electrospun energy harvesting materials convert mechanical vibrations from the environment into electrical energy that can be used to power micro/nanodevices. Among the different materials, the piezoelectric polymer PVDF has been particularly studied due to its high flexibility and long-term physical/chemical stability. Furthermore, electrospun ceramic materials, such as $BaTiO_3$ and PZT, have been developed.

Despite the findings reported on the developments of piezoelectric electrospun materials for energy harvesting, there are relevant issues that need to be addressed. Those issues are mainly concerned with improving materials performance for meeting a large spectrum of application demands, but also time stability and simplicity in device integration to meet manufacturing demands. In this sense, polymer composite electrospun materials can play a relevant role in the near future, allowing us to combine the low cost and higher versatility, stability, and flexibility of the polymers with the higher response of inorganic piezoelectric materials.

Table 2.3. Summary of electrospun piezoelectric energy harvesting systems.

Piezoelectric material	Output performance	Measurement conditions	Ref.
PVDF	5 V/400 nA/200 nW	Load resistance 2 MΩ	[146]
PVDF	0.52 V/18.76 nA/5.54 μW cm^{-3}	Periodic pressure of 0.02 MPa at 1.85 Hz	[147]
PVDF	1.6 V	Frequency of 2 Hz	[148]
PVDF	76 mV/39 nA	Strain of 0.05% at 7 Hz	[149]
PVDF–TrFE	0.4 V	Cantilever pressure 8 mN at 3 Hz	[150]
PVDF–TrFE	14.5 V/8.5 μA/306.5 μW cm^{-3}	Sound pressure 115 dB at 210 Hz	[151]
PVDF–TrFE	57.1 V/2.95 μA/0.85 μW	Load resistance 600 MΩ	[152]
PVDF–TrFE	0.02 μW–25 μW	Load resistance 1 MΩ and displacement of 7 mm	[153]
PVDF/Au	4.1 V/295 nA1.8 V/140 nA	Strain of 0.5% at 5 Hz	[154]
PVDF/MWCNTs	6 V/81.8 nW	—	[155]
PVDF/$BaTiO_3$	0.48 V	Displacement of 6 mm at 0.007 Hz	[156]
PVDF–TrFE/$BaTiO_3$	25 V	Walking frequency of 0.6 Hz	[157]
PVDF–TrFE/NKN	0.98 V/78 nA	—	[158]
PVDF–TrFE/PZN–PZT	3.4 V/240 nA	—	[159]
PVDF/graphene	3.8 V–7.9 V/2 μA–4.5 μA	Pressure of 0.2 MPa at 1 Hz	[160]
PVDF/$BaTiO_3$/graphene	11 V/4.1 μW	Load frequency of 2 Hz and strain of 4 mm	[161]
PVDF/Pt	30 V/6 mA cm^{-2}/22 μW cm^{-2}	Load resistance of 1 MΩ	[162]
PVDF–HFP/Ag	3 V/0.9 μA cm^{-2}	Periodic pressure of 15 kPa at 2 Hz	[163]
PVDF–HFP/graphene/Eu^{3+}	9 V/5.8 μW4.5 V/2.4 μW	Pressure of 5.6 kPa Wind velocity of 8.5 m s^{-1} and incident angle of 45°	[164]
PVA/ZnS	1.7 nW mm^{-3}4 V	Load resistance 40 MΩ Sound pressure of 100 dB	[165]
PLLA	250 mV/96 nA0.55 V/230 nA/9.5 nW	Displacement of 1 mm Deformation of 28.9°	[166]
$BaTiO_3$	0.45 V/60 nW	Load resistance of 1 MΩ at 45 Hz	[167]
BNT–BKT	3.5 V/280 nA/37.2 nW	Load resistance 60 MΩ	[168]
PZT	1.63 V/0.03 μW	Load resistance of 6 MΩ at 40 Hz	[169]
PZT	3.2 V/50 nA/70 μW/cm^{-3}	—	[170]
PZT	6 V/45 nA/0.12 μW	Load resistance 100 MΩ	[171]
PZT	198 V/17.8 μA209 V/53 μA	Load resistance of 100 MΩ	[172]
NKN	0.3 V/50 nA	—	[173]

Finally, due to the increasingly important environmental concerns, more environmentally friendly solvents and recyclable materials must be used in this field. Piezoelectric electrospun materials offer relevant scientific challenges and present large technological interest, playing an increasing and needed role in key areas such as sensing, energy harvesting, and the biomedical field.

Acknowledgments

This work was supported by the Portuguese Foundation for Science and Technology (FCT) in the framework of the Strategic Funding UID/FIS/04650/2013 and projects NORTE-01-0145-FEDER-028237, POCI-01-0145-FEDER-028159, PTDC/EEI-SII/5582/2014 and PTDC/CTM-ENE/5387/2014. TM, ACL, PC, and PLM give thanks to the FCT for its financial support through Grant Nos. SFRH/BD/140242/2018, SFRH/BD/132624/2017, SFRH/BPD/110914/2015, SFRH/BPD/96227/2013, and SFRH/BD/98219/2013. Financial support from the Spanish Ministry of Economy and Competitiveness (MINECO) through the project MAT2016-76-039-C4-3-R (AEI/FEDER, UE) (including the FEDER financial support) and from the Basque Government Industry and Education Departments under the ELKARTEK, HAZITEK, and PIBA (PIBA-2018-06) programs are also acknowledged.

Nomenclature			
D	Electric displacement vector	s	Elastic compliance matrix
d	Piezoelectric strain coefficient	S	Strain tensor
E	Electric field vector	T	Stress tensor
k	Electromechanical coupling factor	T_C	Curie temperature

Symbols			
α	Alpha	Γ	Gamma
β	Beta	ε	Dielectric permittivity
δ	Delta	Ω	Omega

Abbreviations			
AlN	Aluminum nitrite	PLLA	Poly(L-lactic acid)
BT, $BaTiO_3$	Barium titanate	PLZT	Lead lanthanum zirconate titanate
BKT	Bismuth potassium titanate	PMN	Lead magnesium niobite
BNKT	Bismuth sodium potassium titanate	PMN–PT	Lead magnesium niobite–lead titanate
BNLT	Bismuth sodium lanthanum titanate	PMN–PZT	Lead magnesium niobite–lead zirconate titanate
BNT	Bismuth sodium titanate	PS	Polystyrene

(*Continued*)

(*Continued*)

Abbreviations			
BNT–BKT	Bismuth sodium titanate–bismuth potassium titanate	$PbTiO_3$; PT	Lead titanate
BNT–BZT	Bismuth sodium titanate–bismuth zirconium titanate	PVA	Poly(vinyl alcohol)
BNT–NN	Bismuth sodium titanate–sodium niobate	PVDCN/VAc	Poly(vinylidene cyanide–vinyl acetate)
CA	Cellulose acetate	PVDF	Poly(vinylidene fluoride)
EVA	Poly(ethylene-*co*-vinyl alcohol)	PVDF–HFP	Poly(vinylidene fluoride-co-hexafluoropropene)
Fe_3O_4	Magnetite	PVDF–TrFE	Poly(vinylidene fluoride trifluoroethylene)
I_{SC}	Short-circuit current	PVP	Polyvinyl phenol
ITO	Indium tin oxide	PZN	Lead zinc niobite
KN	Potassium niobate	PZN–PT	Lead zinc niobite–lead titanate
KNN	Potassium sodium niobite	PZN–PZT	Lead zinc niobite–lead zirconate titanate
KNN–KCT–Mn	KNN–potassium copper titanate–manganese	PZT	Lead zirconate titanate
$LiNbO_3$	Lithium niobite	SiO_2	Quartz
$LiTaO_3$	Lithium tantalite	TiO_2	Titanium dioxide
MEMS	Micro-electromechanical systems	V_{oc}	Open-circuit voltage
$MgNb_2O_9Pb_3$; PMN	Lead magnesium niobate	ZnO	Zinc oxide
MoS_2	Molybdenum disulfide	ZnS	Zinc sulfide
MPB	Morphologic phase boundary		
MWCNT	Multiwalled carbon nanotubes		
$NaKC_4H_4O_6$–$4H_2O$	Rochelle salt		
NKN	Sodium potassium niobate		
PA	Polyamide		
PAA–PM	Poly(acrylic acid)–poly(pyrene methanol)		
PAN	Poly(acrylonitrile)		
PANI	Polyaniline		
PDMS	Polydimethylsiloxane		
PEO	Poly(ethylene oxide)		
PET	Polyethylene terephthalate		
PIN–PMN–PT	Lead indium niobate		
PLA	Polylactic acid		
PLGA	Polylactic-*co*-glycolic acid		

References

[1] Murphy E B and Wudl F 2010 The world of smart healable materials *Prog. Polym. Sci.* **35** 223–51
[2] Ribeiro C *et al* 2018 Electroactive poly(vinylidene fluoride)-based structures for advanced applications *Nat. Protoc.* **13** 681–704
[3] Hinchet R, Seung W and Kim S W 2015 Recent progress on flexible triboelectric nanogenerators for selfpowered electronics *Chem. Sus. Chem.* **8** 2327–44
[4] Jain A *et al* 2015 Dielectric and piezoelectric properties of PVDF/PZT composites: A review *Polym. Eng. Sci.* **55** 1589–616
[5] Silva M *et al* 2013 Optimization of the magnetoelectric response of poly(vinylidene fluoride)/epoxy/vitrovac laminates *ACS Appl. Mater. Interfaces* **5** 10912–9
[6] Phan H *et al* 2015 The piezoresistive effect of SiC for MEMS sensors at high temperatures: a review *J. Microelectromech. Syst.* **24** 1663–77
[7] Song I S *et al* 2015 Sensitivity enhancement of a UV photo-sensor based on a fiber Bragg grating coated by a photomechanical functional polymer *Sens. Actuators,* A **232** 223–8
[8] Romero Gómez J *et al* 2013 Magnetocaloric effect: A review of the thermodynamic cycles in magnetic refrigeration *Renew. Sustain. Energy Rev.* **17** 74–82
[9] Khan A *et al* 2016 Piezoelectric thin films: an integrated review of transducers and energy harvesting *Smart Mater. Struct.* **25** 053002
[10] Jung J *et al* 2017 Review of piezoelectric micromachined ultrasonic transducers and their applications *J. Micromech. Microeng.* **27** 113001
[11] Tiersten H F 1995 *Linear Piezoelectric Plate Vibrations: Elements of the Linear Theory of Piezoelectricity and the Vibrations of Piezoelectric Plates* (Berlin: Springer)
[12] Narita F and Fox M 2018 A review on piezoelectric, magnetostrictive, and magnetoelectric materials and device technologies for energy harvesting applications *Adv. Eng. Mater.* **20** 1700743
[13] Martins P and Lanceros-Méndez S 2013 Polymer-based magnetoelectric materials *Adv. Funct. Mater.* **23** 3371–85
[14] Martins P, Lopes A C and Lanceros-Mendez S 2014 Electroactive phases of poly(vinylidene fluoride): Determination, processing and applications *Prog. Polym. Sci.* **39** 683–706
[15] Tiwari S *et al* 2019 Enhanced piezoelectric response in nanoclay induced electrospun PVDF nanofibers for energy harvesting *Energy* **171** 485–92
[16] Erturk A and Inman D J 2011 *Piezoelectric Energy Harvesting* (New York: Wiley)
[17] Pradeep Fulay J-K L 2010 *Electronic, Magnetic, and Optical Materials* (Boca Raton, FL: United States: Taylor & Francis)
[18] Nunes-Pereira J, Costa P and Lanceros-Mendez S 2018 Piezoelectric energy production *Comprehensive Energy System* ed I Dincer (Amsterdam: Elsevier), ch 3.9, pp 380–415
[19] Jiří E 2013 Experiments to demonstrate piezoelectric and pyroelectric effects *Phys. Educ.* **48** 438
[20] Zhan S and Li F 2012 High performance ferroelectric relaxor-$PbTiO_3$ single crystals: status and perspective *J. Appl. Phys.* **111** 031301
[21] Khaled S R, Sameoto D and Evoy S 2014 A review of piezoelectric polymers as functional materials for electromechanical transducers *Smart Mater. Struct.* **23** 033001
[22] Harb A 2011 Energy harvesting: state-of-the-art *Renew. Energy* **36** 2641–54
[23] Vullers R J M *et al* 2009 Micropower energy harvesting *Solid-State Electron.* **53** 684–93

[24] Proto A *et al* 2017 Nanogenerators for human body energy harvesting *Trends Biotechnol.* **35** 610–24
[25] Wang H, Jasim A and Chen X 2018 Energy harvesting technologies in roadway and bridge for different applications—A comprehensive review *Appl. Energy* **212** 1083–94
[26] Shanmugam V, Selvakumar S and Yeh C S 2014 Near-infrared light-responsive nanomaterials in cancer therapeutics *Chem. Soc. Rev.* **43** 6254–87
[27] Shi H *et al* 2014 Preparation of functionalized graphene/SEBS-g-MAH nanocomposites and improvement of its electrical, mechanical properties *Mater. Lett.* **133** 200–3
[28] Sridhar R and Ramakrishna S 2013 Electrosprayed nanoparticles for drug delivery and pharmaceutical applications *Biomaterials* **3** e24281
[29] Dewangan K *et al* 2012 Synthesis and characterization of self-assembled nanofiber-bundles of V2O5: their electrochemical and field emission properties *Nanoscale* **4** 645–51
[30] Lakshmi B B, Patrissi C J and Martin C R 1997 Sol–gel template synthesis of semiconductor oxide micro- and nanostructures *Chem. Mater.* **9** 2544–50
[31] Badrossamay M R *et al* 2010 Nanofiber assembly by rotary jet-spinning *Nano Lett.* **9** 2257–61
[32] Beachley V and Wen X 2010 Polymer nanofibrous structures: Fabrication, biofunctionalization, and cell interactions *Prog. Polym. Sci.* **1** 868–92
[33] Dersch R *et al* 2007 Electrospinning of nanofibres: towards new techniques, functions, and applications *Aust. J. Chem.* **60** 719
[34] Yoo H S, Kim T G and Park T G 2009 Surface-functionalized electrospun nanofibers for tissue engineering and drug delivery *Adv. Drug Deliv. Rev.* **6** 1033–42
[35] Liu X and Ma P X 2009 Phase separation, pore structure, and properties of nanofibrous gelatin scaffolds *Biomaterials* **30** 4094–103
[36] Ellison C J *et al* 2007 Melt blown nanofibers: Fiber diameter distributions and onset of fiber breakup *Chem. Eng. Mater. Sci.* **48** 3306–16
[37] Leach M K *et al* 2011 Electrospinning fundamentals: optimizing solution and apparatus parameters *J. Vis. Exp.* **47** e2494
[38] Koštáková E *et al* 2012 Electrospinning and electrospraying of polymer solutions with spherical fullrenes *4th Int. Conf. Proc. of NANOCON 2012*
[39] Bhushania J A and Anandharamakrishnana C 2014 Electrospinning and electrospraying techniques: Potential food based applications *Trends Food Sci. Technol.* **38** 21–33
[40] Kessick R, Fenn J and Tepper G 2004 The use of AC potentials in electrospraying and electrospinning processes *Polymer* **45** 2981–9
[41] Reneker D H and Yarin A L 2008 Electrospinning jets and polymer nanofibers *Polymer* **49** 2387–425
[42] Dersch R *et al* 2005 Nanoprocessing of polymers: applications in medicine, sensors, catalysis, photonics *Polym. Adv. Technol.* **16** 276–82
[43] Tapia-Hernandez J A *et al* 2015 Micro- and nanoparticles by electrospray: advances and applications in foods *J. Agric. Food Chem.* **63** 4699–707
[44] Vonnegut B and Neubauer R L 1952 Production of monodisperse liquid particles by electrical atomization *J. Colloid Sci.* **7** 616–22
[45] Pillay V *et al* 2013 A review of the effect of processing variables on the fabrication of electrospun nanofibers for drug delivery applications *J. Nanomater.* **2013** 789289

[46] Chen J-Y *et al* 2015 Electrospun nanofibers with dual plasmonic-enhanced luminescent solar concentrator effects for high-performance organic photovoltaic cells *J. Mater. Chem.* A **3** 15039–48

[47] Thavasi V, Singh G and Ramakrishna S 2008 Electrospun nanofibers in energy and environmental applications *Energy Environ. Sci.* **1** 205–21

[48] Sankaran K K *et al* 2014 Development and evaluation of axially aligned nanofibres for blood vessel tissue engineering *J. Tissue Eng. Regen. Med.* **8** 640–51

[49] Sorayani Bafqi M S, Bagherzadeh R and Latifi M 2017 Nanofiber alignment tuning: an engineering design tool in fabricating wearable power harvesting devices *J. Ind. Text.* **47** 535–50

[50] Kim K W *et al* 2004 The effect of molecular weight and the linear velocity of drum surface on the properties of electrospun poly(ethylene terephthalate) nonwovens *Fibers Polym.* **5** 122–7

[51] Dersch R *et al* 2003 Electrospun nanofibers: internal structure and intrinsic orientation *J. Polym. Sci. Part A: Polym. Chem.* **41** 545–53

[52] Li D, Wang Y and Xia Y 2003 Electrospinning of polymeric and ceramic nanofibers as uniaxially aligned arrays *Nano Lett.* **3** 1167–71

[53] Teo W E *et al* 2007 A dynamic liquid support system for continuous electrospun yarn fabrication *Polymer* **48** 3400–5

[54] Huanga Z-M *et al* 2003 A review on polymer nanofibers by electrospinning and their applications in nanocomposites *Compos. Sci. Technol.* **63** 2223–53

[55] Dagdeviren C *et al* 2016 Recent progress in flexible and stretchable piezoelectric devices for mechanical energy harvesting, sensing and actuation *Extreme Mech. Lett.* **9** 269–81

[56] Bognitzki M *et al* 2001 Nanostructured fibers via electrospinning *Adv. Mater.* **13** 70–2

[57] MacDiarmid A G *et al* 2001 Electrostatically-generated nanofibers of electronic polymers *Synth. Met.* **119** 27–30

[58] Buchko C J *et al* 1999 Processing and microstructural characterization of porous biocompatible protein polymer thin films *Polymer* **40** 7397–407

[59] Koombhongse S, Liu W X and Reneker D H 2001 Flat polymer ribbons and other shapes by electrospinning *J. Polym. Sci., Part B: Polym. Phys.* **39** 598–606

[60] Ribeiro C *et al* 2010 Influence of processing conditions on polymorphism and nanofiber morphology of electroactive poly(vinylidene fluoride) electrospun membranes *Soft Mater.* **8** 274–87

[61] Liu H Q and Hsieh Y L 2002 Ultrafine fibrous cellulose membranes from electrospinning of cellulose acetate *J. Polym. Sci., Part B: Polym. Phys.* **40** 2119–29

[62] Bergshoef M M and Vancso G J 1999 Transparent nanocomposites with ultrathin, electrospun Nylon-4,6 fiber reinforcement *Adv. Mater.* **11** 1362–5

[63] Kenawy E R *et al* 2003 Electrospinning of poly (ethylene-covinyl alcohol) fibers *Biomaterials* **24** 907–13

[64] Kenawy E R and Abdel-Fattah Y R 2002 Antimicrobial properties of modified and electrospn poly (vinyl phenol) *Macromol. Biosci.* **2** 261–6

[65] Wang X Y *et al* 2002 Highly sensitive optical sensors using electrospun polymeric nanofibrous membranes *Mater. Res. Soc. Symp. Proc.* **708** 397–402

[66] Torres B 2001 Ultrafine fibers of polystyrene dissolved in tetrahydrofuran prepared using the electrospinning method *Proc. National Conf. on Undergraduate Research* vol 15 pp 1–5

[67] Theron A, Zussman E and Yarin A L 2001 Electrostatic field-assisted alignment of electrospun nanofibres *Nanotechnology* **12** 384–90
[68] Wua C M *et al* 2015 Sound absorption of electrospun polyvinylidene fluoride/carbon nanotube membranes *20th Int. Conf. on Composite Materials Copenhagen*
[69] Xia W and Zhang Z 2018 PVDF-based dielectric polymers and their applications in electronic materials *IET Nanodielectr.* **1** 17–31
[70] Li L *et al* 2014 Studies on the transformation process of PVDF from α to β phase by stretching *RSC Adv.* **4** 3938–43
[71] Sencadas V, Gregorio Jr R and Lanceros-Méndez S 2009 α to β phase transformation and microestructural changes of PVDF films induced by uniaxial stretch *J. Macromol. Sci., Part B: Phys.* **48** 514–25
[72] Gregorio R Jr and Borges D S 2008 Effect of crystallization rate on the formation of the polymorphs of solution cast poly(vinylidene fluoride) *Polymer* **49** 4009–16
[73] Sharma M, Madras G and Bose S 2014 Process induced electroactive β-polymorph in PVDF: effect on dielectric and ferroelectric properties *Phys. Chem. Chem. Phys.* **16** 14792–9
[74] Okada D *et al* 2015 Colloidal crystallization and ionic liquid induced partial β-phase transformation of poly(vinylidene fluoride) nanoparticles *Macromolecules* **48** 2570–5
[75] Martins P *et al* 2012 On the origin of the electroactive poly(vinylidene fluoride) β-phase nucleation by ferrite nanoparticles via surface electrostatic interactions *Cryst. Eng. Comm.* **14** 2807–11
[76] Kim G H, Hong S M and Seo Y 2009 Piezoelectric properties of poly(vinylidene fluoride) and carbon nanotube blends: β-phase development *Phys. Chem. Chem. Phys.* **11** 10506–12
[77] Miranda D *et al* 2009 Influence of silver nanoparticles concentration on the α- to β-phase transformation and the physical properties of silver nanoparticles doped poly(vinylidene fluoride) nanocomposites *J. Nanosci. Nanotechnol.* **9** 2910–6
[78] Ouyang Z-W, Wu T-M and Chen E-C 2015 Enhanced piezoelectric and mechanical properties of electroactive polyvinylidene fluoride/iron oxide composites *Mater. Chem. Phys.* **1** 149–50
[79] Nandi A K and Mandelkern L 1991 The influence of chain structure on the equilibrium melting temperature of poly(vinylidene fluoride) *J. Polym. Sci., Part B: Polym. Phys.* **29** 1287–97
[80] Wang Y R *et al* 2011 A flexible piezoelectric force sensor based on PVDF fabrics *Smart Mater. Struct.* **20** 045009
[81] Liu Z H *et al* 2014 Direct-write PVDF nonwoven fiber fabric energy harvesters via the hollow cylindrical near-field electrospinning process *Smart Mater. Struct.* **23** 025003
[82] Yu L *et al* 2013 Piezoelectric performance of aligned PVDF nanofibers fabricated by electrospinning and mechanical spinning *2013 13th IEEE Int. Conf. on Nanotechnology Beijing*
[83] Yun J S *et al* 2016 The fabrication and characterization of piezoelectric PZT/PVDF electrospun nanofiber composites *Nanomater. Nanotechnol.* **6** 20
[84] Shashank P *et al* 2009 Piezoelectric and magnetoelectric thick films for fabricating power sources in wireless sensor nodes *Sensors* **9** 6362–84
[85] Li D and Xia Y 2004 Electrospinning of nanofibers: reinventing the wheel? *Adv. Mater.* **16** 1151–70
[86] Kumar P S *et al* 2014 Hierarchical electrospun nanofibers for energy harvesting, production and environmental remediation *Energy Environ. Sci.* **7** 3192–222

[87] Wan T, Ramakrishna S and Yong L 2018 Recent progress in electrospinning TiO_2 nanostructured photo-anode of dye-sensitized solar cells *J. Appl. Polym. Sci.* **135** 45649
[88] Luan J and Li Y 2014 Photocatalytic water splitting for hydrogen production with Gd_2MSbO_7 (M = Fe, In, Y) photocatalysts under visible light irradiation *Materials* **8** 16–30
[89] Ico G *et al* 2016 Size-dependent piezoelectric and mechanical properties of electrospun P (VDF-TrFE) nanofibers for enhanced energy harvesting *J. Mater. Chem.* A **4** 2293–304
[90] Zhu J, Jia L and Huang R 2017 Electrospinning poly(l-lactic acid) piezoelectric ordered porous nanofibers for strain sensing and energy harvesting *J. Mater. Sci. Mater. Electron.* **28** 12080–5
[91] Liu K *et al* 2018 Piezoelectric energy harvesting and charging performance of $Pb(Zn_{1/3}Nb_{2/3})O_3$–$Pb(Zr_{0.5}Ti_{0.5})O_3$ nanoparticle-embedded P(VDF-TrFE) nanofiber composite sheets *Compos. Sci. Technol.* **168** 296–302
[92] Sahatiya P, Kannan S and Badhulika S 2018 Few layer MoS_2 and *in situ* poled PVDF nanofibers on low cost paper substrate as high performance piezo-triboelectric hybrid nanogenerator: Energy harvesting from handwriting and human touch *Appl. Mater. Today* **13** 91–9
[93] Dineva P S *et al* 2014 *Piezoelectric Materials* (Switzerland: Springer), p 119
[94] Gerhard-Multhaupt R 2002 Less can be more. Holes in polymers lead to a new paradigm of piezoelectric materials for electret transducers *IEEE Trans. Dielectr. Electr. Insul.* **9** 850–9
[95] Ounaies Z, Young J A and Harrison J S 1999 An overview of the piezoelectric phenomenon in amorphous polymers *Field Responsive Polymers* (Washington, DC: American Chemical Society), pp 88–103
[96] Mark H F 2013 *Encyclopedia of Polymer Science and Technology, Concise* (New York: Wiley)
[97] Naohiro M and Hiroshi O 1983 Piezoelectric polymers and their applications *Jpn. J. Appl. Phys.* **22** 3
[98] Nakamura K 2012 *Ultrasonic Transducers: Materials and Design for Sensors, Actuators and Medical Applications* (Amsterdam: Elsevier)
[99] Mendes S F *et al* 2012 Effect of filler size and concentration on the structure and properties of poly(vinylidene fluoride)/$BaTiO_3$ nanocomposites *J. Mater. Sci.* **47** 1378–88
[100] Patro T U *et al* 2008 Studies on poly(vinylidene fluoride)–clay nanocomposites: effect of different clay modifiers *Polymer* **49** 3486–99
[101] Dias J C *et al* 2016 Electromechanical actuators based on poly(vinylidene fluoride) with $[N_{1112(OH)}][NTf_2]$ and $[C_2mim]$ $[C_2SO_4]$ *J. Mater. Sci.* **51** 9490–503
[102] Martins P *et al* 2011 Optimizing piezoelectric and magnetoelectric responses on $CoFe_2O_4$/P (VDF-TrFE) nanocomposites *J. Phys. D: Appl. Phys.* **44** 495303
[103] Xu Y 2013 *Ferroelectric Materials and Their Applications* (Amsterdam: Elsevier)
[104] Sousa R E *et al* 2014 Microstructural variations of poly(vinylidene fluoride co-hexafluoropropylene) and their influence on the thermal, dielectric and piezoelectric properties *Polym. Test.* **40** 245–55
[105] Sousa R E *et al* 2015 Tailoring poly(vinylidene fluoride-co-chlorotrifluoroethylene) microstructure and physicochemical properties by exploring its binary phase diagram with dimethylformamide *J. Polym. Sci., Part B: Polym. Phys.* **53** 761–73
[106] Cardoso V F *et al* 2018 Fluorinated polymers as smart materials for advanced biomedical applications *Polymers* **10** 161

[107] Dong R *et al* 2016 First-principles simulations of PVDF copolymers with high dielectric energy density: PVDF-HFP and PVDF-BTFE *Phys. Rev.* B **94** 014210
[108] Fukada E 2005 Recent developments of piezoelectric polymers – polyurea, odd-nylon, vinylidenecyanide copolymer, and polylactic acid *12th Int. Symp. on Electrets*
[109] Ribeiro C *et al* 2015 Piezoelectric polymers as biomaterials for tissue engineering applications *Colloids Surf.* B **136** 46–55
[110] Li Y *et al* 2004 Dielectric and piezoelecrtic properties of lead-free $(Na_{0.5}Bi_{0.5})TiO_3$–$NaNbO_3$ ceramics *Mater. Sci. Eng.:* B **112** 5–9
[111] Petzelt J 2010 Dielectric grain-size effect in high-permittivity ceramics *Ferroelectrics* **400** 117–34
[112] Hu S *et al* 2019 Multifunctional piezoelectric elastomer composites for smart biomedical or wearable electronics *Compos. Part B: Eng.* **160** 595–604
[113] Saha-Dasgupta T 2014 Ferroic properties in bi-component perovskites: artificial superlattices and naturally forming compounds *J. Phys. Condens. Matter.* **26** 193201
[114] Kang M-G *et al* 2016 Recent progress on PZT based piezoelectric energy harvesting technologies *Actuators* **5** 5
[115] Cheng L-Q and Li J-F 2016 A review on one dimensional perovskite nanocrystals for piezoelectric applications *J. Materiomics* **2** 25–36
[116] Waser R 1992 Ceramic materials for electronics; processing, properties, and applications 2nd edn; R C Buchanan *Marcel Dekker Inc.* New York 1991; XII, 532 pp., ISBN 0-8247-8194-5 *Adv. Mater.* **4** 311
[117] Kenji U 2015 Glory of piezoelectric perovskites *Sci. Technol. Adv. Mater.* **16** 046001
[118] Yao Z *et al* 2010 Morphotropic phase boundary of $(Bi_{0.9}La_{0.1})ScO_3$-$PbTiO_3$ piezoelectric ceramics for high-temperature application *Ferroelectrics* **409** 21–6
[119] Pardo L and Ricote J 2011 *Multifunctional Polycrystalline Ferroelectric Materials: Processing and Properties* (Berlin: Springer)
[120] Reichmann K, Feteira A and Li M 2015 Bismuth sodium titanate based materials for piezoelectric actuators *Materials* **8** 8467–95
[121] Dunmin L *et al* 2008 Lead-free piezoelectric ceramic $(K_{0.5}Na_{0.5})NbO_3$ with MnO_2 and $K_{5.4}Cu_{1.3}Ta_{10}O_{29}$ doping for piezoelectric transformer application *Smart Mater. Struct.* **17** 035002
[122] Panda P K 2009 Review: environmental friendly lead-free piezoelectric materials *J. Mater. Sci.* **44** 5049–62
[123] Pan Z *et al* 2015 Enhanced piezoelectric properties of tetragonal $(Bi_{1/2}K_{1/2})TiO_3$ lead-free ceramics by substitution of pure Bi-based $Bi(Mg_{2/3}Nb_{1/3})O_3$ *J. Am. Ceram. Soc.* **98** 104–8
[124] Ye Z G 2008 *Handbook of Advanced Dielectric, Piezoelectric and Ferroelectric Materials: Synthesis, Properties and Applications* (Amsterdam: Elsevier)
[125] Korir K K, Cicero G and Catellani A 2013 Piezoelectric properties of zinc oxide nanowires: an *ab initio* study *Nanotechnology* **24** 475401
[126] Wang X *et al* 2007 Direct-current nanogenerator driven by ultrasonic waves *Science* **316** 102–5
[127] Mansingh A and Bawa S S 1974 Dielectric properties of compressed Rochelle-salt powders *Phys. Status Solidi.* A **21** 725–31
[128] Wei C and Jing X 2017 A comprehensive review on vibration energy harvesting: Modelling and realization *Renew. Sustain. Energy Rev.* **74** 1–18

[129] Kong L B *et al* 2002 Lead zinc niobate (PZN)–barium titanate (BT) ceramics from mechanochemically synthesized powders *Mater. Res. Bull.* **37** 2491–8

[130] Thorp J S *et al* 1990 The dielectric properties of aluminium nitride substrates for microelectronics packaging *J. Mater. Sci.* **25** 4965–71

[131] Mortet V *et al* 2004 Physical properties of polycrystalline aluminium nitride films deposited by magnetron sputtering *Diamond Relat. Mater.* **13** 1120–4

[132] Jacob M V *et al* 2003 Lithium tantalate – a high permittivity dielectric material for microwave communication systems *TENCON 2003. Conf. on Convergent Technologies for Asia-Pacific Region*

[133] Zhang S J *et al* 2008 High-performance, high-TC piezoelectric crystals *Handbook of Advanced Dielectric, Piezoelectric and Ferroelectric Materials* ed Z-G YeZ-G (Cambridge: Woodhead), ch 5, pp 130–57

[134] Yue W and Yi-jian J 2003 Crystal orientation dependence of piezoelectric properties in $LiNbO_3$ and $LiTaO_3$ *Opt. Mater.* **23** 403–8

[135] Han P, Tian J and Yan W 2008 Bridgman growth and properties of PMN–PT-based single crystals *Handbook of Advanced Dielectric, Piezoelectric and Ferroelectric Materials* ed Z-G YeZ-G (Cambridge: Woodhead), ch 1, pp 3–37

[136] Guo R *et al* 2018 Comprehensive analysis of Mn:PIN-PMN-PT single crystals for Class IV flextensional transducer *Ceram. Int.* **44** 2864–8

[137] Lee H Y 2008 Development of high-performance piezoelectric single crystals by using solid-state single crystal growth (SSCG) method *Handbook of Advanced Dielectric, Piezoelectric and Ferroelectric Materials* ed Z-G YeZ-G (Cambridge: Woodhead), ch 6, pp 158–72

[138] Lim L C 2008 Flux growth and characterization of PZN–PT and PMN–PT single crystals *Handbook of Advanced Dielectric, Piezoelectric and Ferroelectric Materials* ed Z-G Ye (Cambridge: Woodhead), ch 2, pp 38–72

[139] Chaudhari V A and Bichile G K 2013 Synthesis, structural, and electrical properties of pure $PbTiO_3$ ferroelectric ceramics *Smart Mater. Res.* **2013** 9

[140] Kurosawa M K 2012 Thin-film PZT-based transducers *Ultrasonic Transducers* ed K Nakamura (Cambridge: Woodhead), ch 4, pp 117–53

[141] Yoo J *et al* 2004 Dielectric and piezoelectric characteristics of lead-free $Bi_{0.5}(Na_{0.84}K_{0.16})_{0.5}TiO_3$ ceramics substituted with Sr *Mater. Lett.* **58** 3831–5

[142] Kantha P *et al* 2014 Structural and electrical properties of BZT-added BNLT ceramics *Ceram. Int.* **40** 4251–6

[143] Elvin N and Erturk A 2013 *Advances in Energy Harvesting Methods* 1st edn (Berlin: Springer)

[144] Lu Q Q *et al* 2016 Dynamic responses of SMA-epoxy composites and application for piezoelectric energy harvesting *Compos. Struct.* **153** 843–50

[145] Koh S J A *et al* 2011 Dielectric elastomer generators: how much energy can be converted? *IEEE-ASME Trans. Mechatronics* **16** 33–41

[146] Fuh Y K *et al* 2017 Self-powered active sensor with concentric topography of piezoelectric fibers *Nanoscale Res. Lett.* **12** 8

[147] Gao H *et al* 2018 High-performance flexible yarn for wearable piezoelectric nanogenerators *Smart Mater. Struct.* **27** 11

[148] Kang S B *et al* 2017 Enhanced piezoresponse of highly aligned electrospun poly(vinylidene fluoride) nanofibers *Nanotechnology* **28** 11

[149] Liu Z H *et al* 2014 Direct-write PVDF nonwoven fiber fabric energy harvesters via the hollow cylindrical near-field electrospinning process *Smart Mater. Struct.* **23** 11

[150] Beringer L T *et al* 2015 An electrospun PVDF-TrFe fiber sensor platform for biological applications *Sens. Actuators,* A **222** 293–300

[151] Lang C H *et al* 2017 High-output acoustoelectric power generators from poly(vinylidenefluoride-co-trifluoroethylene) electrospun nano-nonwovens *Nano Energy* **35** 146–53

[152] Chen X X *et al* 2017 Flexible fiber-based hybrid nanogenerator for biomechanical energy harvesting and physiological monitoring *Nano Energy* **38** 43–50

[153] Nunes-Pereira J *et al* 2013 Energy harvesting performance of piezoelectric electrospun polymer fibers and polymer/ceramic composites *Sens. Actuators,* A **196** 55–62

[154] Fuh Y K *et al* 2016 All-fiber transparent piezoelectric harvester with a cooperatively enhanced structure *Nanotechnology* **27** 10

[155] Yu H *et al* 2013 Enhanced power output of an electrospun PVDF/MWCNTs-based nanogenerator by tuning its conductivity *Nanotechnology* **24** 9

[156] Lee C *et al* 2016 Electrospun uniaxially-aligned composite nanofibers as highly-efficient piezoelectric material *Ceram. Int.* **42** 2734–40

[157] Siddiqui S *et al* 2016 A durable and stable piezoelectric nanogenerator with nanocomposite nanofibers embedded in an elastomer under high loading for a self-powered sensor system *Nano Energy* **30** 434–42

[158] Kang H B *et al* 2015 (Na,K)NbO_3 nanoparticle-embedded piezoelectric nanofiber composites for flexible nanogenerators *Compos. Sci. Technol.* **111** 1–8

[159] Liu K *et al* 2018 Piezoelectric energy harvesting and charging performance of $Pb(Zn_{1/3}Nb_{2/3})O_3$–$Pb(Zr_{0.5}Ti_{0.5})O_3$ nanoparticle-embedded P(VDF-TrFE) nanofiber composite sheets *Compos. Sci. Technol.* **168** 296–302

[160] Abolhasani M M, Shirvanimoghaddam K and Naebe M 2017 PVDF/graphene composite nanofibers with enhanced piezoelectric performance for development of robust nanogenerators *Compos. Sci. Technol.* **138** 49–56

[161] Shi K M *et al* 2018 Synergistic effect of graphene nanosheet and $BaTiO_3$ nanoparticles on performance enhancement of electrospun PVDF nanofiber mat for flexible piezoelectric nanogenerators *Nano Energy* **52** 153–62

[162] Ghosh S K and Mandal D 2018 Synergistically enhanced piezoelectric output in highly aligned 1D polymer nanofibers integrated all-fiber nanogenerator for wearable nano-tactile sensor *Nano Energy* **53** 245–57

[163] Mandal D, Henkel K and Schmeisser D 2014 Improved performance of a polymer nanogenerator based on silver nanoparticles doped electrospun P(VDF-HFP) nanofibers *Phys. Chem. Chem. Phys.* **16** 10403–7

[164] Adhikary P, Biswas A and Mandal D 2016 Improved sensitivity of wearable nanogenerators made of electrospun Eu^{3+} doped P(VDF-HFP)/graphene composite nanofibers for self-powered voice recognition *Nanotechnology* **27** 11

[165] Sultana A *et al* 2016 Fabrication of wearable semiconducting piezoelectric nanogenerator made with electrospun-derived zinc sulfide nanorods and poly(vinyl alcohol) nanofibers *Transl. Mater. Res.* **3** 11

[166] Zhu J X, Jia L Y and Huang R 2017 Electrospinning poly(L-lactic acid) piezoelectric ordered porous nanofibers for strain sensing and energy harvesting *J. Mater. Sci.-Mater. Electron.* **28** 12080–5

[167] Wang F F *et al* 2015 High quality barium titanate nanofibers for flexible piezoelectric device applications *Sens. Actuators,* A **233** 195–201
[168] Yang Q *et al* 2016 Lead-free $(Na_{0.83}K_{0.17})_{0.5}Bi_{0.5}TiO_3$ nanofibers for wearable piezoelectric nanogenerators *J. Alloys Compd.* **688** 1066–71
[169] Chen X *et al* 2010 1.6 V nanogenerator for mechanical energy harvesting using PZT Nanofibers *Nano Lett.* **10** 2133–7
[170] Cui N Y *et al* 2012 Magnetic force driven nanogenerators as a noncontact energy harvester and sensor *Nano Lett.* **12** 3701–5
[171] Wu W W *et al* 2012 Lead zirconate titanate nanowire textile nanogenerator for wearable energy-harvesting and self-powered devices *ACS Nano* **6** 6231–5
[172] Gu L *et al* 2013 Flexible fiber nanogenerator with 209 V output voltage directly powers a light-emitting diode *Nano Lett.* **13** 91–4
[173] Kang H B *et al* 2014 High quality Mn-doped $(Na,K)NbO_3$ nanofibers for flexible piezoelectric nanogenerators *ACS Appl. Mater. Interfaces* **6** 10576–82

Chapter 3

Triboelectric effect and triboelectric energy generators

Yuying Cao, Hao Shao, Hongxia Wang, Tong Lin and Jian Fang

More than 25 centuries ago, the Greek scientist Thales of Miletus first observed electrostatic charging by rubbing amber with animal furs. The rubbed amber could attract small particles of dust due to the electrostatic force, and this was the first record on electricity. This phenomenon was later known as a surface charging process induced by contact electrification.

Static charges exist everywhere in our daily life and in different industrial activities. We have a long history of observing and understanding electrostatic charge generation processes, as well as utilizing and avoiding these static charges. Electrostatic induction and contact electrification are two main principles in generating static charges. We have achieved a clear understanding of the mechanism of electrostatic induction, which is a process of redistributing charges under the interaction of an external electric field. However, there are still debates about the mechanism of contact electrification.

Along with the advancements of theoretical understanding, a large number of practical applications of static charges have been explored. On one hand, we try to avoid static electricity because the electrostatic discharge can cause many safety issues like the accidental explosion of explosives and electromagnetic interference of electronic devices. Approaches such as reducing contact area, choosing neutral materials on triboelectric series, and improving the electrical conductivity of materials have been widely used to reduce static charge accumulation. On the other hand, lots of efforts have also been made to benefit human activities using static electricity. For example, electrostatics have been widely used in printing, painting, air cleaning, and various types of power generators. However, low current output, bulky device structure, and low energy conversion efficiency have hindered the

doi:10.1088/978-0-7503-2005-4ch3

practical application of triboelectric power generators using conventional materials for high-power electricity.

In 2012, the triboelectric nanogenerator (TENG) was invented by coupling electrostatic induction and contact electrification. After that, a considerable amount of research efforts has been made to harvest small mechanical energies from our daily activities or surrounding environment, and convert them into useful triboelectric signals for powering low energy consumption portable and wearable electronics [1–6].

At present, there are four fundamental device structures for triboelectric power generation: the vertical contact–separation mode, lateral sliding mode, single-electrode mode, and freestanding triboelectric layer mode. Theoretically, the materials at the two ends of a triboelectric series are more promising in generating a high level of triboelectric outputs. For example, PVDF [7, 8], PTFE [2, 9], and PDMS [1] are ideal negative triboelectric materials due to their strong ability of gaining electrons. Meanwhile, electron-losing polyamide (PA), nylon [2], glass [2, 10], and PMMA [10] are commonly chosen as positive triboelectric materials.

For obtaining high-density static charge generation, different microstructures have been introduced onto the triboelectric material surface to increase the specific surface area and enhance surface interaction during triboelectric charge generation. One of the promising materials that attracted a lot of research interest is electrospun nanofibers. Nanofibrous materials prepared by electrospinning processes have a very high specific surface area. Electrospinning is the most promising method to fabricate continuous nanofibers to form nanofibrous mats with many unique features such as naturally carried electric charges, controllable fiber surface and morphology, high porosity, and wide range of suitable material selection [11]. In this chapter, the fundamentals of triboelectric charge generation will be discussed in detail, along with the recent developments of TENGs.

3.1 Static electricity

Materials keep electrical neutrality because of the equal quantity of positive and negative charges they normally carry. Static electricity is a phenomenon of charge imbalance within or on the surface of materials. Charge imbalance can result from the separation of two contacted materials or even non-contact processes. The imbalanced charges will flow under the electric potential if the materials are conductive, and then the materials return back to the electrically neutral state. If a material is electrically insulating, the generated charges will be preserved at the place of generation and show static electricity.

Charge accumulation on a material surface will build an electric field. Generally, static electricity can result in a high potential (0.1–100 kV) but the electric current is normally low (0.01–100 μA) [12]. The electric field strength rises with the increasing surface charge density. This process will continue until the breakdown strength is reached at surrounding atmosphere. The strong electric field can cause ionization of the atmosphere and the flow of ions will generate a short-live current flow, which is the typical process of electrostatic discharge.

3.1.1 History of electrostatics

In ancient Greece, scientist Thales of Miletus was the first to observe electrostatic charging by rubbing amber with animal fur. The rubbed amber can attract small particles of dust due to electrostatic force. The Greek word 'elektron', which means amber, was used to name the negatively charged elemental particles [13]. Later on, William Gilbert found that in addition to amber, many other materials could also attract small particles after rubbing with different materials. The attracting force was named 'via electrica', and the materials were distinguished as 'electriks', which, like amber, could be charged, and 'noelectriks', which could not be charged. Even though the theory of electrostatic attraction had not been understood at that time, it was clearly noticed that the electrostatic attraction was not the same phenomenon as magnetic attraction. These observations attracted the interest of many scientists, and a large number of experiments were carried out to prove the existence of electrica. Around 1663, Otto von Guericke invented a primitive form of a frictional electric generator using a rotated and rubbed sulfur ball to generate static electricity. Both attractive and repulsive forces were observed via placing different bodies close to the sulfur ball. Grericke was also considered as the first person to produce artificial electroluminescence by wrapping the sulfur ball with dry hands.

English physicist Stephen Gray was the first person to make the distinction between conduction and insulation. In 1729, Gray rubbed a glass tube with dry paper to make it electrified, and then placed two tampions close to the tube. He found the tampions could also attract fur so he believed that the electricity was transferable from the glass tube to the tampions. After testing more materials using the same method, he found that materials such as iron, copper (Cu), and silver could convey electricity, and therefore he named them conductors. The materials that could not convey electricity were insulators, and included silk and stone.

Benjamin Franklin proved that sparks are similar to lightning in his famous kite experiment. Both Benjamin Franklin and William Watson considered that it was electric matter that caused the electrical phenomenon. They further named the body with an excess of electric matter as a 'plus electric', and the body with little electric matter as a 'minus electric'. In 1759, Robert Symmer found that electricity was not based on one electric matter but two. In a normal condition, a body contains equal quantities of two different electric matters. The body will show electricity once one electric matter is in a greater quantity than the other. Based on this fundamental theory, J C Wilcke found that two bodies repelled each other when they carried the same type of charges, but attracted each other when they showed opposite types of charges. Later on, Georg Christoph Lichtenberg introduced the terms 'positive' and 'negative' to replace 'plus electric' and 'minus electric'. Joseph Priestley described the electrical attractive and repulsive forces, which showed similar formulations as gravity; this was later proven by Charles Coulomb in his Coulomb's law, which was the first qualitative law of electricity.

3.1.2 Electrostatic induction

Electrostatic induction is a process of redistributing charges under the interaction of an external electric field. When a charged body approaches an uncharged conductor, the charges within the conductor will redistribute. The opposite charges are attracted to the vicinity of the charged body, and the same charges will be repelled to the opposite direction, that is, away from the charged body. This principle has been used to design electrostatic generation devices such as the Wimshurst machine, Van de Graaff generator, and electrophorus.

Not only conductors can be electrified by charged objects; dielectric objects also have an electrostatic induction effect. For insulators, electrons are bonded by atoms or molecules and not free to move within the material. However, the electrons can move slightly within a microscale space. If a positively charged object is approaching a dielectric object, the electrons in each atom will be attracted and move slightly, while the atomic nuclei will be repelled and move in the opposite direction. This phenomenon is named polarization. Since the negative charges within the object are closer to the charged body than the positive charges, the attractive force is slightly greater than the repulsive force. Even though polarization happens at a microscopic level, the accumulative force of a large number of atoms is enough to move light objects like shredded paper. Electrostatic induction will cause charge accumulation on a material surface, which can be hazardous to human health and general safety.

3.2 Contact electrification

Contact electrification is a surface charging process. When the distance of two materials is close to the atom level, charge transfer can be induced at the contact surface by a complicated process consisting of electron transfer, ion transfer, or material transfer, which electrifies the materials. Contact electrification can lead to a very high electric potential between two surfaces. When the potential is high enough to reach the breakdown potential during separation, dielectric discharge occurs. Since the sparks during a discharge can cause safety issues [12, 14], static electricity is normally considered as a hazard that should be avoided in our daily life and industry activities. The triboelectric effect is a special contact electrification process. It is usually induced by frictional contact between different materials, which is the general reason for daily static electricity. Even though the contact electrification phenomenon has been known for more than 2000 years, there are still many debates about its mechanism. One critical point is the charging media: electron transfer, ion transfer, or material transfer.

3.2.1 Electron transfer

Electron transfer is normally considered to happen in all kinds of contact electrifications, and it can play a dominant role in charge transfer (especially when involving metals or inorganic insulators) during contact electrification.

3.2.1.1 Metal–metal charge generation

The triboelectric effect is often treated as a phenomenon occurring between insulators. In fact, the contact between metals can also lead to a great charge transfer between them. This charge transfer is normally neglected because metals have good conductivity and a weak capability of accumulating charges. Excess charges on metals are easily transferred unless protective measures are used.

It was proposed that the electron transfer induced by contact electrification between metals is based on the premise of thermodynamic equilibrium. That is, the charge transfer between two metals can unite the electrochemical potentials at the same level.

When two different metals are brought close together, electrons are exchanged because of the difference in their work functions. A thermodynamic equilibrium for the contact can consequently be achieved. This process can be described in detail by a basic physical model. As depicted in figure 3.1, for an individual metal, the work function $\varnothing$ is the work to bring an electron from vacuum to the inside of a metal, and eVs is the amount of work which must be done to remove an electron from just outside the metal surface to infinity. The electrochemical potential ξ is the minimum energy that must be given to an electron to take it from the Fermi level to infinity. From figure 3.1, the electrochemical potential ξ is the sum of work function $\varnothing$ and eVs.

When two metals are contacting, the difference between the surface potentials of the metals (V_c) is

$$V_c = (\varnothing_B - \varnothing_A)/e.$$

The amount of charge Q at contact state can be further expressed as

$$Q = V_c C_0$$

where C_0 is the effective capacitance between two metals.

When the metals are separated, an electric potential is built between the separating surfaces due to the charge transfer in the contact state. When the separating distance reaches a required limit, the electrons will flow back to the source metal. In this way, the two metals go back to their initial thermodynamic equilibrium.

3.2.1.2 Metal–insulator charge generation

The contact electrification-induced electron transfer can also happen between a metal and an insulator. Different from metals, insulators have discontinuous electron density of states in the range of energy levels. An energy band gap exists between the filled valence band and unfilled conductive band. Such a different electronic structure makes metal–insulator transfer operate on a slightly different electron transfer mechanism than metal–metal transfer.

Insulators have an energy level E_0 because of the thermodynamic equilibrium, which is a similar concept to the Fermi level of metals. In a typical case, E_0 lays in the energy gap. When an insulator contacts a metal, electron transfer happens only when E_f lays upon E_0. If the difference in surface potential V_c is large enough, the electrons of the metal can be easily transferred to the unfilled conduction band of the

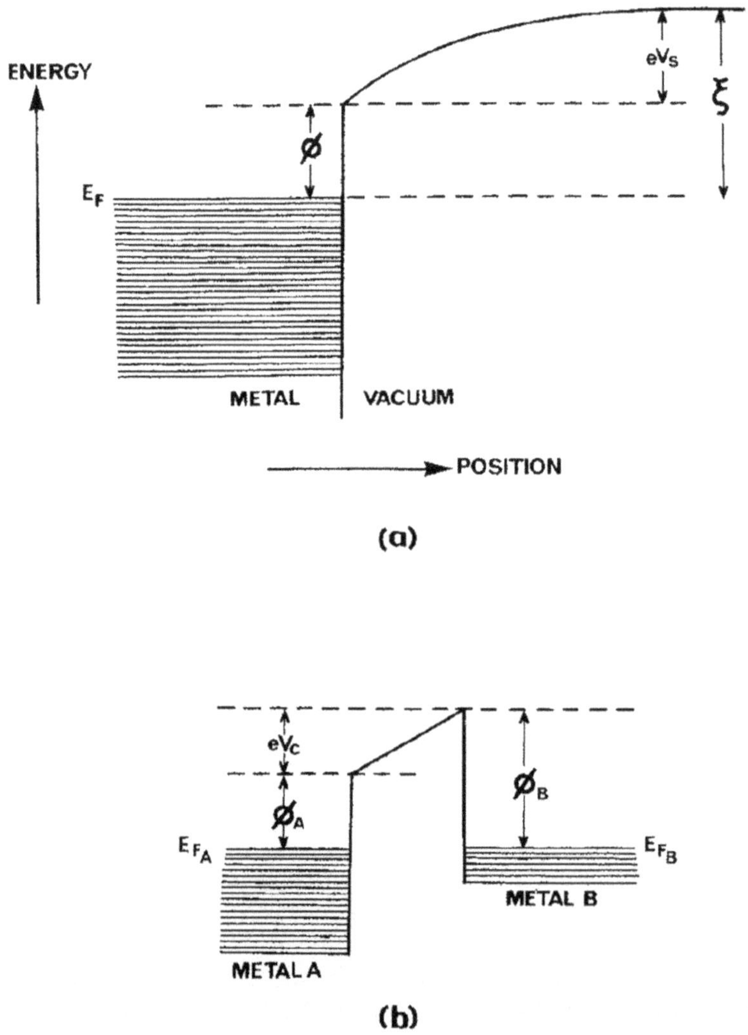

Figure 3.1. The proposed mechanism of electron transfer between metals. Reproduced with permission from Taylor & Francis [15].

insulator. Therefore, we will mainly discuss the case of a small V_c, where electrons are not easy to transfer to the conduction band of the insulator. In reality, there are many defects on the insulator surface. Affected by these defects, there are electron trap states in the band gap, as shown in figure 3.2(a). Lowell and Truscott [16] indicated that there are nonequilibrated electrons in the defects or 'traps'. Although the nonequilibrated electrons are not in their lowest energy state, they are responsible for the finite conductivity of insulators, and also provide 'accepters' for contact charging. It was reported that charges transferred to an insulator from a voltage-biased metal can be 'stored' at the surface and retrieved later when an unbiased metal is then contacted with the surface [17].

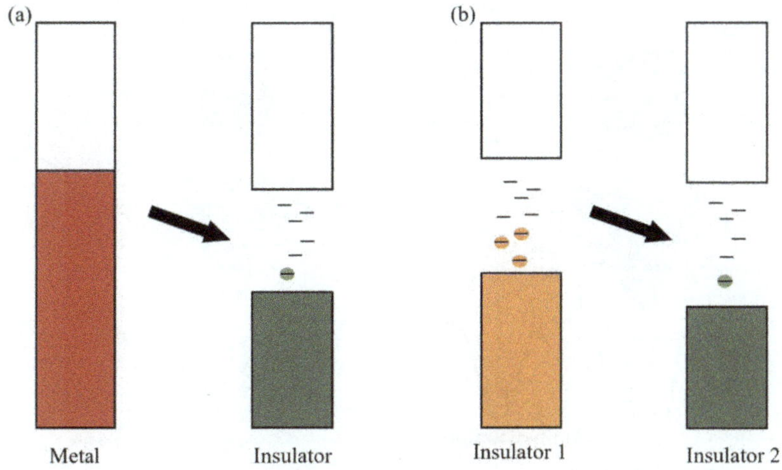

Figure 3.2. The electron transfer models of (a) metal to insulator, in which the free electrons of the metal transfer to the traps of the insulator, and (b) insulator to insulator, in which the trapped electrons in mid-gap defects of insulator 1 transfer to lower energy state traps of insulator 2 [14].

3.2.1.3 Insulator–insulator charge generation

The insulator and insulator charge generation can be generally divided into two cases: (1) an inorganic insulator contacting with another inorganic insulator, and (2) a polymer contacting with either an inorganic insulator or another polymer. In the former case, the charge generation is dominated by electron transfer. The previously discussed thermodynamic equilibrium of surface state theory can be applied. Figure 3.2(b) presents the electron transfer model. Due to the nonequilibrated electrons trapped in the insulators, electron transfer happens when insulators contact with each other. The trapped electrons in the surface state can also transfer to a lower energy state. However, this process is time-consuming, and could take years or centuries. The precise nature of these states is still unknown; however, they are attributed to conformational and chemical defects through the interaction between the two materials [18–21].

In 2018, a paper proposed an electron cloud–potential well model to explain electron transfer between two inorganic materials. It suggests that the dominant process for contact electrification between different inorganic solids is associated with different surface potentials [22].

Figure 3.3 shows the electron cloud–potential well model based on the electron emission-dominated charge transfer mechanism. The electron cloud is used to describe the probability of an electron appearing around the nuclear space. A potential well represents an atom with electrons bounded to form an electron cloud. E_A and E_B are the energy levels of electrons around the atoms of material A and B, and E_1 and E_2 are the potential energies of the electrons escaped from the atoms. When two materials come into contact with each other, the electron clouds overlap and this leads to a shielding effect of two contacting atoms. The original potential well becomes an asymmetric double-well potential where an electron could hop

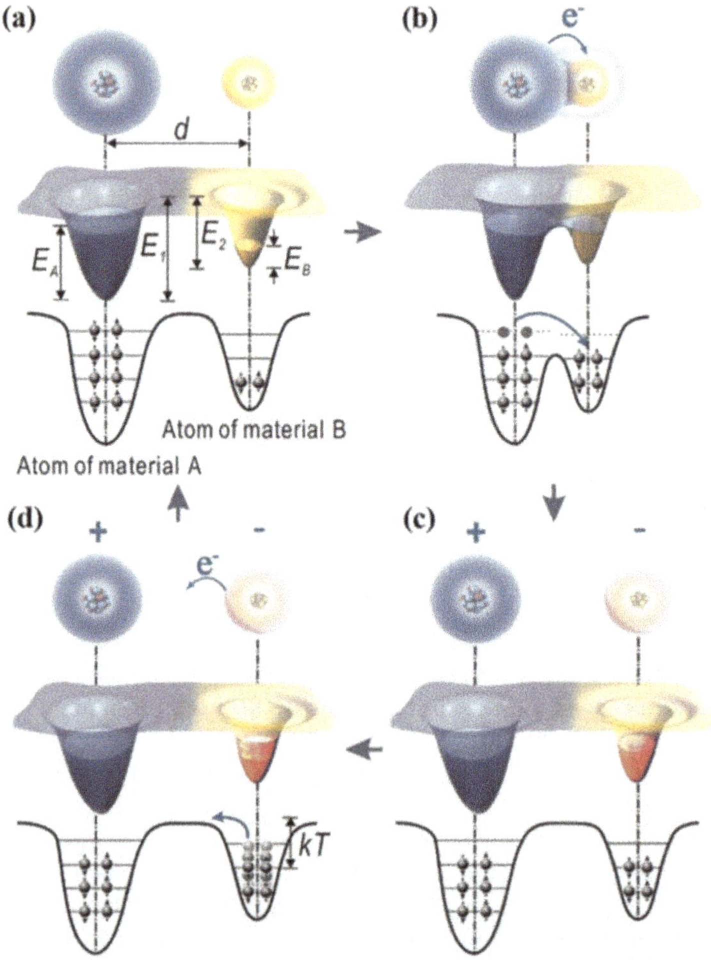

Figure 3.3. The electron cloud–potential well model of contact electrification (a) before contact, (b) during contact, and (c) after contact; (d) charge release from the atom at an elevated temperature T once kT approaches the barrier height. Reproduced with permission from Wiley [22].

between two atoms, as shown in figure 3.3(b). The electron transfer occurs from atom A to atom B. When they are separated at a normal temperature, most of the electrons will not flow back due to the existence of the energy barrier E_2 in material B (figure 3.3(c)). Material A is positively charged and material B is negatively charged due to the contact electrification. At a high temperature, the electrons will flow back, as explained in figure 3.3(d). This model can explain contact electrification between different types of materials.

Charge transfer between different polymers may consist of electron transfer, ion transfer, and material transfer, depending on the polymer types. However, the dominating process varies. Here, we focus our discussion on ion transfer and

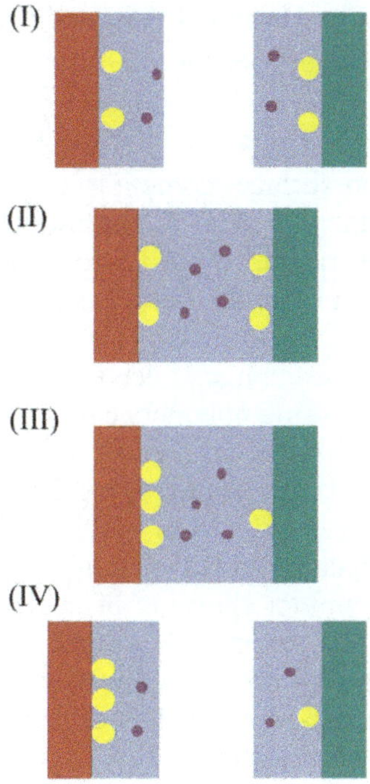

Figure 3.4. Ion transfer model, in which charge transfer is due to the redistribution of ions in adsorbed water layers on the material surface [14].

material transfer because charge transfer has been well explained in the discussion above.

3.2.2 Ion transfer

The ion transfer theory explains the charge generation of insulating materials with mobile ions: ionomers or polymers with molecular salts on their surface. They consist of strongly bound ions and loosely bound counter ions. Two type of ions show opposite charge polarities to keep the material electrically neutral. When these types of materials make contact with other materials, the loosely bound ions are easily transferred to the other surface while the strongly bound ions are left on the original surface.

This mechanism has been extended to materials without mobile ions. Generally, ion transfer happens in a water layer on the material surface. The model in figure 3.4 shows that the contact between two surfaces will lead to the redistribution of adsorbed hydroxide ions. This produces a chance to charge. When two materials separate at stage I (figure 3.4), the mobile ions in the water layer are absorbed. When two materials are moved to come into contact with each other, the two hydration layers merge with each other; as a result, ions redistribute in the layer, as shown in stages II and III in figure 3.4. When the materials are moved apart from each other, they bring extra opposite charges (stage IV). In this case, factors like humidity [23], atmosphere [23], surrounding pressure [24, 25], and surface characteristics [26] can all influence the contact-charging process.

3.2.3 Material transfer

In some cases, when two solid surfaces come into contact with each other, material transfer happens in the interfacial layers. If there are electric charges carried by the transferred materials, charge transfer occurs across the two surfaces. Actually, the materials are not only the impurities, but also the parts of the materials themselves. For example, physical contact between two polymers can lead to a breakage of chemical bonds that causes the exchange of deeper-layer materials. This is one of the reasons why the contact of the same polymers can sometimes cause charge transfer.

Generally, it is considered that within a given pair of contact-charging materials, the charge in the two materials are homogeneously positive and negative, respectively. The material transfer was ignored in previous research due to its capricious property. A recent study has provided strong evidence for material transfer. Baytekin *et al* reported the contact charging of polymers, and imaged the surface potentials by Kelvin force microscopy. They indicated that the contact materials comprised a mosaic of (+) and (−) regions. Further coherent Raman spectroscopy and x-ray photoelectron spectroscopy results verified that material transfer indeed takes place during contact electrification [27].

3.3 Triboelectric materials and triboelectric series

The relative charge generating capacity of different materials for a series of triboelectric materials are listed in table 3.1. The materials at a higher position in the table are more easily lose electrons and become positively charged; meanwhile, the materials at a lower position in the table are more likely to gain electrons and carry negative charges. The first triboelectric series was proposed by J C Wilcke in 1757 [13], and since then, various versions of the triboelectric series have been produced by different researchers. There are minor differences in the order of materials in these series because the triboelectric effect is dramatically affected by the surface condition of materials, such as impurities and surface roughness. The triboelectric effect is also sensitive to contact area, rubbing intensity, and environmental conditions (e.g. humidity and temperature).

Table 3.1 also shows the tendency of triboelectric materials to lose or gain electrons [28]. Polyformaldehyde at the top of the table is the easiest material to lose electrons while silicon rubber (at the bottom of the table position) is the easiest to gain electrons. The longer the distance between two coupled materials in the triboelectric series, the stronger triboelectric effect.

3.4 Antistatics

Static charges exist everywhere in our daily life and industrial activities, and they can cause many issues such as the accidental explosion of explosives and electric pyrotechnics. Electromagnetic radiation generated from electrostatic discharge can cause electromagnetic interference to electronic devices and information systems [29].

Table 3.1. The triboelectric series.

Positive	Polyformaldehyde	(continued)	
	Etylcellulose	Polyester	
	Polyamide 11	Polyisobutylene	
	Polyamide 6-6	Polyurethane flexible sponge	
	Melanime formol	Polyethylene terephthalate	
	Wool, knitted	Polyvinyl butyral	
	Silk, woven	Polychlorobutadiene	
	Aluminum	Natural rubber	
	Paper	Polyacrylonitrile	
	Cotton, woven	Acrylonitrile-vinyl chloride	
	Steel	Polybisphenol carbonate	
	Wood	Polychloroether	
	Scaling wax	Polyvinylidene chloride	
	Hard rubber	Polystyrene	
	Nickel, copper	Polyethylene	
	Sulfur	Polypropylene	
	Brass, silver	Polyimide	
	Acetate, rayon	Polyvinyl chloride	
	Polymethyl methacrylate	Polyvinylidene fluoride	
	Polyvinyl alcohol	Polytrifluorochloroethylene	
	Polyester	Polytetrafluoroethylene	
	(continued)	Silicone rubber	Negative

To avoid the hazardous effects of electrostatics, several methods have been applied to reduce static charge accumulation such as reducing contact area, choosing neutral materials based on triboelectric series, and improving the electrical conductivity of materials. The first two methods are aiming to control the generation of charges, and the third one accelerates charge transfer and thus reduces charge intensity.

In the textile industry, electrostatic charges can be produced at every step of industrial production since the majority of textile materials are nonconductive. So far, three main approaches have been developed for antistatic treatment in textile manufacturing: hydrophilic treatment, preparing intrinsically antistatic fibers, and adding conductive fibers.

Hydrophilic treatment of fibers or fabrics with surfactants can increase the hygroscopicity of the textile materials, thereby reducing their electrical resistivity and accelerating the dissipation of charges. This method was firstly introduced in the 1950s and is closely related to the relative humidity of the surrounding atmosphere. In order to prepare intrinsically antistatic fibers, fiber-forming polymers are blended or copolymerized with functional groups or materials to carry hydrophilic polar groups. Fabrics made from antistatic fibers or blended in high proportions to common synthetic fibers can eliminate electrostatic problems during processing and use; however, a highly humid environment is still necessary for charge dissipation. Alternatively, many conductive materials, such as metals, carbon, and conducting polymers, can be fabricated into fibrous structures or onto a conventional fiber surface. To add some conductive fibers into traditional textiles is also an effective method to increase electrical conductivity for antistatic purposes.

3.5 Applications for energy generation

Static charges are not always a cause for concern; they also benefit human society. For example, electrostatics have been widely used in printing, painting, and air cleaning industries. Electrospraying and electrospinning, which are driven by static charges, are effective tools in the fabrication of thin films and nanofibers. Some early studies were also conducted to utilize charges for electricity generation.

The first electrostatic power generator was invented in 1663 via rubbing a sulfur globe by hand to produce electricity [30]. In 1929, American physicist Robert J Van de Graaff invented an electrostatic generator to generate a high electrical potential; this device was named a Van de Graaff generator [31]. The high electrical potential was used as a particle accelerator. As shown in figure 3.5, a conductive comb was charged by a high-voltage power supply and transferred charges to the moving belt. The moving belt continuously redeposits charges on the metal sphere through a collecting comb. A very high potential difference has been created between the metal sphere and ground. Van de Graaff generators can generate up to 10 million volts.

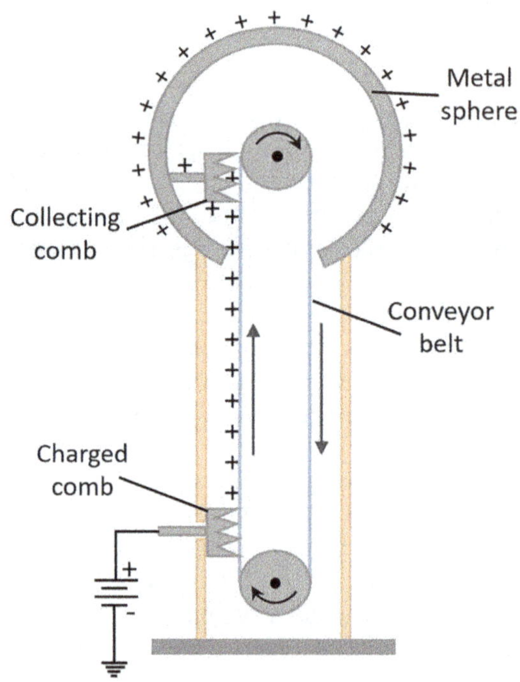

Figure 3.5. Schematic illustration of a Van de Graaff generator.

Aside from the Van de Graaff generator [32], other types of generators have also been designed and fabricated, such as the Holtz machine, Wimshurst machine [33], and Pidgeon machine [34]. However, low current output, bulky device structure, and low energy conversion efficiency have hindered the wide development of triboelectric power generators.

In recent years, after the pioneering work by Prof Zhonglin Wang at the Georgia Institute of Technology in 2012 [1], considerable research has been devoted to using nanomaterials and versatile devices to fabricate TENGs [1–3]. TENGs can efficiently convert small mechanical energies into useful electricity to meet the power generation requirements.

The function of TENGs relies on the coupling of electrostatic induction and contact electrification. In their initial work, polyethylene terephthalate and Kapton® films were selected as the triboelectric materials in a sandwiched structure [1]. Metal layers were deposited on the back surfaces of the triboelectric materials as electrodes. A voltage output of 3.3 V was generated to drive small electronic components, with an output power density of 10.4 mW cm^{-3}. Ever since this pioneering study, there have been enormous amounts of studies on TENGs for effectively harnessing mechanical energies from ambient vibrations, wind, ocean waves, human motion, and body movement, and then converting these energies to useable electricity. So far, many different applications have been demonstrated with TENGs, including but not limited to self-powered sensors, and power supplies for electrospinning systems and drug delivery system.

3.5.1 Fundamental modes of TENGs

At present, four TENGs have been reported. According to the fundamental structure, these TENGs are classified into the vertical contact–separation mode, lateral sliding mode, single-electrode mode, and freestanding triboelectric layer mode, as shown in figure 3.6.

3.5.1.1 Vertical contact–separation mode

The vertical contact–separation mode is a basic TENG mode and is also the first reported TENG. In a typical TENG structure (figure 3.7(a)), two dissimilar dielectric layers are used. Two metal layers are deposited or laminated onto the triboelectric layers as electrodes. When two triboelectric materials are brought into contact with each other, the interaction between two surfaces will generate opposite charges due to contact electrification. Induced electrons will flow through the extra circuit to balance the surface charges. The electrons will flow back once the triboelectric materials are completely separated [10].

Figure 3.7(a) describes the voltage generated from a vertical contact–separation mode TENG. In its origin state, the triboelectric layers (Kapton and poly(methyl methacrylate) (PMMA)) are placed away from each other, and therefore no electrical signals can be recorded. Driven by an external force, once the triboelectric layers make contact as shown in stage II, charge transfer occurs on surfaces due to contact electrification. Based on the positions in the triboelectric series, electron

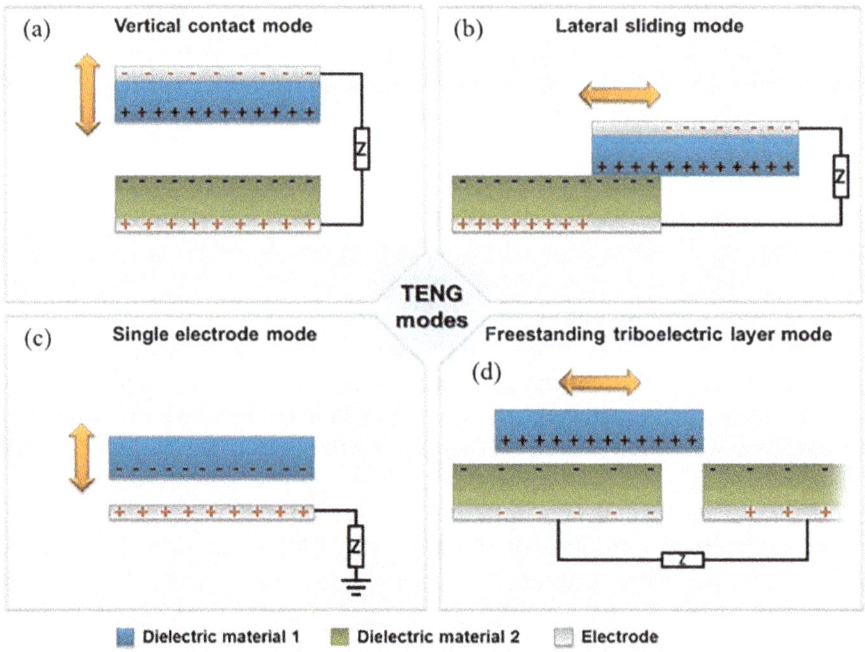

Figure 3.6. Four fundamental modes of the TENG. Reproduced with permission from Wiley [35].

transfers from the PMMA layer to the Kapton layer, leaving the PMMA surface with positive charges and the Kapton surface with negative charges. When these two surfaces are separated again, surface charges will produce an electric potential between them. With increasing distance between two triboelectric layers, the electrical potential rises until the maximum distance is reached (stages III and V). Another contact will gradually reduce the electrical potential between the triboelectric layers (stage VI). The overall voltage output signal during a pressing–releasing cycle is shown in figure 3.7(a) IV.

Figure 3.7(b) illustrates the principle of current generation of a vertical contact–separation mode TENG. There is no charge flow through the extra circuit in the origin state (I). Once the surfaces of two triboelectric materials make contact (II), the releasing force generates a growing electric potential difference that induces electron transfer through the external circuit (III). The electrons flow from the top electrode to the bottom electrode until the charges are balanced (V). When the materials are pressed together again, the electrons will flow in the reverse direction. The typical current signals are shown in figure 3.7(b) IV. In this mode, the separation distance between triboelectric layers and the speeds of contact and separation are critical factors in electric signal generation.

3.5.1.2 Lateral sliding mode

The basic structure of the lateral sliding mode is similar to that of the vertical contact–separation mode, except that it uses frictional movement for charge

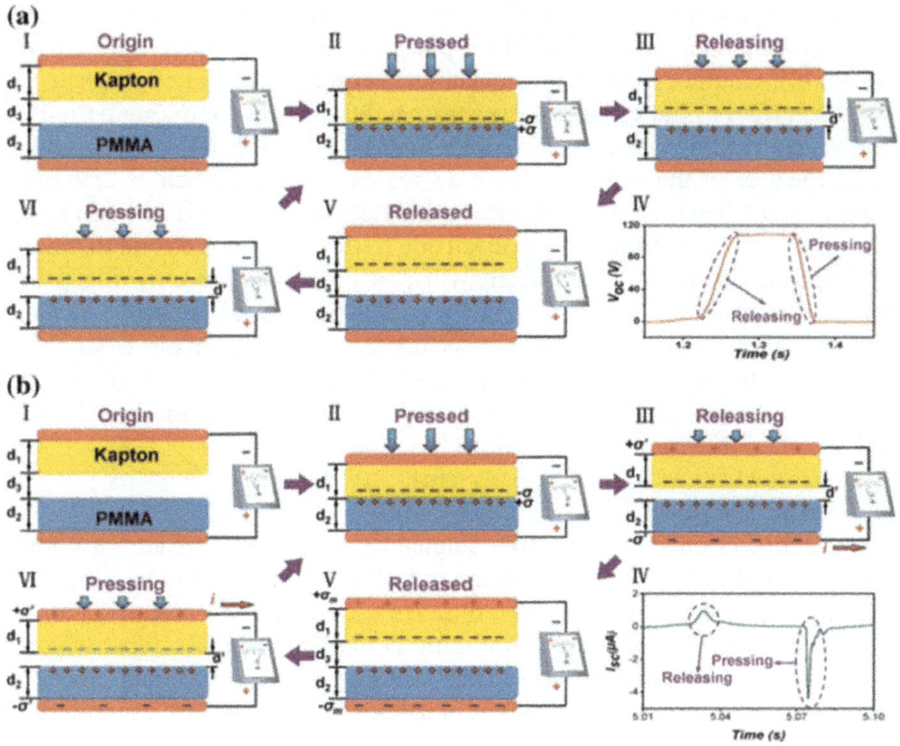

Figure 3.7. Operating principle of a vertical contact–separation mode TENG. Reproduced with permission from the American Chemical Society [10].

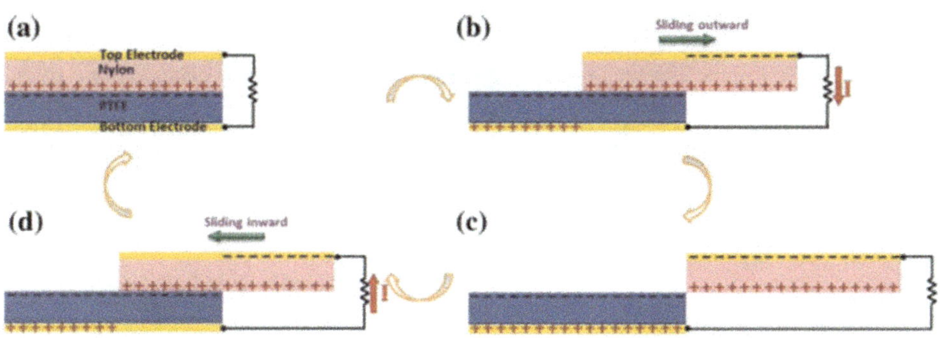

Figure 3.8. Working mechanism of TENG based on the lateral sliding mode. Reproduced with permission from the American Chemical Society [2].

generation. Both modes consist of a pair of triboelectric materials in the middle and two electrodes as the top and bottom layers. They are both considered as having an attached-electrode structure. The working mechanism of the lateral sliding mode TENG is illustrated in figure 3.8.

In a typical structure, nylon and polytetrafluoroethylene (PTFE) can be used as triboelectric materials. Nylon and PTFE plates have close contact in the origin state (figure 3.8(a)). Contact electrification generates equal quantities of opposite charges on the two surfaces. These charges can stay on the surfaces for a long period of time due to the electrical insulation of both nylon and PTFE. When the nylon plate slides outward, the area of the contact surface decreases. An electric field is then produced from left to right. This electric field induces an electrical potential between two plates that drives current flow from the top electrode to the bottom electrode to balance the charge-induced potential. The current will continue with the on-going process of sliding until the contact area of two surfaces reaches the minimum level. Once the nylon plate slides back, the current flows back from the bottom electrode to the top electrode due to reversed electrostatic induction. Unlike the vertical contact–separation mode, no air gap is required in the lateral sliding mode, which makes TENG easier to integrate in some cases.

3.5.1.3 Single-electrode mode

Unlike attached-electrode modes, the single-electrode mode includes a conducting material as the triboelectric layer. As seen in figure 3.9, an aluminum (Al) plate and a fluorinated ethylene propylene (FEP) film are used as the triboelectric pair [36]. A Cu plate is connected to the external circuit as the reference electrode, while the Al plate acts as the primary electrode. The FEP and Al plates contact closely at the original state, and a equal quantities of opposite charges are generated on the

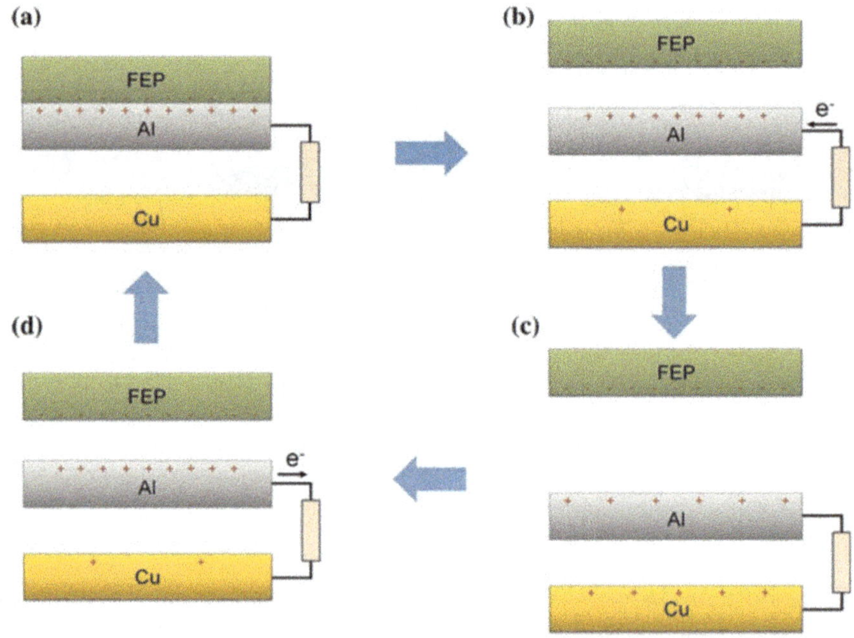

Figure 3.9. Working mechanism of the single-electrode mode TENG. Reproduced with permission from Springer Nature [36].

contact surfaces. Because both electrodes (Al and Cu plates) are fixed and connected to the external circuit, when the FEP plate moves away from the Al plate, the weakened electric field between the Al plate and the FEP film leads to an increased potential difference between the Al plate and the Cu plate. This will drive electrons to flow from the reference electrode to the primary electrode. When the FEP plate moves back to the Al plate, the electrons flow back from the primary electrode to the reference electrode to balance the potential change.

Because of the wireless feature of the moving part in the single-electrode mode TENG, this structure has been widely utilized in self-powered sensors such as displacement sensors, touch sensors, and biosensors.

3.5.1.4 Freestanding triboelectric layer mode

The freestanding triboelectric layer mode always consists of a freestanding triboelectric layer and a pair of stationary electrodes. The movement of the freestanding layer between two stationary electrodes can induce a potential difference that drives the current flow through external circuit. The freestanding triboelectric layer mode has two typical structures: a sliding freestanding triboelectric layer TENG and a contact freestanding triboelectric layer TENG. Similar to in the lateral sliding mode, the potential difference of the sliding freestanding TENG is generated via sliding the triboelectric layer on the electrodes.

Figure 3.10 describes the structure and principle of a sliding freestanding triboelectric layer TENG [37]. In this case, the FEP film is used as the freestanding negative triboelectric layer, and the Al plate acts as both the positive triboelectric material and electrode due to its top position in the triboelectric series and good electrical conductivity. In the original state, the freestanding FEP film contacts closely with one of the Al plates, and opposite charges are generated on two contact surfaces because of contact electrification. When the FEP layer slides from one electrode to another, an electric current is induced in the same direction as the sliding to balance the negative charges. Another electric current with reversed direction can be observed when the FEP layer slides back.

The working principle of the contact freestanding TENG is similar to the contact–separation mode, as shown in figure 3.11. A freestanding FEP plate is placed in the middle of two electrodes, and contact and separation between the triboelectric layer and two electrodes induce the electron flow through the external circuit. In the original state, equal quantities of opposite charges are induced on the top electrode layer and the freestanding layer. The movement of the freestanding layer will lead to a potential difference between the two electrodes, which will drive electron flow through the external circuit.

In summary, the contact–separation mode has the advantages of simple device structure, high instantaneous power density, and easy integration of multilayer devices. Different from the contact–separation mode, no air gap is required in the lateral sliding mode, which makes device encapsulation easier. Moreover, device structure can be designed as planar or rotation-disk. Both contact–separation and lateral sliding modes require external circuit connection to move triboelectric layers, which can be an obstacle in some application conditions. The single-electrode mode

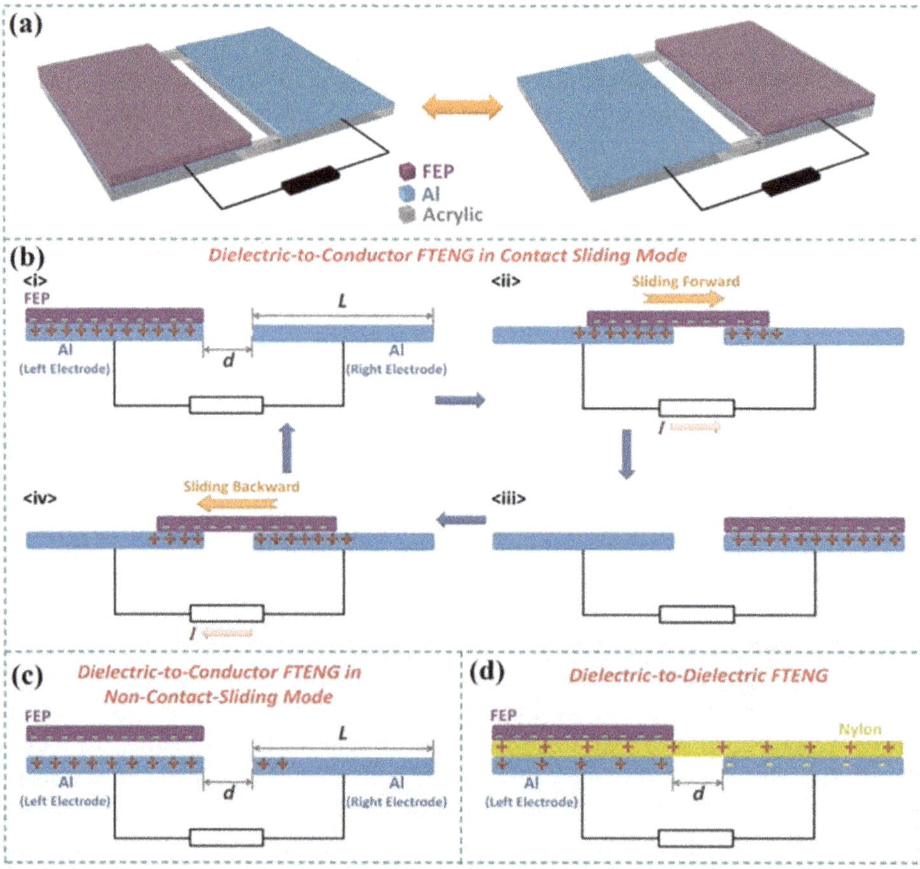

Figure 3.10. Working principle of the sliding freestanding triboelectric layer TENG. Reproduced with permission from Wiley [37].

and freestanding triboelectric layer mode can effectively solve this problem. There is no requirement of electrode connection to any moving part. However, in the single-electrode mode, the electrostatic screening effect of the primary electrode limits the efficiency of electron transfer. The freestanding triboelectric layer mode combines the advantages of the contact–separation mode and single-electrode mode. No requirement of electrodes on moving objects and no interference of the electrostatic screening effect make it easy to fabricate and operate, and thus high efficiency is achieved.

3.5.2 Improvement of TENG performances

Since the invention of the first TENG in 2012, the research interests in TENG have grown significantly. Most of the studies have focused on improving TENG outputs [1, 2, 10, 39–42], enhancing theoretical understanding [4, 43–46], achieving novel generator devices, and exploring new practical applications [47–53].

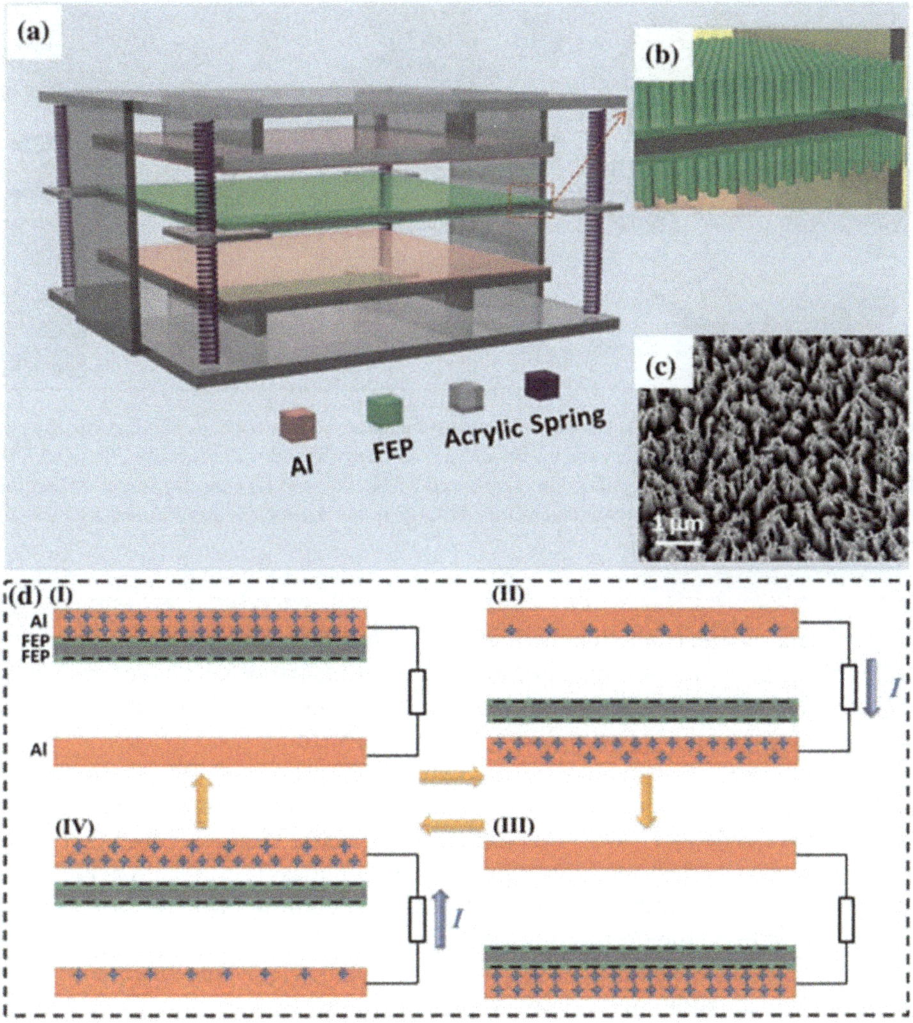

Figure 3.11. Working mechanism of the contact freestanding TENG. Reproduced with permission from the American Chemical Society [38].

3.5.2.1 High-output TENGs

As an energy harvester, high energy conversion is a significant factor in TENGs. Several high-output TENGs were designed as shown in figure 3.12. Three approaches have been used so far: triboelectric material selection, material surface engineering, and modulating bulk friction materials.

Any material that is capable of gaining or losing electrons can be considered for use in TENGs. However, better-paired triboelectric materials can lead to higher intensity output signals. Theoretically, the materials at the two ends of the triboelectric series are most promising. For example, PVDF [7, 8], PTFE [2, 9], PDMS [1], and FEP [39, 54] are ideal negative triboelectric materials due to their strong

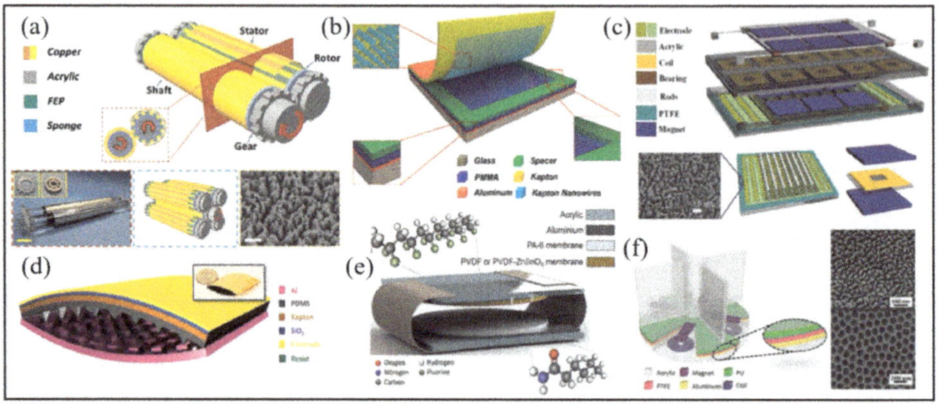

Figure 3.12. High-output TENGs: (a) nanowire array-enhanced rolling friction TENG [54]; (b) polymer nanowire-enhanced TENG [10]; (c) nanowire-enhanced blue energy hybrid nanogenerator [9]; (d) arch-shaped TENG with nanostructure [1]; (e) polarization-enhanced TENG [8]; and (f) nanostructure-modified rotating disc TENG [55]. Reproduced with permission from Wiley [54], the American Chemical Society [1, 9, 10, 55], and Elsevier [8].

ability of gaining electrons; meanwhile, electron-losing PA, nylon [2], glass [2, 10], and PMMA [10] are commonly chosen as the positive triboelectric material. Al [1, 7, 8, 41] and Cu [9, 39, 54] are also chosen because of their simultaneous strong ability of losing electrons and high conductivity as an electrode.

Surface engineering has also been used to increase the output performance of TENGs. Fan *et al* [56] reported three micro-pattern arrays (micro-line, micro-cube, and micro-pyramid) on PDMS thin surfaces to increase the area of contact surfaces. This research clearly demonstrates that surface roughness is critical in achieving high electrical outputs. Compared with flat surfaces without any pattern, patterned surfaces can increase the output by about five times. To obtain different micro/nano-structures, plasma treatment, nanostructure templating, and ion beam etching have been actively used. Inductively coupled plasma reactive ion etching has been used to fabricate various polymeric nanowires, such as FEP nanowires [39, 54, 57], PTFE nanowires [2, 55, 58], and Kapton nanowires [59]. Sandpaper-based microstructures [60] were also prepared to enhance surface roughness. Besides paper [7] and graphene oxide sheets [41] with micro surface features and high surface area, electrospun nanofibers have also been investigated for use as triboelectric layers. Electrospinning is the most promising method to fabricate continuous nanofibers to form nanofibrous mats with many unique features such as high specific surface area, controllable fiber surface and morphology, high porosity, and wide range of suitable material selection [11]. In addition, high-voltage DC power is the used during the electrospinning process, and most of the electrospun nanofibers can carry a certain amount of residual electric charges. In 2015, Cui *et al* [61] used electrospun PVDF nanofibers to fabricate a TENG for harvesting sound vibrations. A maximum current of 0.45 mA and charging rate of 61 $\mu C\ s^{-1}$ were achieved in this research. Since then, many other polymers have been electrospun into nanofibers for triboelectric power generation, such as PVDF–trifluoroethylene [62], PVDF–HFP [63], polyimide [64], PAN [65], POM [66], and PEO/PTFE [67]. With the wide material and fibrous structure selection of the electrospinning process, electrospun nanofibers for building high-performance TENGs are attracting more interest.

Enhancing the polarity of triboelectric layers is another method to improve TENG outputs. In 2017, Yang *et al* reported that polydopamine-modified paper can be easily polarized when paired with a PVDF triboelectric layer [7]. After dopamine modification, the electric outputs of paper-based TENG can be enhanced by more than 3.5 times. Doping nanomaterials and adding sublayers into bulk friction materials are the main ways to modulate bulk friction materials. Molybdenum disulfide monolayer, titania monolayers, SiO_2, TiO_2, $BaTiO_3$, and $SrTiO_3$ nanoparticles have been doped onto triboelectric layers. Many results showed that such TENGs have higher output compared with nondoped TENGs.

3.5.2.2 Applications of TENGs

TENGs have been found to have many applications in different fields such as in human motion, mechanical triggering and blowing wind. As shown in figure 3.13, the TENG is used to power cell phones [81] and electrospinning systems [90], to detect bio signals as a self-powered sensor [53, 71] and to catalyze chemical reactions [88, 89]. There has been rapid development in flexible energy harvesters for wearable electronics. The most common structures of flexible TENGs are flexible films [47, 50, 52, 56, 71–74], fibers [3, 5, 48, 49, 57, 75–78], and fabrics [51, 79–87].

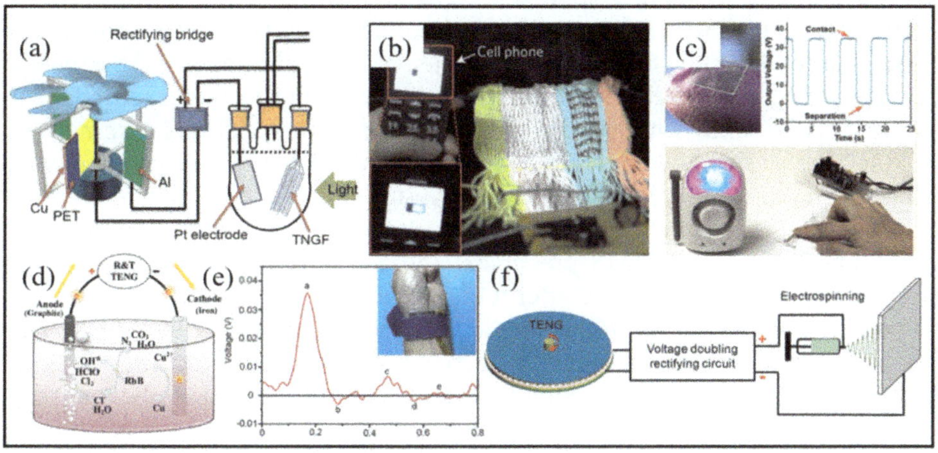

Figure 3.13. Widespread application of the TENG: (a) wind driven TENG power-assisted photocatalytic H_2 production [88]; (b) TENG continuously powering a cell phone in a wearable manner [81]; (c) TENG triggering a wireless alarm by gentle finger tapping on a triboelectric sensor [71]; (d) TENG removing Cu^{2+} and RhB from waste water [89]; (e) TENG real-time monitoring of physiological signals [53]; and (f) TENG driving an electrospinning system [90]. Reproduced with permission from Elsevier [53, 88], Springer Nature [81], the American Chemical Society [71, 89], and Wiley [90].

Film-based TENGs could achieve relatively higher electrical outputs compared with other kinds of flexible TENGs. In this regard, PDMS [56, 74], nylon [71], FEM [50, 71], PTFE [50, 73], and silicone rubber [52] are commonly used in flexible TENGs not only because of their promising triboelectric property, but also because of their excellent flexibility, and sometimes biocompatibility. Metal thin film [47, 72, 73], carbon nanotube (CNT), graphene, and indium tin oxide (ITO) [50, 56, 71] are common electrode materials.

Textiles and fibers are more suitable for wearable electronics because they have good breathability and flexibility compared with films. Generally, there are three methods in fabricating triboelectric fabrics: conductive coating on textiles, conductive textiles with a triboelectric coating, and weaving triboelectric fibers/yarns/stripes into textile structures.

Pu *et al* used common polyester fabric as the starting substrate, and coated it with a conductive Ni film to fabricate a wearable TENG [80]. A 20 μA short-circuit current was generated by the flexible TENG in the contact–separation mode. Zhang *et al* reported a conductive fabric-based TENG using silver nanowire-coated glass fabric as the electrode and a PDMS film as the triboelectric layer. Another triboelectric layer was a FEP film attached to an ITO electrode. This conductive fabric-based TENG can generate 125 V of output with a 100 MΩ resistance and 48 μA short-circuit current. Furthermore, it was integrated with a bus grip to harvest energy from human body movement [51].

The triboelectric yarns mainly include two parts, the triboelectric material coating and the core electrode. Similar to film-based TENGs, polymers with strong losing or gaining capability are commonly used as coating materials, such as PTFE [5, 48], PDMS [3], PVDF [49, 76], silicone rubber, etc [77]. Several studies used conductive fibers as the core material coated with triboelectric materials to fabricate triboelectric fibers. Metal wire [77], metal-coated polymer yarn [49], CNT-coated yarn [5, 57], and carbon fiber [3, 48] have been reported as the central part of the triboelectric fiber and acting as the electrode. In triboelectric yarn fabrication, methods like dip coating [83] and sandwiching core yarn with polymer materials are commonly used [81, 82]. The chemical vapor deposition method was used to coat triboelectric material (parylene) onto conductive Ni–polyester yarn [80, 85]. Dong *et al* reported a silicone rubber coating method, in which the core yarn was first inserted into a plastic tube filled with PDMS precursor, and then the PDMS-coated conductive yarn was pulled out of the tube after a curing process [91].

In addition to woven TENGs, there are also some single fiber-based triboelectric devices. Gong *et al* designed a sheath–core-structured triboelectric fiber that offered high sensitivity in response to stretching, bending, and pressing [77]. Nylon wrapped conductive metal fiber in the core was used as both the positive triboelectric layer and the electrode. The sheath part of the fiber was made of bamboo fiber twined silicone rubber tube, with silver nanowire and PDMS coatings. This fiber-based TENG has good flexibility and can generate a maximum 4.16 V under 100% tensile strain. Another fiber-like TENG was designed by silicone rubber and Cu wire [57]. The CNT coating layer on a silicon rubber fiber and wrapped Cu wire were used as two electrodes of the fibrous device. The releasing and stretching process of the fiber-like TENG can be simplified as a contact–separation movement between the inner fiber and the Cu coils. The stretching mode can generate a maximum V_{OC} of 142.8 V and a maximum Q_{SC} of 61 nC per stretching cycle in a single fiber-like TENG.

3.6 Summary

Static electricity has been known for a long time, and the static charge generation can be easily found in our daily life and many industrial activities. The most common way to generate static charges is through electrostatic induction and contact electrification (triboelectric effect). Along with the progress in eliminating static charges in unwanted situations, electrostatics have been widely used in printing, painting, air cleaning and various types of power generators. The invention of TENGs has significantly expanded the application of static electricity. The optimization of triboelectric materials, surface functionalization, and device structure has largely enhanced the output performance of TENGs. With the growing number of practical applications, TENGs are expected to make more important contributions to the development of microelectronic systems and wearable electronics.

References

[1] Wang S, Lin L and Wang Z L 2012 Nanoscale triboelectric-effect-enabled energy conversion for sustainably powering portable electronics *Nano Lett.* **12** 6339–46

[2] Wang Z L 2013 Triboelectric nanogenerators as new energy technology for self-powered systems and as active mechanical and chemical sensors *ACS Nano* **7** 9533–57

[3] Li X H, Lin Z H, Cheng G, Wen X N, Liu Y, Niu S M and Wang Z L 2014 3D fiber-based hybrid nanogenerator for energy harvesting and as a self-powered pressure sensor *ACS Nano* **8** 10674–81

[4] Tang W, Zhou T, Zhang C, Fan F R, Han C B and Wang Z L 2014 A power-transformed-and-managed triboelectric nanogenerator and its applications in a self-powered wireless sensing node *Nanotechnology* **25** 225402

[5] Zhong J, Zhang Y, Zhong Q, Hu Q, Hu B, Wang Z L and Zhou J 2014 Fiber-based generator for wearable electronics and mobile medication *ACS Nano* **8** 6273–80

[6] Zhou T, Zhang C, Han C B, Fan F R, Tang W and Wang Z L 2014 Woven structured triboelectric nanogenerator for wearable devices *ACS Appl. Mater. Interfaces* **6** 14695–701

[7] Feng Y, Zheng Y, Rahman Z U, Wang D, Zhou F and Liu W 2016 Paper-based triboelectric nanogenerators and their application in self-powered anticorrosion and antifouling *J. Mater. Chem.* A **4** 18022–30

[8] Soin N *et al* 2016 High performance triboelectric nanogenerators based on phase-inversion piezoelectric membranes of poly(vinylidene fluoride)-zinc stannate (PVDF-$ZnSnO_3$) and polyamide-6 (PA6) *Nano Energy* **30** 470–80

[9] Wang X, Wen Z, Guo H, Wu C, He X, Lin L, Cao X and Wang Z L 2016 Fully packaged blue energy harvester by hybridizing a rolling triboelectric nanogenerator and an electromagnetic generator *ACS Nano* **10** 11369–76

[10] Zhu G, Pan C, Guo W, Chen C Y, Zhou Y, Yu R and Wang Z L 2012 Triboelectric-generator-driven pulse electrodeposition for micropatterning *Nano Lett.* **12** 4960–5

[11] Thenmozhi S, Dharmaraj N, Kadirvelu K and Kim H Y 2017 Electrospun nanofibers: New generation materials for advanced applications *Mater. Sci. Eng.:* B **217** 36–48
[12] Britton L G 2010 *Avoiding Static Ignition Hazards in Chemical Operations: A CCPS Concept Book* vol 20 (New York: Wiley)
[13] Wilcke J C 1757 *Disputatio physica experimentalis, de electricitatibus contrariis* (Rostochii: Typis Ioannis Iacobi Adleri)
[14] Lacks D J and Sankaran R M 2011 Contact electrification of insulating materials *J. Phys. D: Appl. Phys.* **44** 453001
[15] Lowell J and Rose-Innes A 1980 Contact electrification *Adv. Phys.* **29** 947–1023
[16] Lowell J and Truscott W 1986 Triboelectrification of identical insulators. II. Theory and further experiments *J. Phys. D: Appl. Phys.* **19** 1281
[17] Labadz A and Lowell J 1991 Charge transfer across metal-SiO_2 interfaces *J. Phys. D: Appl. Phys.* **24** 1416
[18] Yu Z-Z and Watson P 1989 Contact charge accumulation and reversal on polystyrene and PTFE films upon repeated contacts with mercury *J. Phys. D: Appl. Phys.* **22** 798
[19] Meunier M and Quirke N 2000 Molecular modeling of electron trapping in polymer insulators *J. Chem. Phys.* **113** 369–76
[20] Cubero D, Quirke N and Coker D F 2003 Electronic transport in disordered n-alkanes: from fluid methane to amorphous polyethylene *J. Chem. Phys.* **119** 2669–79
[21] Yu Z-Z G and Watson K 2001 Two-step model for contact charge accumulation *J. Electrostat.* **51** 313–8
[22] Xu C *et al* 2018 On the electron-transfer mechanism in the contact-electrification effect *Adv. Mater.* **30** e1706790
[23] Ducati T R, Simões L H and Galembeck F 2010 Charge partitioning at gas–solid interfaces: humidity causes electricity buildup on metals *Langmuir* **26** 13763–6
[24] Hogue M, Buhler C, Calle C, Matsuyama T, Luo W and Groop E 2004 Insulator–insulator contact charging and its relationship to atmospheric pressure *J. Electrostat.* **61** 259–68
[25] Hogue M, Mucciolo E, Calle C and Buhler C 2005 Two-phase equilibrium model of insulator–insulator contact charging with electrostatic potential *J. Electrostat.* **63** 179–88
[26] Friedle S and Thomas S W III 2010 Controlling contact electrification with photochromic polymers *Angew. Chem. Int. Ed.* **49** 7968–71
[27] Baytekin H, Patashinski A, Branicki M, Baytekin B, Soh S and Grzybowski B A 2011 The mosaic of surface charge in contact electrification *Science* **333** 308–12
[28] Davies D 1969 Charge generation on dielectric surfaces *J. Phys. D: Appl. Phys.* **2** 1533
[29] Saini P and Choudhary V 2013 *Electrostatic Charge Dissipation and Electromagnetic Interference Shielding Response of Polyaniline Based Conducting Fabrics*
[30] de V. Heathcote N 1950 Guericke's sulphur globe *Ann. Sci.* **6** 293–305
[31] Löfhede J, Seoane F and Thordstein M 2012 Textile electrodes for EEG recording—A pilot study *Sensors* **12** 16907–19
[32] Van de Graaff generator n.d. Retrieved May 28th, from http://yourdictionary.com/van-de-graaff-generator
[33] McNaughton D 1898 Collectors for Wimshurst machines *Sci. Am.* **79** 247
[34] Maver W 1918 Electricity, its history and progress *The Encyclopedia Americana* 172
[35] Hinchet R, Seung W and Kim S W 2015 Recent progress on flexible triboelectric nanogenerators for selfpowered electronics *Chem. Sus. Chem.* **8** 2327–44

[36] Niu S, Liu Y, Wang S, Lin L, Zhou Y S, Hu Y and Wang Z L 2014 Theoretical investigation and structural optimization of single-electrode triboelectric nanogenerators *Adv. Funct. Mater.* **24** 3332–40

[37] Wang S, Xie Y, Niu S, Lin L and Wang Z L 2014 Freestanding triboelectric-layer-based nanogenerators for harvesting energy from a moving object or human motion in contact and non-contact modes *Adv. Mater.* **26** 2818–24

[38] Wang S, Niu S, Yang J, Lin L and Wang Z L 2014 Quantitative measurements of vibration amplitude using a contact-mode freestanding triboelectric nanogenerator *ACS Nano* **8** 12004–13

[39] Niu S, Wang X, Yi F, Zhou Y S and Wang Z L 2015 A universal self-charging system driven by random biomechanical energy for sustainable operation of mobile electronics *Nat. Commun.* **6** 8975

[40] Zheng Q *et al* 2016 *In vivo* self-powered wireless cardiac monitoring via implantable triboelectric nanogenerator *ACS Nano* **10** 6510–8

[41] Guo H, Li T, Cao X, Xiong J, Jie Y, Willander M, Cao X, Wang N and Wang Z L 2017 Self-sterilized flexible single-electrode triboelectric nanogenerator for energy harvesting and dynamic force sensing *ACS Nano* **11** 856–64

[42] Chen B *et al* 2018 Three-dimensional ultraflexible triboelectric nanogenerator made by 3D printing *Nano Energy* **45** 380–9

[43] Zi Y, Wang J, Wang S, Li S, Wen Z, Guo H and Wang Z L 2016 Effective energy storage from a triboelectric nanogenerator *Nat. Commun.* **7** 10987

[44] Kim J, Lee J H, Ryu H, Lee J-H, Khan U, Kim H, Kwak S S and Kim S-W 2017 High-performance piezoelectric, pyroelectric, and triboelectric nanogenerators based on P(VDF-TrFE) with controlled crystallinity and dipole alignment *Adv. Funct. Mater.* **27** 1700702

[45] Xi F, Pang Y, Li W, Jiang T, Zhang L, Guo T, Liu G, Zhang C and Wang Z L 2017 Universal power management strategy for triboelectric nanogenerator *Nano Energy* **37** 168–76

[46] Cheng G, Zheng H, Yang F, Zhao L, Zheng M, Yang J, Qin H, Du Z and Wang Z L 2018 Managing and maximizing the output power of a triboelectric nanogenerator by controlled tip–electrode air-discharging and application for UV sensing *Nano Energy* **44** 208–16

[47] Fan F-R, Tian Z-Q and Lin Wang Z 2012 Flexible triboelectric generator *Nano Energy* **1** 328–34

[48] Wang J, Li X, Zi Y, Wang S, Li Z, Zheng L, Yi F, Li S and Wang Z L 2015 A flexible fiber-based supercapacitor-triboelectric-nanogenerator power system for wearable electronics *Adv. Mater.* **27** 4830–6

[49] Sim H J, Choi C, Kim S H, Kim K M, Lee C J, Kim Y T, Lepro X, Baughman R H and Kim S J 2016 Stretchable triboelectric fiber for self-powered kinematic sensing textile *Sci. Rep.* **6** 35153

[50] Yao C, Hernandez A, Yu Y, Cai Z and Wang X 2016 Triboelectric nanogenerators and power-boards from cellulose nanofibrils and recycled materials *Nano Energy* **30** 103–8

[51] Zhang K, Wang Z L and Yang Y 2016 Conductive fabric-based stretchable hybridized nanogenerator for scavenging biomechanical energy *ACS Nano* **10** 4728–34

[52] Ahmed A, Zhang S L, Hassan I, Saadatnia Z, Zi Y, Zu J and Wang Z L 2017 A washable, stretchable, and self-powered human-machine interfacing triboelectric nanogenerator for wireless communications and soft robotics pressure sensor arrays *Extreme Mech. Lett.* **13** 25–35

[53] Chen X, Song Y, Su Z, Chen H, Cheng X, Zhang J, Han M and Zhang H 2017 Flexible fiber-based hybrid nanogenerator for biomechanical energy harvesting and physiological monitoring *Nano Energy* **38** 43–50

[54] Yeh M-H, Guo H, Lin L, Wen Z, Li Z, Hu C and Wang Z L 2016 Rolling friction enhanced free-standing triboelectric nanogenerators and their applications in self-powered electrochemical recovery systems *Adv. Funct. Mater.* **26** 1054–62

[55] Zhang B, Chen J, Jin L, Deng W, Zhang L, Zhang H, Zhu M, Yang W and Wang Z L 2016 Rotating-disk-based hybridized electromagnetic-triboelectric nanogenerator for sustainably powering wireless traffic volume sensors *ACS Nano* **10** 6241–7

[56] Fan F R, Lin L, Zhu G, Wu W, Zhang R and Wang Z L 2012 Transparent triboelectric nanogenerators and self-powered pressure sensors based on micropatterned plastic films *Nano Lett.* **12** 3109–14

[57] He X, Zi Y, Guo H, Zheng H, Xi Y, Wu C, Wang J, Zhang W, Lu C and Wang Z L 2017 A highly stretchable fiber-based triboelectric nanogenerator for self-powered wearable electronics *Adv. Funct. Mater.* **27** 1604378

[58] Fan X, Chen J, Yang J, Bai P, Li Z and Wang Z L 2015 Ultrathin, rollable, paper-based triboelectric nanogenerator for acoustic energy harvesting and self-powered sound recording *ACS Nano* **9** 4236–43

[59] Chen X, Jiang T, Yao Y, Xu L, Zhao Z and Wang Z L 2016 Stimulating acrylic elastomers by a triboelectric nanogenerator — toward self-powered electronic skin and artificial muscle *Adv. Funct. Mater.* **26** 4906–13

[60] Rasel M S U and Park J-Y 2017 A sandpaper assisted micro-structured polydimethylsiloxane fabrication for human skin based triboelectric energy harvesting application *Appl. Energy* **206** 150–8

[61] Cui N, Gu L, Liu J, Bai S, Qiu J, Fu J, Kou X, Liu H, Qin Y and Wang Z L 2015 High performance sound driven triboelectric nanogenerator for harvesting noise energy *Nano Energy* **15** 321–8

[62] Jang S, Kim H, Kim Y, Kang B J and Oh J H 2016 Honeycomb-like nanofiber based triboelectric nanogenerator using self-assembled electrospun poly(vinylidene fluoride-co-trifluoroethylene) nanofibers *Appl. Phys. Lett.* **108** 143901

[63] Lin M-F, Xiong J, Wang J, Parida K and Lee P S 2018 Core–shell nanofiber mats for tactile pressure sensor and nanogenerator applications *Nano Energy* **44** 248–55

[64] Gu G Q, Han C B, Lu C X, He C, Jiang T, Gao Z L, Li C J and Wang Z L 2017 Triboelectric nanogenerator enhanced nanofiber air filters for efficient particulate matter removal *ACS Nano* **11** 6211–7

[65] Sun N *et al* 2017 All flexible electrospun papers based self-charging power system *Nano Energy* **38** 210–7

[66] Zhang F, Li B, Zheng J and Xu C 2015 Facile fabrication of micro-nano structured triboelectric nanogenerator with high electric output *Nanoscale Res. Lett.* **10** 1001

[67] Zhao P *et al* 2018 Emulsion electrospinning of polytetrafluoroethylene (PTFE) nanofibrous membranes for high-performance triboelectric nanogenerators *ACS Appl. Mater. Interfaces* **10** 5880–91

[68] Song P *et al* 2017 Drug-delivery system using biokinetic energy *Adv. Mater.* **29** 1605668

[69] Ouyang H *et al* 2017 Self-powered pulse sensor for antidiastole of cardiovascular disease *Adv. Mater.* **29** 1703456

[70] Khan Y, Ostfeld A E, Lochner C M, Pierre A and Arias A C 2016 Monitoring of vital signs with flexible and wearable medical devices *Adv. Mater.* **28** 4373–95

[71] Zhu G, Yang W Q, Zhang T, Jing Q, Chen J, Zhou Y S, Bai P and Wang Z L 2014 Self-powered, ultrasensitive, flexible tactile sensors based on contact electrification *Nano Lett.* **14** 3208–13

[72] Jie Y, Zhu H, Cao X, Zhang Y, Wang N, Zhang L and Wang Z L 2016 One-piece triboelectric nanosensor for self-triggered alarm system and latent fingerprint detection *ACS Nano* **10** 10366–72

[73] Li J and Wang X 2017 Research update: materials design of implantable nanogenerators for biomechanical energy harvesting *APL Mater.* **5** 073801

[74] Pu X, Liu M, Chen X, Sun J, Du C, Zhang Y, Zhai J, Hu W and Wang Z L 2017 Ultrastretchable, transparent triboelectric nanogenerator as electronic skin for biomechanical energy harvesting and tactile sensing *Sci. Adv.* **3** e1700015

[75] Huang T, Wang C, Yu H, Wang H, Zhang Q and Zhu M 2015 Human walking-driven wearable all-fiber triboelectric nanogenerator containing electrospun polyvinylidene fluoride piezoelectric nanofibers *Nano Energy* **14** 226–35

[76] Lu C, Lee W Y, Shih C C, Wen M Y and Chen W C 2017 Stretchable polymer dielectrics for low-voltage-driven field-effect transistors *ACS Appl. Mater. Interfaces* **9** 25522–32

[77] Gong W, Hou C, Guo Y, Zhou J, Mu J, Li Y, Zhang Q and Wang H 2017 A wearable, fibroid, self-powered active kinematic sensor based on stretchable sheath-core structural triboelectric fibers *Nano Energy* **39** 673–83

[78] Park S, Ryu H, Park S, Hong H, Jung H Y and Park J-J 2017 Rotating triboelectric generator using sliding contact and noncontact from 1D fiber friction *Nano Energy* **33** 184–94

[79] Jung S, Lee J, Hyeon T, Lee M and Kim D H 2014 Fabric-based integrated energy devices for wearable activity monitors *Adv. Mater.* **26** 6329–34

[80] Pu X, Li L, Song H, Du C, Zhao Z, Jiang C, Cao G, Hu W and Wang Z L 2015 A self-charging power unit by integration of a textile triboelectric nanogenerator and a flexible lithium-ion battery for wearable electronics *Adv. Mater.* **27** 2472–8

[81] Chen J, Huang Y, Zhang N, Zou H, Liu R, Tao C, Fan X and Wang Z L 2016 Micro-cable structured textile for simultaneously harvesting solar and mechanical energy *Nature Energy* **1** 16138

[82] Pu X, Song W, Liu M, Sun C, Du C, Jiang C, Huang X, Zou D, Hu W and Wang Z L 2016 Wearable power-textiles by integrating fabric triboelectric nanogenerators and fiber-shaped dye-sensitized solar cells *Adv. Energy Mater.* **6** 1601048

[83] Zhao Z, Yan C, Liu Z, Fu X, Peng L M, Hu Y and Zheng Z 2016 Machine-washable textile triboelectric nanogenerators for effective human respiratory monitoring through loom weaving of metallic yarns *Adv. Mater.* **28** 10267–74

[84] Xiong J, Lin M-F, Wang J, Gaw S L, Parida K and Lee P S 2017 Wearable all-fabric-based triboelectric generator for water energy harvesting *Adv. Energy Mater.* **7** 1701243

[85] Pu X, Li L, Liu M, Jiang C, Du C, Zhao Z, Hu W and Wang Z L 2016 Wearable self-charging power textile based on flexible yarn supercapacitors and fabric nanogenerators *Adv. Mater.* **28** 98–105

[86] Wen Z, Yeh M-H, Guo H, Wang J, Zi Y, Xu W, Deng J, Zhu L, Wang X and Hu C 2016 Self-powered textile for wearable electronics by hybridizing fiber-shaped nanogenerators, solar cells, and supercapacitors *Sci. Adv.* **2** e1600097

[87] Liu M, Pu X, Jiang C, Liu T, Huang X, Chen L, Du C, Sun J, Hu W and Wang Z L 2017 Large-area all-textile pressure sensors for monitoring human motion and physiological signals *Adv. Mater.* **29**

[88] Yu X, Han X, Zhao Z, Zhang J, Guo W, Pan C, Li A, Liu H and Wang Z L 2015 Hierarchical TiO_2 nanowire/graphite fiber photoelectrocatalysis setup powered by a wind-driven nanogenerator: a highly efficient photoelectrocatalytic device entirely based on renewable energy *Nano Energy* **11** 19–27

[89] Li C, Yin Y, Wang B, Zhou T, Wang J, Luo J, Tang W, Cao R, Yuan Z and Li N 2017 Self-powered electrospinning system driven by a triboelectric nanogenerator *ACS Nano* **11** 10439–45

[90] Chen S, Wang N, Ma L, Li T, Willander M, Jie Y, Cao X and Wang Z L 2016 Triboelectric nanogenerator for sustainable wastewater treatment via a self-powered electrochemical process *Adv. Energy Mater.* **6** 1501778

[91] Dong K, Wang Y-C, Deng J, Dai Y, Zhang S L, Zou H, Gu B, Sun B and Wang Z L 2017 A highly stretchable and washable all-yarn-based self-charging knitting power textile composed of fiber triboelectric nanogenerators and supercapacitors *ACS Nano* **11** 9490–9

Chapter 4

Piezoelectric energy conversion performance of electrospun nanofibers

Hao Shao, Hongxia Wang and Jian Fang

Piezoelectric materials that can convert mechanical forces or vibrations into electricity have attracted increasing attention due to the energy crisis and problems of environmental pollution. Electrospinning is an effective technique to produce piezoelectric nanofibers for energy conversion purposes. In the electrospinning process, the dipoles in materials tend to be oriented within the high electrical field, resulting in piezoelectricity. PVDF is the most widely studied piezoelectric polymer because of its high piezoelectric coefficient and easy electrospinning preparation. Besides, other electrospun piezoelectric polymers, piezoelectric ceramics, and composites with nanofibrous structures have also been prepared for energy harvesting. In this chapter, a detailed introduction of electrospun piezoelectric materials and their piezoelectricity will be provided.

4.1 Introduction

Mechanical-to-electrical energy conversion can be achieved based on different principles such as piezoelectric, triboelectric, and electromagnetic effects, and the dynamic Schottky contact effect. These different principles can lead to great differences in device structure and electric output features. Among them, the piezoelectric effect, which converts mechanical forces or vibration into electric signals, has attracted the most attention owing to the large amount of existing piezoelectric materials around our life.

Piezoelectric materials are conventionally divided into two main classes, inorganic ceramics and organic polymers, and piezoelectric polymers are further divided into three subclasses, bulk polymers, composite, and voided charged polymers (VCPs). Most of the piezoelectric materials except VCPs have a dense thin film structure. Piezoelectric ceramics normally have higher piezoelectric coefficients and higher piezo responses. Piezoelectric polymers with nontoxic, biocompatibility,

doi:10.1088/978-0-7503-2005-4ch4

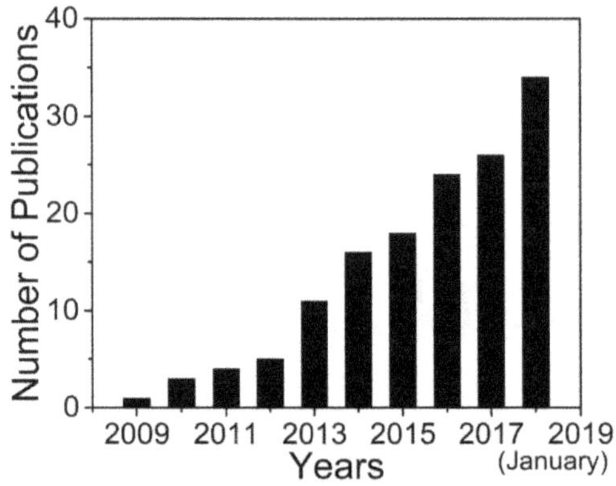

Figure 4.1. Number of scientific publications on electrospun nanofibers for piezoelectric energy conversion application between 2009–18. Source: SciFinder; January 2019.

biodegradability, and high-flexibility properties have exhibited great significance in many application fields. The fabrication of piezoelectric films often requires harsh and costly conditions. Recently, electrospinning has been found as a simple and effective tool to process piezoelectric materials into nanofibrous structures, especially for piezoelectric polymers. Piezoelectric nanofibers from electrospinning have shown many unique features apart from high surface area to volume ratios, high porosity, good flexibility, and light weight [1]. As shown in figure 4.1, since the first piezoelectric nanogenerator (PENG) prepared by electrospinning was reported in 2009, the research has continuously expanded, as indicated by the increased number of publications. In this chapter, the details of research findings about electrospun nanofibers for piezoelectric energy conversion will be summarized.

4.2 Brief history

The word 'piezo' is derived from the Greek word 'piezoin', which means to press or squeeze, and 'piezoelectricity' means electricity resulting from pressing. Since it was discovered in 1880, piezoelectricity has been investigated for over one century. It has two effects: the direct piezoelectric effect and the converse piezoelectric effect. These are shown in figure 4.2. When a force is applied to a piezoelectric material, it changes the dipole moment, resulting in electric outputs (the direct effect). Meanwhile, piezoelectric material deforms in the direction of the poling voltage when it is charged by an external electrical voltage (the converse effect).

Lots of natural materials were found to have piezoelectric properties, such as quartz, DNA, and bone [3]. Artificial piezoelectric materials that have been developed include PZT, $BaTiO_3$, ZnO, and polymers. PZT is ranked to have the highest piezoelectric coefficient among the piezoelectric ceramics. However, the toxicity originating from lead has restricted its wide application. Thus, environmentally friendly piezoelectric ceramics, such as $BaTiO_3$ and ZnO, have been developed.

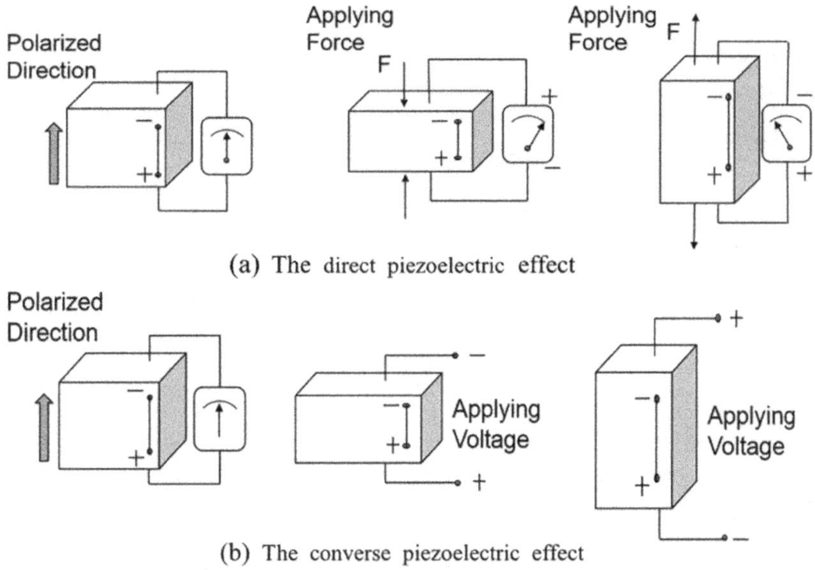

Figure 4.2. A schematic of (a) direct and (b) converse effects of piezoelectric materials [2].

PVDF was the first reported piezoelectric polymer [4]. After that, more piezoelectric polymers, such as PVDF copolymers, nylon-11, and polyrurea-9, were discovered [5]. Currently, piezoelectric polymers normally show a lower piezoelectric coefficient than piezoelectric ceramics, but they are much more flexible and can undergo a higher level of physical deformation. Recently, electrospinning was found to be an effective method to prepare piezoelectric webs from polymers, ceramics, and their combinations.

4.3 Piezoelectricity of PVDF nanofibers

PVDF is a multi-ferric material with superior piezoelectric, pyroelectric, and ferroelectric properties [6]. PVDF has a simple monomer structure of $-(C_2H_2F_2)-$, which includes atoms of carbon, hydrogen, and fluorine. In the solid-state, PVDF is a semicrystalline polymer that can have five different crystal phases: α, β, γ, δ, and ε. These crystal phases contain three different conformations: an all-trans (TTT) planar zigzag for the β phase, trans-gauche–trans-gauche (TGTG′) for α and δ phases, and TTTGTTTG′ for γ and ε phases, as shown in figure 4.3. The β crystal phase has the highest electrical dipole moment per unit, about $5–8 \times 10^{-30}$ C · m, which is the main reason for PVDF piezoelectricity.

The α crystal phase shows no piezoelectric property, but it is the most thermodynamically stable phase. Conventionally, piezoelectric PVDF films are prepared by conducting a series of post treatments to PVDF cast films, which are dominated by the α crystal phase. Post stretching enables a PVDF film to have a β crystal phase, and electrical poling at an elevated temperature assists in improving the alignment of dipole moments in the film.

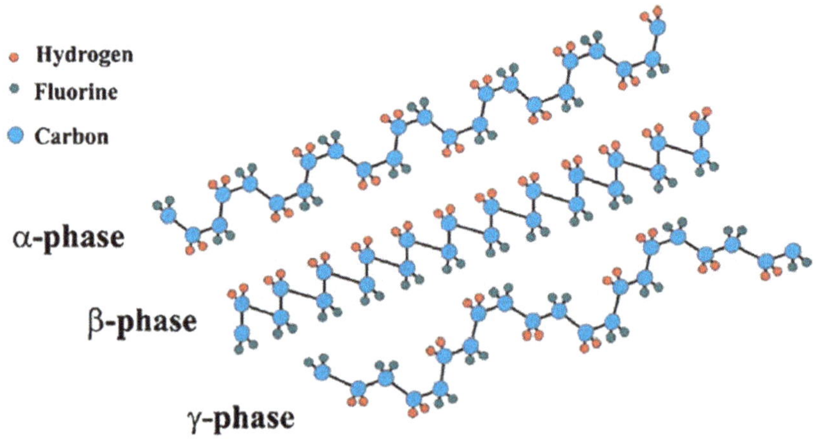

Figure 4.3. Chain conformation for the α, β, and γ phases of PVDF [7]. Reproduced with permission. Copyright 2014, Elsevier.

Electrospinning is able to directly produce PVDF nanofibers with high β crystal phase content without any post treatments [8]. However, in 2009, the direct piezoelectric effect was reported on a single electrospun PVDF fiber. Chang *et al* [8] used a near-field electrospinning technique to prepare a single PVDF fiber. The single PVDF fiber device showed a piezoelectric property with pulse outputs of 2.5 mV when an external strain of 0.092% was applied. Slightly later, a strong piezoelectric property was reported on aligned nanofiber web [9] and randomly orientated PVDF nanofibers webs, i.e. nonwovens [10].

According to fiber structure and electric output, the studies on the piezoelectric effect of electrospun PVDF nanofibers can be separated into a few different groups: single fiber/multiple fibers, randomly orientated fiber webs (nonwoven), and aligned nanofiber webs. A summary of neat PVDF piezoelectric devices are provided in table 4.1.

4.3.1 Single fiber/multiple fibers

Near-field electrospinning is an effective method to prepare a single fiber. In the electrospinning process, the stretching forces and strong electric fields (greater than 10^7 V m^{-1}) naturally align dipole moments in the fiber to form a polar β phase, determining the polarity of the electrospun nanofiber (figure 4.4). The piezoelectric device fabricated based on a single electrospun PVDF nanofiber with a diameter of 700 nm could produce an electric voltage output up to 2.5 mV. Limited electric outputs were generated by a single fiber. Later, the same group found that the single PVDF nanofiber device could generate the electrical outputs of 32 mV and 4.5 nA under low stretch strain (around 0.08%) [9]. Figures 4.4(c) and (d) indicate the typical electrical voltage and current outputs. A pulse electrical signal accompanied by an opposite pulse electrical output was generated. The first signal was derived from the deformation of nanofiber, while the subsequent opposite output was

Table 4.1. A summary of piezoelectric electrospun neat PVDF fiber devices.

PVDF	Testing mode	Fiber diameter	Web thickness	Working area (cm^2)	Strain/Force/ Frequency	Output types	Poling	V	I	Ref
Single fiber	Stretching/ Releasing	700 nm	—	—	0.092% strain	d_{33}	No	2.5 mV	—	[8]
	Stretching/ Releasing	500 nm–6.5 µm	—	—	0.085% strain, 2 Hz	d_{33}	No	32 mV	4.5 nA	[9]
Multiple fibers	Vibration	0.9 µm–5 µm	~50 parallel fibers	9.75	Wind speed 11.3 m s^{-1}	d_{33} and d_{31}	No	2.5 V	100 nA	[11]
	Vibration	0.9 µm–2.5 µm	~500 fibers	5	Wind speed 19.8 m s^{-1}	d_{33} and d_{31}	No	2.0 V	—	[12]
	Bending/Releasing	0.9 µm–2.5 µm	~500 fibers	3.5	0.5% strain, 10 Hz	d_{33} and d_{31}	No	4 V	75 nA	[13]
	Stretch/Release	—	~120 buckled fibers	—	30% strain, 0.5 Hz	d_{33}	No	40 mV	1.2 nA	[14]
	Stretch/Release	—	40 fibers	—	70% strain, 0.8 Hz	d_{33}	No	—	2.8 nA	[15]
	Stretch/Release	—	20–150 fibers	—	160 mm s^{-1}, 120% strain	d_{33} and d_{31}	No	—	20 nA	[16]
	Bending	400 nm	—	—	0.05% strain, 1.67% strain rate	d_{33}	20 kV mm^{-1}, 15 min	20 mV	0.3 nA	[17]
Nonwoven web	Compression/ Decompression	183 ± 37 nm	140 µm	2	5 Hz	d_{33}	No	2.21 V	4.5 µA	[10]
	Compression/ Decompression	—	20 µm	2.5	0.2 MPa, 3 Hz	d_{33}	No	175mV	—	[18]

(*Continued*)

Table 4.1. (*Continued*)

PVDF	Testing mode	Fiber diameter	Web thickness	Working area (cm^2)	Strain/Force/ Frequency	Output types	Poling	V	I	Ref
	Drop stainless steel	812 ± 123 nm	310 μm	2	5 g	d_{33}	12.6 kV mm^{-1}, 4h room temperature	4.0 V	—	[19]
	Compression/ Decompression	284 ± 70 nm	70 μm	4	10 N, 1 Hz	d_{33}	No	2.2 V	2.3 μA	[20]
	Finger compression/ decompression	124 nm	120 μm	2	—	d_{33}	No	1 V	—	[21]
	Vibration	300 ± 70 nm	—	2	1.54 Hz	d_{33}	20 kV mm^{-1}, 30 min	1.3 V	13 nA	[22]
	Compression/ Decompression	—	—	2.25	4 Hz, 15 N	d_{33}	No	0.97 V	—	[23]
	Periodic bending	500 ± 220 nm	50 μm	3	—	d_{33} and d_{31}	No	2.8 V	1.32 μA	[24]
	Compression/ Decompression	300 nm	100 μm	4	10 N, 1 Hz	d_{33}	No	1.0 V	1.2 μA	[25]
	Compression/ Decompression	539 ± 175 nm	100 μm	2	10 N, 5 Hz	d_{33}	No	2.6 V	4.5 μA	[26]
	Compression/ Decompression	—	—	—	4.44 N, 20 Hz	d_{33}	No	0.28 V	—	[27]
	Sound wave reacted vibration	150–180 nm	—	—	100 Hz, 100 dB	d_{33}	No	0.3 V	—	[28]
	Bending/Releasing	—	200 μm	8.4	37 N, 1.4 Hz	d_{33} and d_{31}	25 kV mm^{-1}	1.5 V	400 nA	[29]
	Bending/Releasing	124 nm	120 μm	16	0.5 Hz	d_{33} and d_{31}	No	0.7 V	—	[30]
	Compression/ Decompression	~100 nm	260 μm	6	8.3 kPa	d_{33}	No	48 V	6 μA	[31]

	Compression/ Decompression	1137 ± 200 nm (Smooth fibers)	100 μm	15	5 Hz, 10 N	d_{33}	No	1 V	1.6 μA	[32]
	Compression/ Decompression	1129 ± 168 nm (Porous fibers)	100 μm	15	5 Hz, 10 N	d_{33}	No	1.3 V	2.1 μA	[32]
	Compression/ Decompression	1130 ± 158 nm (Wrinkled fibers)	100 μm	15	5 Hz, 10 N	d_{33}	No	1.8 V	3.2 μA	[32]
	Bending	350 ± 23 nm	60 μm	1.2	2 Hz,	d_{33} and d_{31}	No	0.5 V	13 nA	[33]
Aligned fiber web	Stretch/Release	1.6 μm	100 fibers	—	0.05% Strain, 7 Hz	d_{33}	No	76 mV	39 nA	[34]
	Vibration	Hollow tube web, 13.18 μm outer, 3.26 μm inner	—	5.25	7 Hz	d_{33}	No	71.66 mV	153 nA	[35]
	Finger press	0.9 μm–3 μm	400 fibers	1	—	d_{31}	No	4 V	100 nA	[36]
	Bending	400 ± 35 nm	60 μm	1.2	2 Hz,	d_{33} and d_{31}	No	1.1 V	40 nA	[33]
	Compression/ Decompression	1052 ± 129 nm (Smooth fibers)	100 μm	15	5 Hz, 10 N	d_{33}	No	1.7 V	2.2 μA	[32]
	Compression/ Decompression	1038 ± 113 nm (Porous fibers)	100 μm	15	5 Hz, 10 N	d_{33}	No	2.2 V	2.7 μA	[32]
	Compression/ Decompression	1047 ± 100 nm (Wrinkled fibers)	100 μm	15	5 Hz, 10 N	d_{33}	No	2.8 V	3.9 μA	[32]

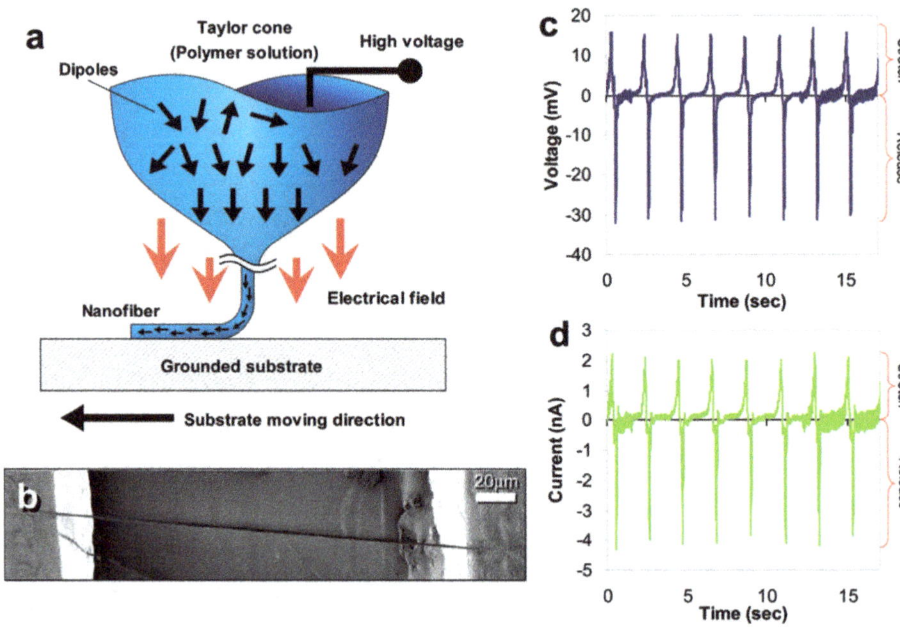

Figure 4.4. (a) Near-field electrospinning combining direct-writing, mechanical stretching, and electrical poling to create piezoelectric nanogenerators on a substrate. (b) SEM image of a single PVDF nanofiber. Output (c) voltage and (d) current measured with respect to time under an applied strain at 2 Hz [9]. Reproduced with permission. Copyright 2010, American Chemical Society.

attributed to the recovery deformation of the nanofiber due to the release of the force.

Fuh *et al* [11–13] developed highly flexible devices with multiple PVDF fibers for sensors and energy generating applications. The sensor device was attached on a flag surface and it produced maximum current and voltage outputs of 100 nA and 2.5 V, respectively, when the wind speed was 11.3 m s^{-1}. For energy harvesting purposes, the output voltage of the device exceeds 4 V while the output current reached 75 nA [12, 13].

Stretchable nanogenerators (NGs) with highly uniform controllable buckled PVDF fiber arrays were prepared by direct-writing electrospinning [14–16]. Figure 4.5(a) shows a piezoelectric device containing about 120 buckled PVDF fibers. Under a periodic stretching (30% strain) and releasing process, the output current and voltage of 1.2 nA and 40 mV, respectively, were observed (figure 4.5(b)).

In addition to near-field electrospinning, conventional electrospinning has also been used to prepare aligned multiple nanofibers. Benjamin *et al* [17] prepared a nanogenerator by using aligned PVDF nanofibers obtained by using two grounded copper pieces with a 2 cm gap as a fiber collector. They indicated that random dipoles of the polar β crystal phase existed in the nanofibers; however, they did not show any piezoelectric property. To obtain the piezoelectricity, a high electrical field poling (20 kV mm^{-1}) was employed to align the dipoles along the in-plane direction.

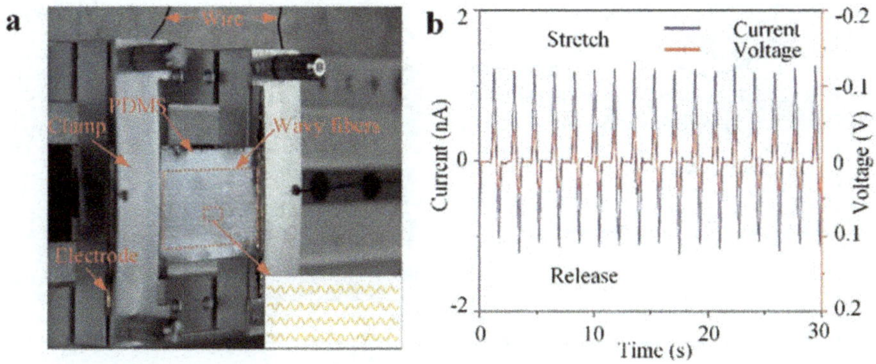

Figure 4.5. (a) Photograph of the tensile test platform. (b) Output current and voltage of the device consisting of 120 PVDF fibers measured under 30% applied strain at 0.5 Hz [14]. Reproduced with permission. Copyright 2014, Royal Society of Chemistry.

The prepared device could generate 20 mV voltage and 0.3 nA current outputs with a maximum bending strain of 0.05%.

4.3.2 Nonwovens

Nanofiber nonwovens are predominant products of conventional electrospinning. In 2011, Jian *et al* [10] first reported that randomly oriented PVDF nanofiber webs prepared by conventional electrospinning showed mechanical-to-electrical energy conversion properties without a post poling process. When an electrospun PVDF fiber nonwoven was sandwiched between two aluminum foils, a voltage and current of 2.4 V and 4.2 μA, respectively, were generated upon compressive impact (figure 4.6).

Lei *et al* [18] compared nanofiber webs produced by a mechanical spinning process without static electric field bias (forcespinning) and an electrospinning technique. It was interesting to find that the two nanofiber webs showed completely difference piezoelectric responses. The electrospun PVDF possesses the piezoelectric effect, whereas the forcespun PVDF does not. The electric field in the electrospinning process induces poling and dipole orientation in PVDF fibers.

The effects of electrospinning parameters and PVDF polymer concentrations on β crystal phase content and mechanical-to-electrical energy conversion of randomly oriented electrospun PVDF nanofiber webs were studied systematically [20–24]. It was found that finer uniform PVDF fibers showed higher β crystal phase content, and hence their energy harvesting devices have higher electrical outputs (figure 4.7). Nanofiber web thickness also plays a role in determining the level of electrical outputs. When the fiber web was thinner than 20 μm, a short circuit occasionally occurred during compression. Upon increasing the web thickness from 20 μm to 70 μm, both voltage and current outputs increased. However, electrical outputs reduced with a further increase in web thickness [20]. It is also found that the residue charges contribute to the mechanical-to-electrical energy conversion of electrospun

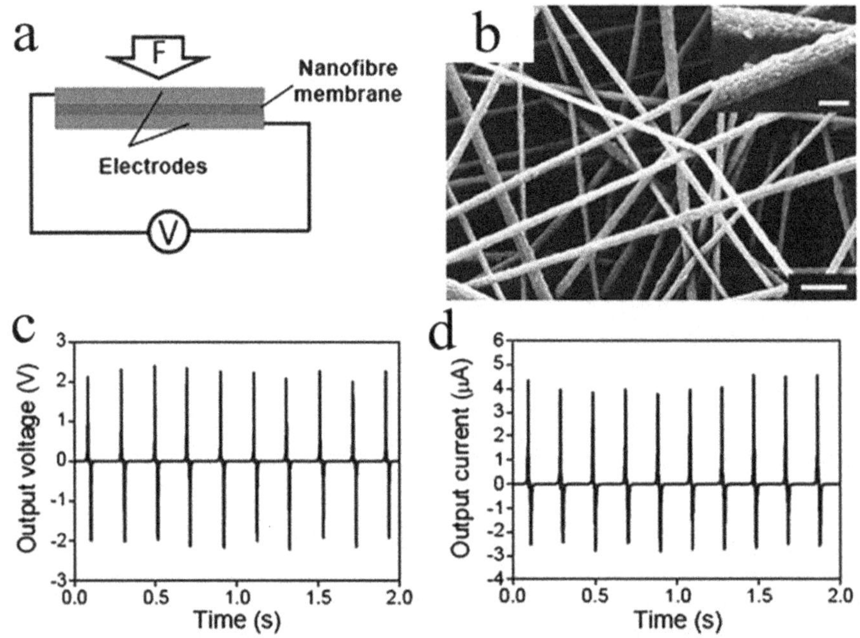

Figure 4.6. (a) Illustration of power generator. (b) SEM images of electrospun PVDF nanofibers (scale bar: 1 mm, inset scale bar: 200 nm). (c) Voltage and (d) current outputs of the same nanofiber device under 5 Hz repeated compressive impacts. (Nanofiber web thickness = 140 mm; working area = 2 cm^2) [10]. Reproduced with permission. Copyright 2011, Royal Society of Chemistry.

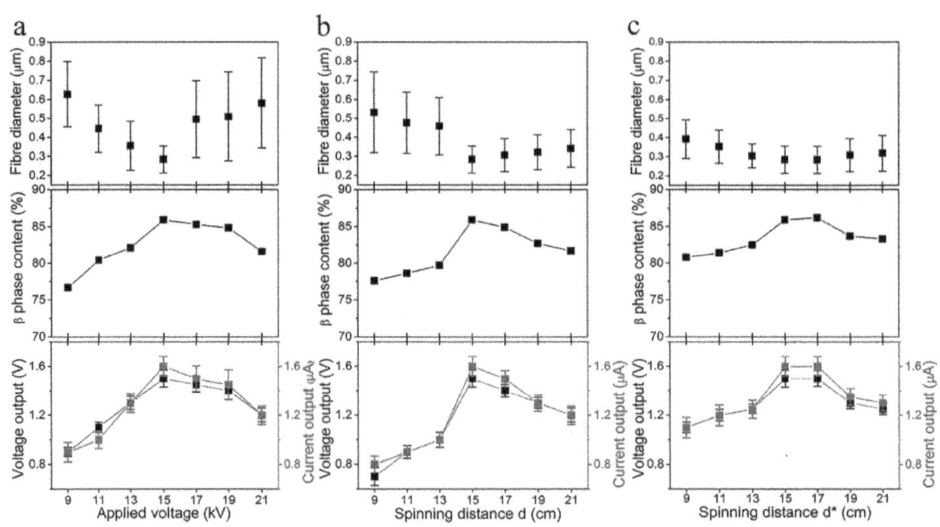

Figure 4.7. Effect of (a) applied voltages, (b) spinning distances, and (c) electric field intensity on PVDF fiber diameters, β crystal phase contents, and electrical outputs of PVDF nanofiber webs (PVDF concentration 20%; nanofiber web thickness 100 mm) [20]. Reproduced with permission. Copyright 2015, Royal Society of Chemistry.

PVDF nanofibers [25]. The electric outputs of the device decreased from 1.0 V and 1.2 μA to 0.45 V and 0.5 μA after the residual charges were removed.

Interconnected PVDF fiber webs were obtained by using short-distance electrospinning [37]. Despite the insufficient fiber whipping movement and limited solvent evaporation, the interconnected fiber webs produced had a comparable β crystal phase content and mechanical-to-electrical energy conversion property to those produced by conventional electrospinning. Moreover, the interconnected fiber webs had higher delamination resistance and tensile strength, and hence were more robust and could maintain structural integrity during compression/decompression.

Fang *et al* [26] examined the effect of electrospinning models on the piezoelectricity of electrospun nanofiber webs. It is reported that PVDF nonwovens prepared by a disc (needleless) electrospinning showed enhanced piezoelectric performance compared to those from conventional needle electrospinning. A device made of needleless spun nanofibers generated 2.6 V and 4.5 μA outputs while the one from needle electrospun nanofibers only produced 2.05 V and 3.12 μA, under the same conditions. This is explained by the higher applied voltage used in needleless electrospinning leading to a higher β crystal phase and better molecular orientation.

Conventional electrospinning was also employed to produce randomly oriented PVDF nanofibers for energy generator and sensor applications [27, 28, 30, 31, 38]. Wang *et al* [27] fabricated piezoelectric force sensors based on random PVDF nanofiber webs. Different PVDF fiber morphologies were obtained by varying the applied voltage and flow rate. As a result, the nanofibers with different β crystal phase contents were obtained. The PVDF nonwoven with a higher β crystal phase content exhibited higher sensitivity. Boongik *et al* [28] integrated randomly oriented electrospun PVDF nanofiber webs with an organic photovoltaic device to fabricate a hybrid energy generator.

In some studies, the electrical poling process was applied to the initial electrospun PVDF webs for obtaining device piezoelectricity [19, 22, 29]. Ali *et al* [19] indicated that the piezoelectric response of initial electrospun PVDF webs was not desirable and detectable due to the random orientation of fibers causing neutralization of electric dipoles. Thus, a high voltage of 3.9 kV was applied through the terminals coming off the electrode faces for a sample with a 310 μm thickness. The poling process was carried out for 4 h. The typical voltage output was 4.0 V when a stainless steel weight (5 g) was dropped from a 15 cm height onto the sample. Liu *et al* [29] polarized electrospun PVDF webs by applying external current electrical fields of 5 kV. The prepared device generated an output voltage and current up to 1.5 V and 400 nA under periodic bending by a line motor.

4.3.3 Aligned fiber web

Aligned PVDF nanofibers were prepared for piezoelectric energy conversion, and it was found that the electrical outputs of the PENG based on the aligned fiber webs were normally higher than those based on the randomly oriented fiber nonwovens [39].

By modifying the near-field electrospinning setup, various PVDF products resulted, as shown in figure 4.8. Liu *et al* [34] prepared aligned PVDF nonwoven fiber web energy harvesting devices by near-field electrospinning with a hollow cylindrical collector. A maximum peak voltage and current of 76 mV and 39 nA were obtained under repeated stretching and releasing of the device with a strain of 0.05% at 7 Hz. Pan *et al* [35] fabricated PVDF hollow fiber tubes with a metallic

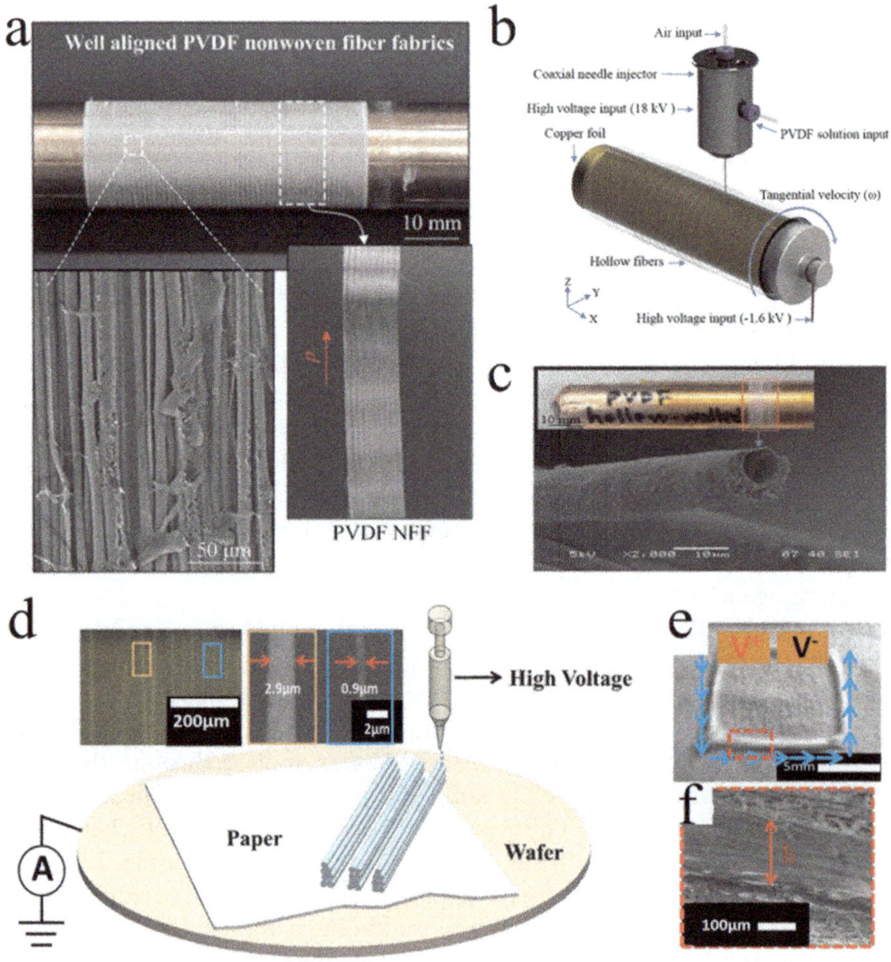

Figure 4.8. (a) The top optical photograph shows well-aligned PVDF fibers on a glass tube collector, and a piece of PVDF nanofiber fabric (NFF) is easily extracted from the glass tube collector. The bottom SEM image shows dense deposition of fibers to form the nonwoven fiber webs. (b) The coaxial needle injector electrospinning and (c) SEM image of the cross-sections of PVDF fiber tubes. (d) Schematic setup of the sequentially stacked 3D fiber structure via near-field electrospinning process. (e) Photograph of a 3D square structure (1×1 cm^2) constructed on a paper substrate. (f) SEM image showing the area of the square and micro-wall structure made of about 600 layers of electrospun fibers stacked on top of each other; the electrospun and stacked fiber bundles are measured to have a width of ~70 μm and height of 300 μm [34–36]. Reproduced with permission. Copyright 2015, Royal Society of Chemistry. Reproduced with permission. Copyright 2016, Elsevier.

coaxial needle injector. The voltage output generated from the PVDF fiber tubes was about 71.66 mV, which was much higher than that of solid PVDF fibers. 3D architectures of PVDF fibers were also prepared by near-field electrospinning with a programmable *x*–*y* translational motion stage [36]. The device with a PVDF stack height of 400 μm (about 800 fibers) generated a voltage of 4 V and a current of 100 nA. The electrical outputs increased with the increase in fiber stack height.

Conventional electrospinning was also used to prepare aligned nanofiber webs. Two types of collectors were reported to prepare an aligned fibrous structure: motion-free separated electrodes, and high-speed rotating collectors. Sung *et al* [33] prepared aligned nanofibers using a metal frame as a collector. Figure 4.9(b) shows

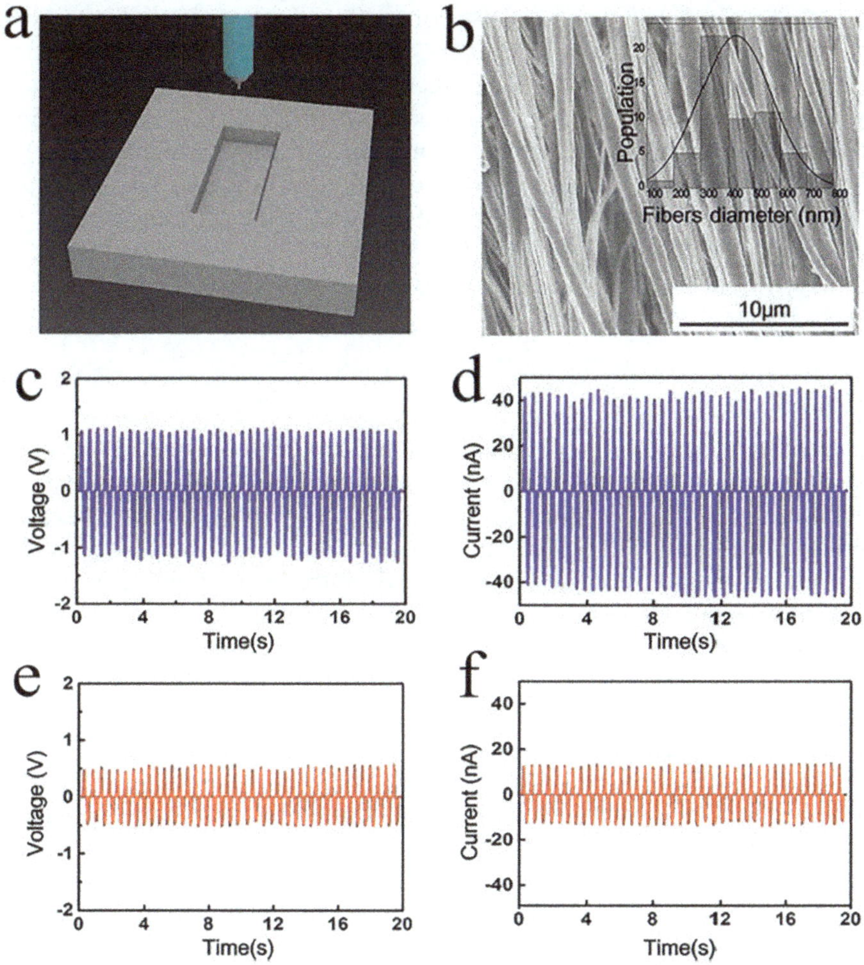

Figure 4.9. (a) Schematic illustration of the electrospinning setup containing a stepped collector, and (b) SEM image of the aligned PVDF nanofibers. Output (c) voltage and (d) current of the piezoelectric device fabricated from the aligned nanofibers. Output (e) voltage and (f) current piezoelectric device fabricated from the randomly oriented nanofibers [33].

the SEM image of prepared PVDF nanofibers. The aligned PVDF nanofibers had a higher content of the β crystal phase compared to those of randomly orientated fibers. The aligned nanofiber-based piezoelectric device could generate an output voltage and current of 1.1 V and 40 nA, respectively, under a bending condition at a frequency of 2 Hz. At the same measurement condition, it exhibited a twofold increase in the output voltage and a threefold increase in the output current compared to the corresponding values obtained for the device manufactured from randomly oriented nanofibers (figure 4.9).

Zaarour *et al* [32] prepared PVDF nanofibers by using a grounded drum collector. Aligned fibers were obtained when the drum rotating speed was 2000 rpm. Fibers with three different morphologies (wrinkled fibers, smooth fibers, and porous fibers) were obtained by controlling the relative humidity, as shown in figure 4.10. The highest β crystal phase content was found on wrinkled fibers while the lowest one was found on porous fibers. Moreover, the β crystal phase content of randomly oriented and aligned PVDF fibers of the same morphology was similar.

For comparison, the piezoelectric property of the randomly oriented fiber web and aligned fiber web was tested under compressive impacts. Figures 4.10(g) and (h) show the schematic structure and photo of the actual device. The outputs of the randomly oriented fiber web were 1 V and 1.6 μA for the smooth fibers, 1.3 V and 2.1 μA for the porous fibers, and 1.8 V and 3.2 μA for the wrinkled fibers (figure 4.10(i)). The output voltage and current values were enhanced for the aligned fiber webs (figure 4.10(j)).

4.3.4 Nanofillers

The nucleation behavior of a polymer depends on the properties of the nucleating agents, such as surface charge, surface area, concentration, and dispersion, as well as the interfacial interactions between the nucleating agents with PVDF chains. Nucleating agents reduce the nucleation energy barrier, and increase crystallization kinetics and the degree of crystallinity [40]. By introducing appropriate nanofillers into the electrospun PVDF, the β crystal phase content in the nanofibers can be significantly increased, which further enhances the mechanical-to-electrical energy conversion property. To get high energy conversion, a variety of nanofillers have been explored, and are described below.

4.3.4.1 Salts

To evaluate the effects of surface charge on nucleation efficiency, various salts have been embedded into electrospun PVDF nanofibers as nucleating agents. Dipti *et al* [41] studied the effect of nickel chloride hexahydrate ($NiCl_2{\cdot}6H_2O$) (NC) on the PVDF phase formation. An addition of 0.5 wt% NC was found to enhance the β crystal phase in the nanofibers by ~30%, and the voltage generated from PVDF–salt nanofiber webs was three times higher than that of neat PVDF nanofiber webs. Mokhtari *et al* [42, 43] added lithium chloride (LiCl) into a PVDF solution to prepare composite nanofibers. The addition of 0.00133 wt% LiCl led to the increase in β crystal phase content from 84% to 93.6%; accordingly, the piezo response

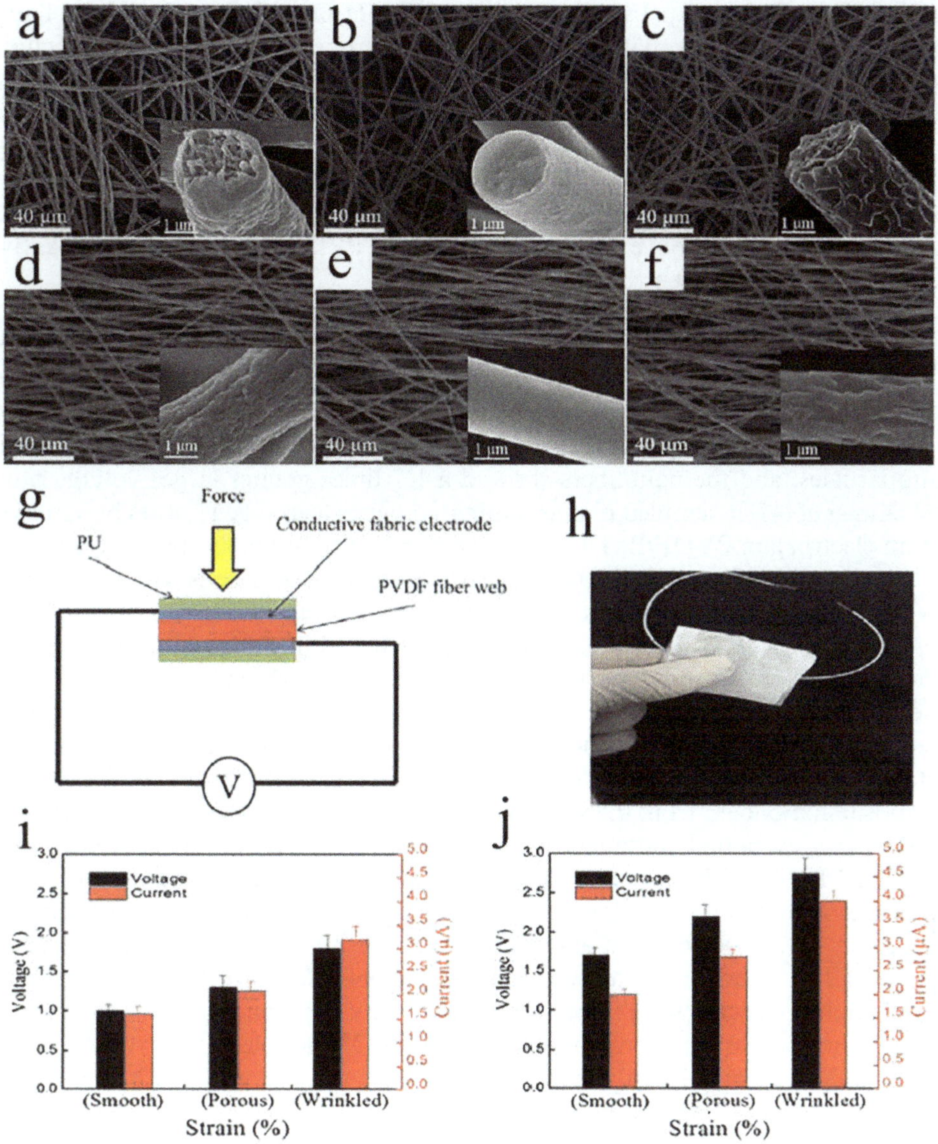

Figure 4.10. Samples prepared by electrospinning of PVDF solutions with different morphologies. (a–c) Randomly oriented fibers: (a) wrinkled, (b) smooth, and (c) porous. (d–f) Aligned fibers: (d) wrinkled, (e) smooth, and (f) porous. (g) Schematic structure of the PENG and (h) photo of the actual device. Summary results of the average voltage and current outputs generated by (i) randomly oriented fiber device and (j) aligned fiber device [32]. Reproduced with permission. Copyright 2019, Wiley.

increased from 1.89 V to 8.0 V when a steel ball (12 g) was dropped from a height of 30 cm through a steel guide pipe to the fiber webs.

In another study, 26 different salts were used to dope PVDF nanofibers to understand their effect on piezoelectric properties [44]. These salts were divided into three groups. The group I salts, such as $SnCl_4 \cdot 5H_2O$ and $Co(Ac)_2 \cdot 4H_2O$, had no obvious influence on the nanofiber piezoelectric property. Group II salts, including $CoCl_2 \cdot 6H_2O$ and $FeCl_2 \cdot 4H_2O$, had a significant enhancement in nanofiber piezoelectric performance, while group III salts ($MgCl_2 \cdot 6H_2O$, $Co(NO_3)_2 \cdot 6H_2O$, etc) resulted in a modest improvement in the piezoelectric property. Furthermore, salts of low concentration induced PVDF transform from amorphous polymer phases to the β crystal phase, resulting in a further increase in piezoelectric outputs. However, the energy conversion property of the PVDF nanofibers was negatively affected if the salt concentration was too high.

4.3.4.2 Ceramic particles

A large number of works were reported on the incorporation of piezoelectric inorganic particles into electrospun PVDF nanofibers to enhance the piezoelectric property [45]. $BaTiO_3$ was added to PVDF nanofibers and with 16 wt% $BaTiO_3$ nanoparticles, and the nanofibers showed a 1.7 times greater larger voltage output [46]. Xin *et al* [47] assembled planar laminated piezoelectric generators by sandwiching an electrospun PVDF/$BaTiO_3$ fiber web with carbon-loaded-polyethylene films. The device with 20 wt% $BaTiO_3$ nanoparticles generated a peak voltage output of 8 V and a current of 50 nA under bend-release deformation.

A vibration energy harvester was prepared by adding piezoelectric $(Na_{0.5}K_{0.5})NbO_3$ (NKN) ceramic nanoparticles into PVDF nanofibers via electrospinning [48]. The PVDF/NKN composite nanofiber web reached the highest piezoelectric coefficient when the NKN concentration was 50 vol%. The largest power output obtained was 145 nW cm^{-2} per vibration cycle at 170 Hz. Sang *et al* [49] synthesized composite nanofibers from $0.78Bi_{0.5}Na_{0.5}TiO_3$–$0.22SrTiO_3$ (BNT–ST) ceramic and PVDF, and examined the effect of BNT–ST concentrations on nanofiber piezoelectricity. BNT–ST has been considered the most promising candidate for lead-free piezoelectric materials, owing to its strong piezoelectric property. It has been found that 60 wt% BNT–ST content allowed the nanofiber web to produce a peak voltage output of 1.31 V under the resonance frequency of 6 MHz.

ZnO nanoparticles were incorporated into PVDF nanofibers to improve the piezoelectric properties. Compared with neat PVDF samples (0.351 V) in the same study, under compression, the composite samples with 15 wt% ZnO nanoparticles generated higher voltage outputs [50].

4.3.4.3 Nanoclay

Nanoclays were used to reinforcement and improving the barrier functions of PVDF. Mohammed *et al* [51] reported that the incorporation of halloysite nanotubes (HNTs) into PVDF not only reduced nanofiber diameter, but also increased β crystal phase content. The optimum HNT concentration to make PVDF nanofibers reach the highest β crystal phase content was 10 wt%; at that time, the β crystal phase content reached 81.1%, in comparison to 34% in the neat PVDF. Meanwhile, its highest voltage output was 0.955 V under a 100 g load, which was three times higher than that of neat PVDF nanofibers. Xin *et al* [52] reported that 5 wt% nanoclay greatly improved the β crystal phase content and piezoelectric output of the PVDF nanofiber web. Hectorite (HRT) clay was also added into electrospun PVDF to enhance the piezoelectric property [53]. The prepared nanofibers with 0.1 wt% HRT clay increased the β crystal phase content to 95% and voltage output to 12 V (neat PVDF nanofibers contained 80% β crystal phase and could generate 5 V outputs).

4.3.4.4 Carbon nanotubes and graphene

MWCNTs and graphene have been used to modify electrospun PVDF nanofibers. Liu *et al* [54–56] prepared PVDF/MWCNT composite nanofiber webs by near-field electrospinning and studied their piezoelectricity. During the electrospinning process, MWCNTs interacted with PVDF, resulting in an obvious increase in β crystal phase content. When the PVDF nanofiber web was doped with 0.03 wt% MWCNTs (the maximum in this study), it showed the highest piezoelectric outputs of about 25 mV and 130 nA under a 15 Hz impact. Hao *et al* [57] examined the effect of MWCNT amount in a high doping level up to 10 wt% on the fiber piezoelectricity. It was found that the β crystal phase content reached the highest level when the MWCNTs amount was 5 wt%. At that time, the generated voltage output was as high as 6 V, which increased by 200% compared with neat PVDF nanofiber webs (figure 4.11). In another study, the effect of graphene on PVDF piezoelectricity was examined [58]. An electrospun PVDF nanofiber web with just 0.1 wt% graphene had a β crystal phase content as high as 83%, and voltage of 7.9 V under compression; this can be compared to the values of 77% and 3.8 V for the neat PVDF nanofiber web.

4.3.4.5 Polymer blending

Some polymers were blended with PVDF to enhance the β crystal phase content and piezoelectricity. Electrospun PVDF nanofibers containing cellulose nanocrystals (CNCs) showed an increase in the β crystal phase content. When the CNC content was increased from 0 wt% to 2 wt% [59, 60], the β crystal phase content increased by 5.7%. Further increasing the CNC content resulted in a decrease in β crystal phase content. The PVDF fiber matrix containing 2 wt% CNCs could produce a voltage output of 60 V, which was almost three times the value for the neat PVDF web NG (figure 4.12) [60].

4.3.4.6 Small molecules

Electron and hole transfer agents, tri-*p*-tolylamine (TTA) and 2-(4-tert-butylphenyl)-5-(4-biphenylyl)-1,3,4-oxadiazole (Butyl-PBD), were added separately into PVDF nanofibers [61]. When the PVDF nanofibers contain 0.5% TTA or 1% Butyl-PBD, they show the highest β phase content and electric outputs (figure 4.13). By applying a 10 N compression force at 1 Hz to the fiber webs, the PVDF/TTA device can generate a maximum voltage of 2.55 V while PVDF/Butyl-PBD generates 2.65 V. These values show increases of 70% and 77%, respectively, compared with those of the neat PVDF nanofiber web.

4.3.5 Device structure

For various applications, devices with specific structures were prepared. For respiration and healthcare monitoring applications, a wearable active sensor was

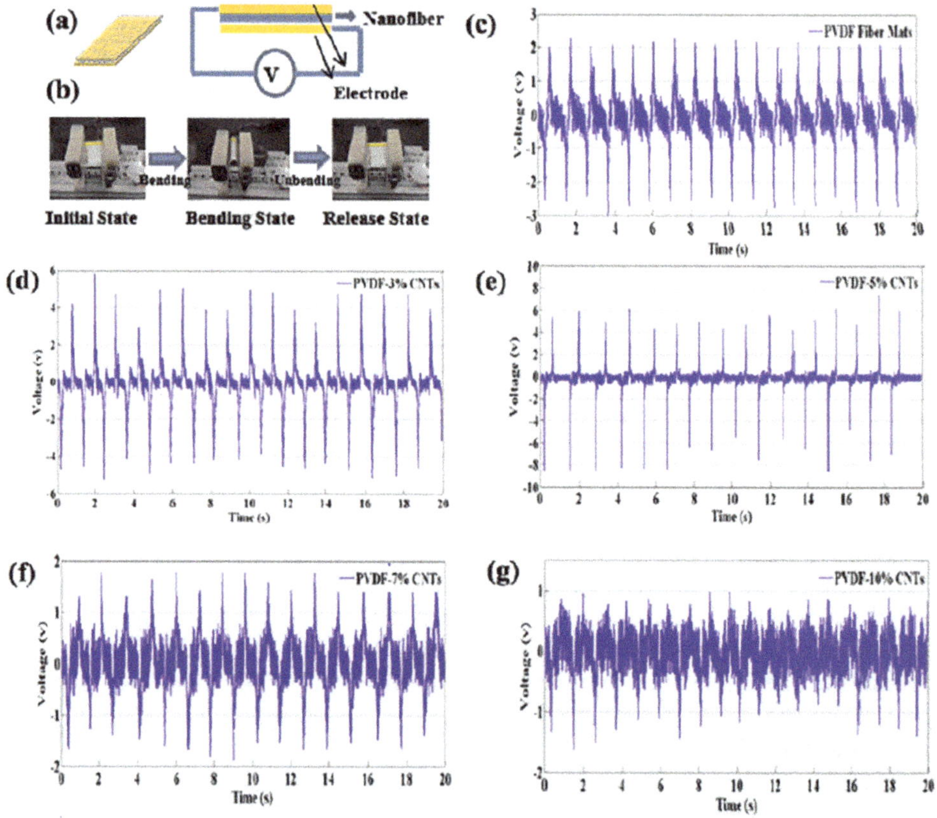

Figure 4.11. (a) Schematics of the appearance and cross-sectional structure of the PVDF power generator; (b) the measured output voltage corresponds to the initial state, bending state, and release state of the power generator. The output voltage of PVDF nanofiber webs with different MWCNT concentrations: (c) neat PVDF fiber webs; (d) PVDF-3% CNTs; (e) PVDF-5% CNTs; (f) PVDF-7% CNTs; and (g) PVDF-10% CNTs [57].

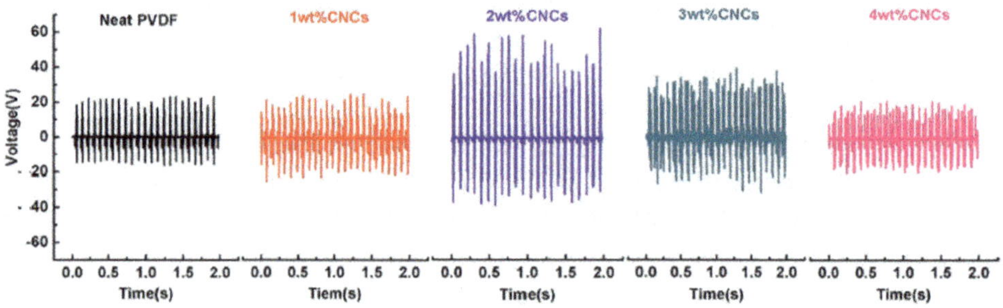

Figure 4.12. The output voltages of electrospun PVDF and PVDF/CNCs-based NGs [60]. Reproduced with permission. Copyright 2016, Elsevier.

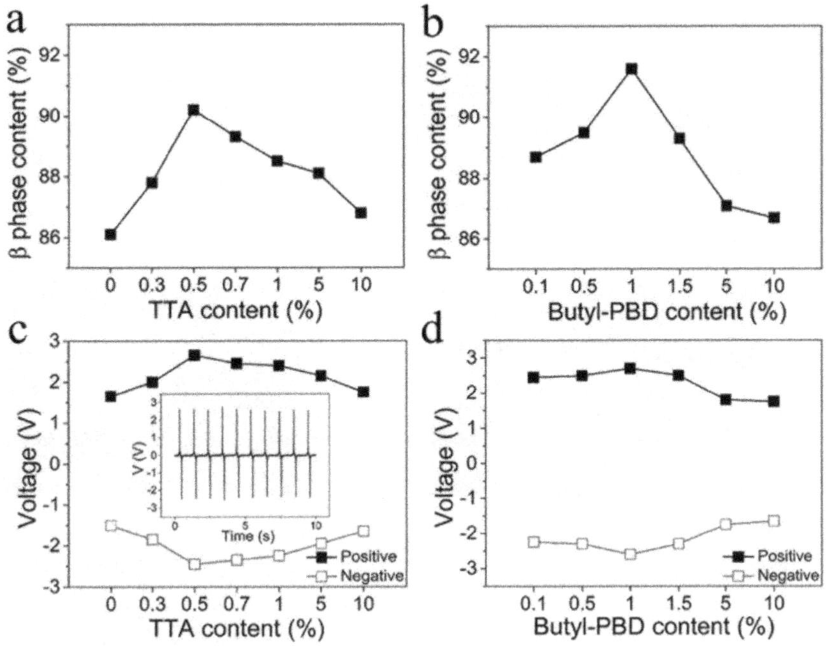

Figure 4.13. Calculated β phase content of PVDF nanofibers with different (a) TTA and (b) Butyl-PBD doping contents. Effect of (c) TTA and (d) Butyl-PBD doping contents on the peak voltage of PVDF nanofibers. Inset in (c): voltage output signals of PVDF nanofibers with 0.5% TTA doping content (web thickness: 100 μm; working area: 4 cm^2) [61]. Reproduced with permission. Copyright 2017, Wiley.

fabricated based on a flexible PENG. Silicone was employed as an elastic substrate to enhance the structure stability of the device [29]. When the device was bent by a cyclic mechanical force, the generated voltage and current were around 1.5 V and 400 nA, respectively. To enhance the electrical outputs, a circuit board substrate with interdigital electrodes (IDEs) was used, as shown in figure 4.14 [22, 62].

Fuh *et al* [63] designed a novel concentric circular topography device, which has spectacular performance in harvesting mechanical deformation in arbitrary directions. As shown in figure 4.15, the PVDF nanofibers were deposited on the electrodes by near-field electrospinning. Later, Cu wires were soldered on two end sides of the electrodes separately and PDMS was coated on the top for device encapsulating purpose. Five positions in the device were tested. The outputs were similar, showing the capability of mechanical energy harvesting in arbitrary directions. The device could be beneficial to the fabrication of electronic sensors.

4.4 Piezoelectricity of other electrospun nanofibers

4.4.1 Polymers

Besides PVDF, some other electrospun nanofibers have also been reported to have piezoelectric properties. These are summarized in table 4.2.

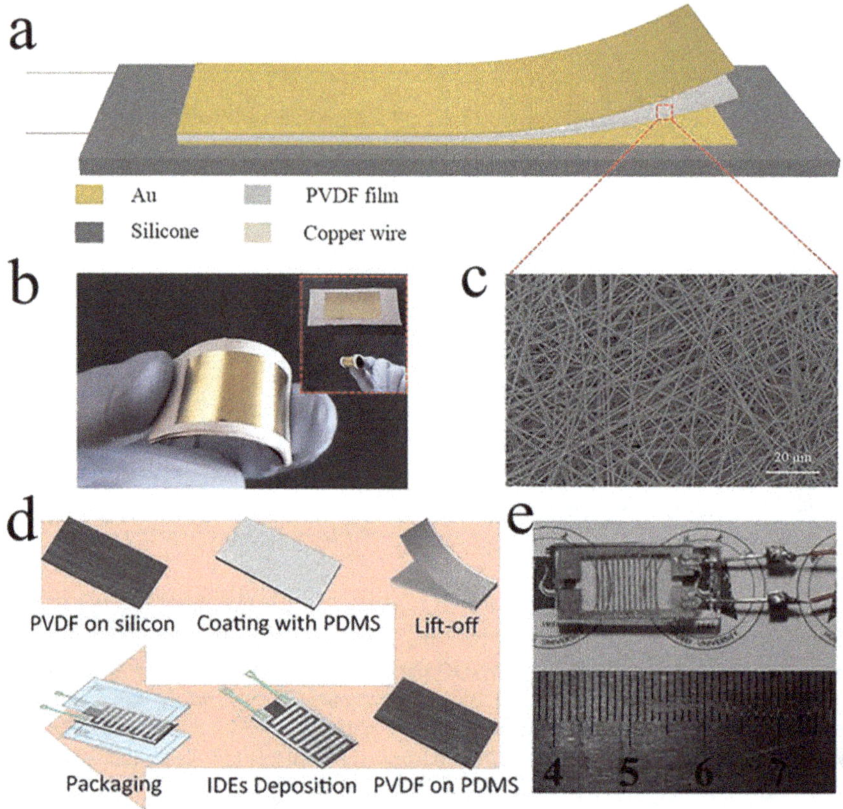

Figure 4.14. (a) Schematic structure of PVDF nanofiber power generator, (b) photographs of the device, and (c) SEM image of the electrospun PVDF nanofibers [29]. (d) Schematic diagram of the device fabrication process and (e) a photo of the device [22].

4.4.1.1 PVDF copolymers

Copolymerization is an effective method to tune the polymorph structure and phase transition behavior of PVDF. The comonomer type and composition ratio determine the crystalline structure of the polymer and degree of crystallinity, which directly affect the piezoelectric properties. So far, a variety of VDF–copolymers have been synthesized by the incorporation of trifluoroethylene (TrFE) and hexafluoropropene (HFP), and have high piezoelectricity [45].

PVDF–TrFE, including α (nonpolar), β (polar), γ (polar), and δ (polar) phases, is one of the most studied copolymers. The β crystal phase, having the antipolar arrangement of fluorine and hydrogen atoms with partial negative and positive charges, respectively, is the most piezoelectrically responsive phase. One advantage of PVDF–TrFE over PVDF is that regardless of the processing method, it always contains the β crystalline phase. This is attributed to the steric hindrance of TrFE in the copolymer, which forces PVDF into the all-trans (TTTT) configuration (β phase). Furthermore, the high degree of crystallinity in the preferred orientation of

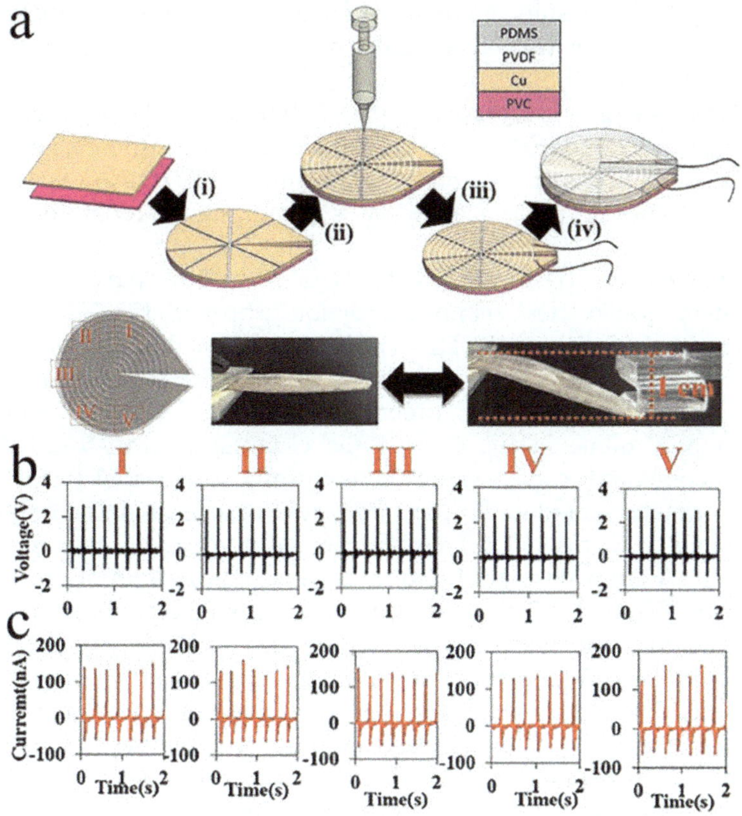

Figure 4.15. (a) Schematic for fabricating PVDF fibers with concentric circle topography, (b) voltage, and (c) current generated by flapping the corresponding position at a constant frequency of approximately 4.5 Hz [63].

Table 4.2. Summary of various electrospun piezoelectric polymer nanofibers (except PVDF).

Polymers	Fiber diameter (nm)	Web thickness (μm)	Working area (cm^2)	Strain/ Force/ Frequency	V	I	Ref
PVDF–TrFE	250–300	—	—	0.3% strain	2.5 V	0.1 μA cm^{-2}	[64]
PVDF–HFP	71	150 ± 20	1.13	2 Hz, 15 kPa pressure	1 V	0.3 μA cm^{-2}	[65]
Poly(γ-benzyl-L-glutamate)	~8000	—	—	30 Hz	14.25 mV	19.00 nA	[66]
Poly(L-lactic acid)	450	40	5	15 Hz, 0.3 MPa	1 V	5 nA	[67]
PAN	610 ± 50	110	5	2 Hz	2.0 V	1.2 μA	[68]
Collagen	210 ± 140	—	3.75	—	0.23 V	—	[69]
Gelatin	170 ± 100	—	3.75	—	0.05 V	—	[69]
Chitin	170 ± 110	—	3.75	—	0.25 V	—	[69]

well-grown crystallites results in a higher remnant polarization compared to that in PVDF, which indicates the higher efficiency in mechanical-to-electrical energy conversion [70].

PVDF–HFP has a lower crystallinity than PVDF due to the presence of the bulky CF_3 groups [71]. PVDF–HFP has an unusual piezoelectric response, i.e. $|d_{31}/d_{33}| > 1$, which is not observed in other piezoelectric polymers [72]. It also has a much higher piezoelectric coefficient (d_{31}), making it a promising material for piezoelectric applications.

Electrospun PVDF–TrFE and PVDF–HFP nanofibers have been reported to have mechanical-to-electrical energy conversion properties [73–77]. Dipti *et al* [64] prepared randomly oriented PVDF–TrFE nanofibers directly on a conducting polyaniline film, which was grown on a flexible PET substrate. Under the application of a flexural strain, orders of magnitude stronger modulation of transport in polyaniline in the case of the electrospun fiber-based device were observed, as compared to the case of the uniformly spin-coated PVDF–TrFE film device. The transport modulation occurred due to strain-induced conformational changes leading to carrier delocalization. Finally, a voltage output of 2.5 V and a current of 0.1 μA cm^{-2} were produced under an applied strain of up to 0.3%.

Electrospun PVDF–TrFE fibers with diameters ranging from 1000 to sub-100 nm were prepared and the piezoelectric performance studied [78]. When the fiber diameter was reduced from 860 nm down to 90 nm, the electric output of nanofiber webs increased. The electric output values were much higher than those of PVDF nanofiber webs. Ma *et al* [79] used a stretching-induced alignment method to achieve highly oriented electrospun PVDF–TrFE fibers on a large scale. These aligned fibers exhibited an enhanced piezoelectric property and high mechanical endurance. The average voltage output of 80% aligned electrospun PVDF–TrFE fibers was 84.96 mV, which was about 226% of the value of similar fibers with random orientation.

Dipankar *et al* [65] prepared PENGs based on electrospun PVDF–HFP nanofiber webs. The fiber webs contained around 77% of the relative proportion of the β crystalline phase. A maximum voltage output of 1 V with a current density of 0.3 μA cm^{-2} was achieved by the generated NG under periodic pressure with a peak amplitude of 15 kPa. Silver nanoparticle doping could increase the β crystalline phase content and further enhance the piezoelectric outputs; this will be described in a later section.

4.4.1.2 Poly(γ-benzyl-L-glutamate) (PBLG)

PBLG, a liquid crystalline rod-like polymer, was reported to have high piezoelectricity when processed into nanofibers. PBLG is a synthetic polypeptide that forms hierarchically ordered structures containing α-helices. The α-helically structured PBLG serves as a rod-like structure in solid state and common organic solvents. PBLG features a great number of directionally aligned hydrogen bonds along the helical axis that form a macroscopic dipole, which can couple synergistically with the external electric field and shear force (figure 4.16) [80].

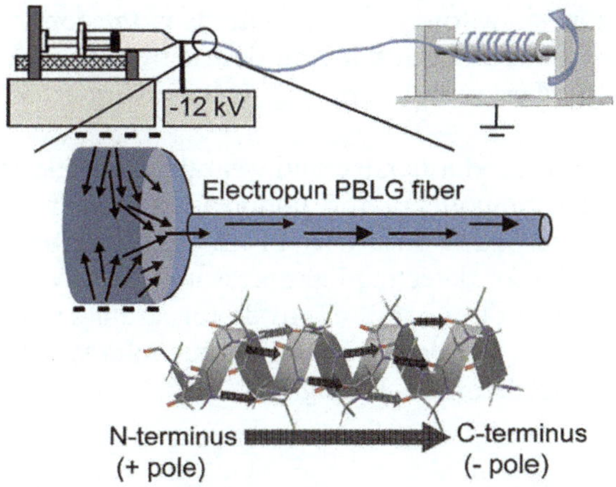

Figure 4.16. Schematic of electrospinning and the PBLG poling process. The α-helical molecular structure of PBLG is shown without side chains for simplicity. Small block arrows in the PBLG helix are individual hydrogen bonds, and long arrows under the PBLG helix and in the expanded portion of needle tip area represent macroscopic dipoles of PBLG molecules [80]. Reproduced with permission. Copyright 2011, Wiley.

Pan *et al* [66] used cylindrical near-field electrospinning to produce PBLG piezoelectric fibers showing permanent piezoelectricity. A single PBLG fiber could generate the maximum voltage output of 33.27 mV with an external resistance of 8 MΩ. The corresponding power output reached 138.42 pW. PBLG fibers were further patterned on a cicada wing with interdigitated electrodes for an energy harvesting application. The voltage output increased with increasing vibration frequency and it ranged from 7.67 mV to 14.25 mV when the strains of 0.04%–0.1% at 10–30 Hz were applied. They also reported that the PBLG fibers had much better mechanical properties than PVDF fibers.

4.4.1.3 Poly(L-lactic acid) (PLLA)

PLLA is a biocompatible, biodegradable, and piezoelectric polymer that has shown great potential for applications in tissue engineering scaffolds, energy harvesting, and sensors [81]. The piezoelectric property of PLLA fibers originates from the C=O dipoles in the polymer chain [82]. The C=O dipoles are located circularly around the helical structure of PLLA with an angle of 125° relative to the C–O–C main chain. As a result, the dipole components in the perpendicular direction of the polymer chain cancel out and the dipole component along the polymer main chain remains. For these reasons, it is difficult to show piezoelectricity when PLLA is electrical poled along the thickness. With the inherent shear force and electric field force in the same direction, the electrospinning process can align the dipole component along the polymer chain. This results in electrospun PLLA fibers showing piezoelectricity along the d_{33} direction [81].

The shear piezoelectric behavior of electrospun PLLA nanofiber web sensors was reported as a function of web stacking arrangement including constructive, destructive, and multilayered folding [83]. Comparing to electrospun PVDF nanofiber web sensors, PLLA sensors show lower voltage outputs in the constructive stacking case and slightly higher voltage outputs in the destructive case. With the increase in stacking of the nanofiber web layers, the voltage output of the PLLA device increases.

Sultana *et al* [67] prepared a flexible and wearable piezoelectric bio-e-skin based on electrospun PLLA nanofibers for noninvasive human physiological signal monitoring. The device had a sensitivity of 22 V N^{-1} stably as well as pressure detection down to 18 Pa for detecting light pressure. Zhu *et al* [84] reported aligned porous PLLA nanofibers on a comb electrode for strain sensing and human joint motion energy harvesting. The device had an output voltage of 0.55 V and current of 230 pA under the strain deformation angle of 28.9°.

4.4.1.4 PAN

PAN is an amorphous polymer comprising a cyano group (–CN) in each repeat unit. The dipole moment of PAN was calculated as nearly 3.5 Debye units, which is much higher than PVDF's (2.1 Debye units) [85]. However, PAN often shows a low piezoelectricity [86].

Wang *et al* [68] reported that PAN nanofibers prepared by electrospinning can have a piezoelectricity even stronger than PVDF. The prepared device consisted of an electrospun nanofiber web, two aluminum foils, and two transparent PET films (figure 4.17(a)). Under the periodic mechanical compression, the PAN nanofiber device generated electric outputs of 2.0 V and 1.2 μA. In the same condition, the device made of PVDF nanofibers had lower electric outputs (figures 4.17(b) and (c)). By adjusting the collector drum's rotating speed, aligned PAN nanofibers were also prepared. The authors further discovered that the aligned PAN nanofiber device could generate much larger electrical outputs, up to 6.0 V and 5.1 μA. In solid state, PAN has two typical conformations: planar zigzag and 3^1-helical. The zigzag conformation with the all-transform (TTT) structure is the reason for PAN's piezoelectricity. A high content of the zigzag conformation (around 80%) exists in electrospun PAN nanofibers, which suggests their higher piezoelectricity.

4.4.1.5 Collagen, gelatin, and chitin

Collagen, gelatin, and chitin are reported to show piezoelectric properties and could be used for making pressure sensors. One advantage of using biological materials is their biocompatibility, allowing for long-term physiological detection without causing any skin infection. Collagen, as one type of protein, its piezoelectricity was reported as early as the 1950s, and the piezoelectricity originated from the oriented crystalline collagen fiber [87]. Gelatin is produced through the partial hydrolyzation of collagen. Chitin is one of the polysaccharides containing β-(1.4) linked *N*-acetylglucosamine monomers.

Street *et al* [69] reported an electrospun chitin nanofiber web with piezoelectricity, and aligned chitin nanofiber webs had a 400% higher piezoelectric response than the random chitin nanofiber webs. The increase in piezoelectricity is a consequence of an increase in α-chitin crystallinity in the nanofibers. They also prepared aligned and

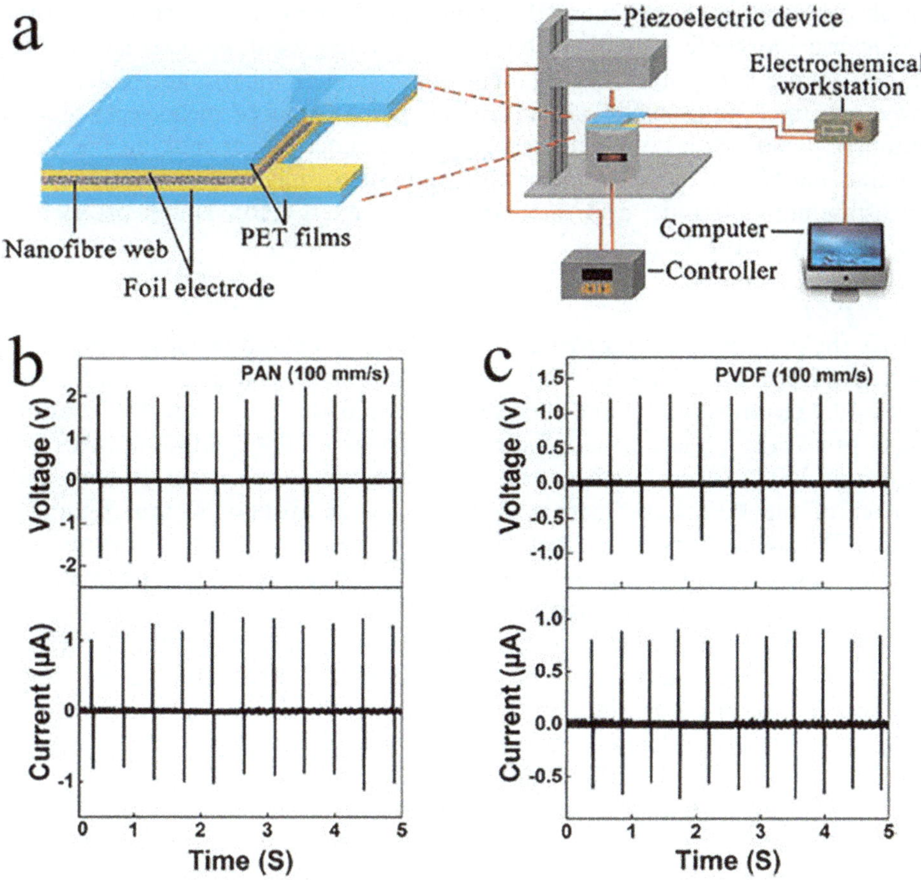

Figure 4.17. (a) Schematic illustration of device structure and energy conversion tests; piezoelectric outputs of (b) electrospun PAN nanofiber device and (c) electrospun PVDF nanofiber device (working area: 5 cm^2; web thickness: 100 μm; frequency: 2 Hz) [68]. Reproduced with permission. Copyright 2019, Elsevier.

random collagen nanofibers, as well as gelatin nanofibers, but found lower piezoelectric responses.

Ghosh *et al* [88] prepared a self-powered pressure sensor based on electrospun gelatin nanofibers for healthcare monitoring. The sensor exhibited high sensitivity (about 0.8 V kPa^{-1}) in a low-pressure region, and is suitable for physiological signal monitoring. It has high durability over 108 000 cycles of piezoelectric output under dynamic tactile stimuli, indicating its potential application in the field of implantable and portable biomedical and personal electronic devices.

4.4.2 Inorganic materials

Piezoelectric inorganic materials normally have a higher piezoelectric coefficient compared to polymer materials. However, they are often rigid and work at a very low strain level. Through processing these piezoelectric inorganic materials into a nanofibrous structure, their work durability can be enhanced. The preparation of piezoelectric inorganic nanofibers by electrospinning contains two steps: electrospinning to prepare polymer/inorganic composite nanofibers, and calcinating to remove organic components. Unlike in the preparation of piezoelectric polymer nanofibers, the electric poling process is always required for making piezoelectric inorganic nanofibers. Various inorganic nanofibers have been prepared by the electrospinning technique, and their mechanical energy conversion properties are detailed below (table 4.3).

4.4.2.1 PZT

In 2010, the first inorganic piezoelectric nanofibers prepared by electrospinning was reported; PZT nanofibers were used [89]. The PVP/PZT composite nanofibers were prepared by electrospinning and then annealed at 650 °C to obtain PZT nanofibers with a perovskite phase. Interdigitated electrodes of fine platinum wires with a distance of 500 μm were used for the fabrication of piezoelectric devices, and the diameter of the PZT nanofibers was controlled at around 60 nm (figure 4.18). Finally, PZT nanofibers were treated within an electric field of 4 kV mm^{-1} at a temperature over 140 °C for about 24 h. The poled nanofibers had orientated electric dipoles along the electric field with an output voltage of 1.63 V. In another study, PZT nanofibers were prepared by the same process, and the effect of the poling process on PZT piezoelectricity was studied [90]. The electromechanical coupling effect was increased by as much as 3.7 times after 90 min of polarization under a 3 kV mm^{-1} electric field. The produced device generated an output voltage of 100 mV in response to the falling ball (8.34 g) impact.

PZT nanowires prepared by electrospinning have also been reported for use in energy harvesting and self-powered devices [91]. PZT precursor salts containing tetrabutyl titanate, zirconium acetylacetonate, lead subacetate, and PVP at a specific ratio were used for making the electrospinning solution. After electrospinning, PZT nanowires were prepared by calcination of the composite fiber at 650 °C for about 3 h. Later, the nanowires were electrically poled in a 4 kV mm^{-1} electric field at a temperature of 130 °C for 15 min. The device could generate electric outputs of 6 V and 45 nA. Electrospinning was also used to prepare vertically aligned $Pb(Zr_{0.52}Ti_{0.48})O_3$ (PZT) nanowire arrays [92]. The voltage output generated by the device was as high as 209 V and it had a current density of 23.5 μA cm^{-2}, which were directly used to stimulate a frog's sciatic nerve and induce contraction of its gastrocnemius.

Andris *et al* [93] produced $PbTiO_3$ nanofibers by electrospinning a solution containing Pb acetate, acetic acid, titanium tetraisopropoxide, and PVP, and by subsequently annealing the composite nanofiber web at an elevated temperature (500 °C for 3 h). Finally, the $PbTiO_3$ nanofiber web was packed into a device and poled in a 0.4 kV mm^{-1} electric field. It could generate a peak voltage output as high as 1 V under compression.

4.4.2.2 Lead-free ceramics

Previously, lead oxides, especially PZT, have been widely used as piezoelectric materials because of their excellent piezoelectric properties. However, their lead

Table 4.3. A summary of inorganic nanofiber energy generators.

Inorganic nanofiber	Fiber diameter (nm)	Precursor polymer	Calcination temperature (°C)	Poling voltage (kV mm^{-1})	Poling temperature/ time	Device size (cm^2)	Force/Strain	Electric outputs	Ref
PZT multifibers	60	PVP	650	4	140 °C, 24 h	—	Finger pressure	1.63 V	[89]
PZT multifibers	50–120	PVP	650	3	90 min	—	8.34 g falling ball impact	0.1 V	[90]
PZT web	370	PVP	650	4	130 °C, 15 min	1.2	Bending/ Releasing	6 V, 45 nA	[91]
PZT nanowire array	370	PVP	650	5	130 °C, 15 min	2.25	0.53 MPa pressure	209 V, 53 μA	[92]
$PbTiO_3$ web	140	PVP	500	0.4	150 °C, 24 h	—	~35 g load	1 V	[93]
Neat ZnO web, doped ZnO web	82.8,61	PVA	550	—	—	1	1.25 kg load	0.12 V, 0.35 V	[94]
$BaTiO_3$ random, horizontal, vertical webs	354.1	PVP	1000	5	120 °C, 12 h	2.21, 1.19, 1.53	0.002 MPa pressure	0.56 V, 57.78 nA, 1.48 V, 103.33 nA, 2.67 V, 261.4 nA	[95]
$BaTiO_3$ nanofibers	45	PVP	750	—	—	5.6	0.16% bending strain	2.7 V	[96]
$BaTi_{0.9}Zr_{0.1}O_3$ nanofibers	300	PVP	800	3	60 °C, 60 min	—	Finger pressure	0.26 V	[97]
$0.5Ba(Zr_{0.2}Ti_{0.8})O_3$–0.5 $(Ba_{0.7}Ca_{0.3})TiO_3$ nanofibers	175	PVP	700	1.5	90 °C, 2 h	1.2	Stretch/ Release	3.25 V, 55 nA	[98]
$(Ba_{0.85}Ca_{0.15})(Ti_{0.9}Zr_{0.1})O_3$, $(Ba_{0.85}Ca_{0.15})(Ti_{0.9}Zr_{0.1})O_3$–Y	300, 350	PVP	700	12.0 kV	25 °C, 30 min	—	Finger pressure	1.9 V, 48 nA, 3 V, 85 nA	[99]

(*Continued*)

Table 4.3. (*Continued*)

Inorganic nanofiber	Fiber diameter (nm)	Precursor polymer	Calcination temperature (°C)	Poling voltage (kV mm^{-1})	Poling temperature/ time	Device size (cm^2)	Force/Strain	Electric outputs	Ref
$(K,Na)NbO_3$ nanofibers	100	PVP	700	2.3	—	1	Compression strain 6.0%, 5 Hz	65 mV	[100]
$NaNbO_3$ nanofibers	78	PVP	700	2	Room temperature	2	Bending	2 V	[101]
$0.96(K_{0.48}Na_{0.52})(Nb_{0.95}Sb_{0.05})O_3$–$0.04Bi_{0.5}(Na_{0.82}K_{0.18})_{0.5}ZrO_3$ nanofibers	250	PVP	700	5	150 °C, 2 h	4	Pressure, 1 Hz	10 V	[102]

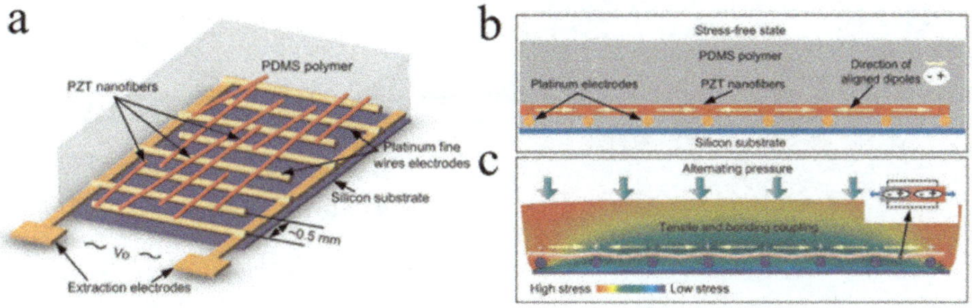

Figure 4.18. (a) Schematic view of the PZT nanofiber generator; (b) cross-sectional view of the poled PZT nanofibers in the generator, and (c) schematic view explaining the mechanism of the PZT nanofibers working in the longitudinal mode [89]. (Reproduced with permission. Copyright 2010, American Chemical Society.)

content raises environmental concerns, which hinders their applications in many fields. Thus, developing a lead-free and environmentally friendly material with a piezoelectric coefficient comparable to that of PZT is highly desirable. In the last decade, many lead-free piezoelectric ceramics, such as ZnO, $BaTiO_3$ (BTO), $BaTi_{0.9}Zr_{0.1}O_3$ (BTZO), $0.5Ba(Zr_{0.2}Ti_{0.8})O_3$–$0.5(Ba_{0.7}Ca_{0.3})TiO_3$ (BZTO–BCTO), $(Ba_{0.85}Ca_{0.15})(Ti_{0.9}Zr_{0.1})O_3$–Y (BCTZY), $(K,Na)NbO_3$ (KNN), and 0.96 $(K_{0.48}Na_{0.52})(Nb_{0.95}Sb_{0.05})$ O_3–$0.04Bi_{0.5}(Na_{0.82}K_{0.18})_{0.5}ZrO_3$ (KNNS–BNKZ), have been processed into nanofibers.

ZnO nanofibers and Al-doped ZnO nanofibers were fabricated by electrospinning at different solution flow rates of 2, 4, and 6 μl min^{-1} followed by annealing [94]. A higher solution flow rate resulted in a larger fiber diameter and worse mechanical-to-electrical conversion performance. The maximum power density of neat ZnO was 17.6 nW cm^{-2}, and for the doped ZnO nanofibers, the device showed a maximum power density of 51.7 nW cm^{-2}.

Yan *et al* [95] studied the energy conversion properties of $BaTiO_3$ nanofibers with different alignment features: aligned vertically, horizontally, and randomly (figure 4.19). The vertically aligned $BaTiO_3$ nanofibers showed the best piezoelectric performance with a maximum voltage output of 2.67 V and a current of 261.40 nA, under a mechanical stress of 0.002 MPa. Comparably, the devices from horizontally aligned nanofibers and randomly aligned nanofibers generated lower voltages, 1.48 V and 0.56 V, and lower currents, 103.33 nA and 57.78 nA, respectively. Another report indicated that electrospun $BaTiO_3$ nanofibers with finer diameters exhibited better piezoelectricity [96]. A device containing nanofibers with an average diameter of 45 nm produced a maximum voltage of 7.94 V_{p-p} under a strain of 0.16%.

$BaTi_{0.9}Zr_{0.1}O_3$ (BTZO) nanofibers with a diameter of nearly 300 nm were prepared by electrospinning [97]. Aligning BTZO nanofibers after poling generated 0.26 V voltage outputs when a finger applied a dynamic load. Lead-free $0.5Ba(Zr_{0.2}Ti_{0.8})O_3$–$0.5(Ba_{0.7}Ca_{0.3})TiO_3$ (BZTO–BCTO) nanowires with a high piezoelectric coefficient were synthesized by electrospinning [98]. This was accomplished through assembling the aligned nanowires into a flexible device, and poling in a 1.5 kV mm^{-1} electric field. The device could harvest weak mechanical energy and generate 3.25 V and 55 nA.

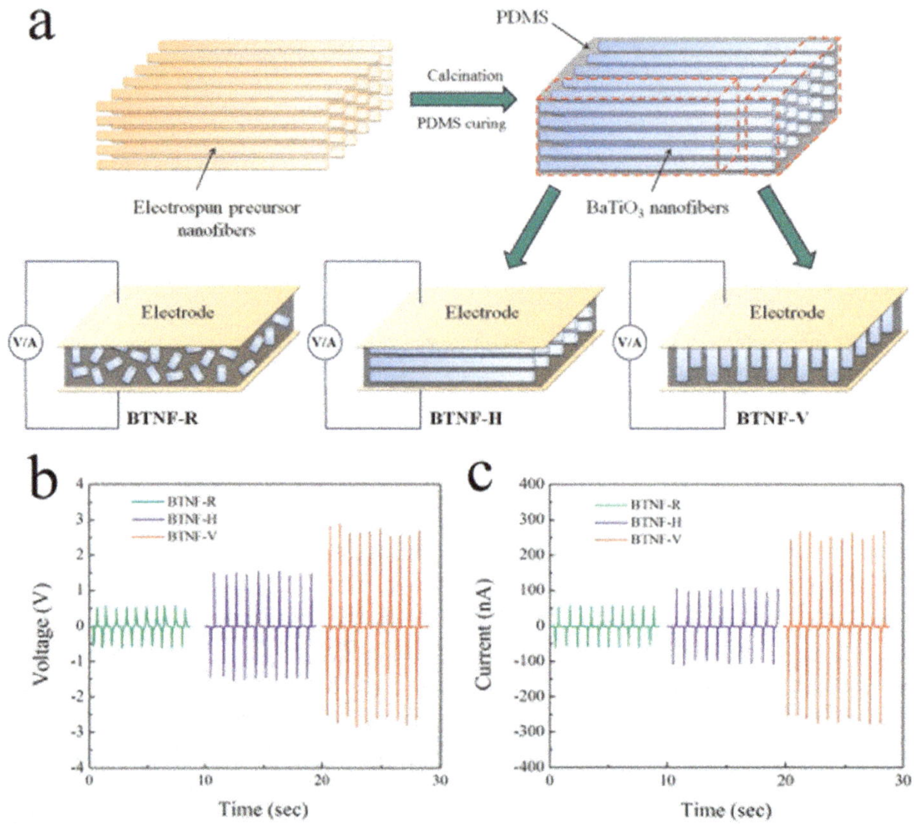

Figure 4.19. (a) Schematic fabrication procedure of the NGs based on $BaTiO_3$ nanofibers in three alignment modes. (b) Voltage and (c) current outputs of the $BaTiO_3$ NGs under a periodic mechanical compression [95]. Reproduced with permission. Copyright 2016, American Chemical Society.

Flexible BCTZY and BCTZ nanofibers were synthesized by electrospinning and the piezoelectric properties of the vertically aligned nanofibers were studied [99]. The introduction of Y^{3+} into BCTZ nanofibers could effectively enhance the nanofiber continuity, flexibility, and stability. Moreover, the Curie temperature of the BCTZY nanofibers is about 280 °C, which is much higher than that of BCTZ nanofibers (about 90 °C). Additionally, the electrical properties such as dielectricity, ferroelectricity, and piezoelectricity were all improved after doping with Y^{3+}. The devices made of vertically aligned BCTZY nanofibers showed an output voltage of 3.0 V and current of 85 nA following finger tapping.

Well-aligned electrospun KNN nanofibers after electric poling showed a fast response to dynamic strain by generating impulsive voltage signals that are dependent on the amplitude of dynamic strain and its frequency [100]. At a 1 Hz repeated impact, the voltage output increased from 1 mV to 40 mV when the strain changed from 1% to 6%. Moreover, the voltage output linearly increased when the input frequency increased from 0.2 to 5 Hz. Electrospun piezoelectric $NaNbO_3$

nanofibers were prepared for humidity sensor applications [101]. When the relative humidity changed from 5% to 80%, the peak-to-peak value of voltage generated by the sensor decreased from 0.40 to 0.07 V. This was because of the increased leakage current in the $NaNbO_3$ nanofibers, which was generated due to proton hopping among the H_3O^+ groups in the absorbed H_2O layers under the driving force of the piezoelectric potential. Piezoelectric KNNS–BNKZ nanofibers were fabricated by electrospinning and calcination [102]. The nanofiber device, after being poled at 150 °C with an electric field of 5 kV mm^{-1} for 2 h, could generate an output voltage up to 10 V under dynamic pressures.

4.4.3 Composites

A number of nanofillers have been added into electrospun PVDF nanofibers for enhancing the piezoelectric property of devices. Besides, some nanofillers were also mixed with other piezoelectric polymers to form composite nanofibers for achieving better energy conversion.

Piezoelectric composite nanofibers based on NKN nanoparticles mixed with PVDF–TrFE were prepared [103]. As expected, the addition of NKN particles in the nanofibers resulted in better energy harvesting performance. It was found that the device made of composite fibers containing 10 vol% NKN had an output voltage and current of 0.98 V and 78 nA, while the neat PVDF–TrFE fiber device only achieved 0.45 V and 56 nA. The improvement in piezoelectricity has been attributed to the contribution of the NKN nanoparticles dispersed intermittently along the polymer nanofibers.

Electrospun PVDF–HFP/Co-doped ZnO (Co–ZnO) composite nanofibers were also prepared for stretchable device fabrication [104]. An increase in β crystal phase concentration in PVDF–HFP nanofibers was observed with the increase in Co–ZnO nanofiller concentration in the PVDF–HFP matrix. Under mechanical compression, the neat PVDF–HFP device had a very low voltage output, about 0.12 V (figure 4.20). Piezoelectric voltages of 2 V, 2.4 V, and 2.8 V were achieved when the devices contained 0.5, 1, and 2 wt% of Co–ZnO, respectively.

Apart from adding piezoelectric ceramic nanoparticles, nanoparticles without a piezoelectric property were also added into piezoelectric polymer nanofibers. PVDF–HFP nanofibers containing silver nanoparticles were prepared by electrospinning [65]. The content of the β phase increased with the addition of silver nanoparticles, resulting in better piezoelectricity. A maximum output voltage of 3.0 V with a current density of 0.9 μA cm^{-2} was achieved under mechanical pressure. Eu^{3+}-doped PVDF–HFP/graphene nanofibers (EuGNFs) were also prepared [72]. The addition of Eu^{3+} and graphene sheets was reported to increase the β crystal phase content and crystallinity. The composite EuGNFs were able to generate a voltage output of 9 V under finger touches (with a pressure of 5.6 kPa), whereas a 4 V voltage was generated from the Eu^{3+}-doped PVDF–HFP EuNFs.

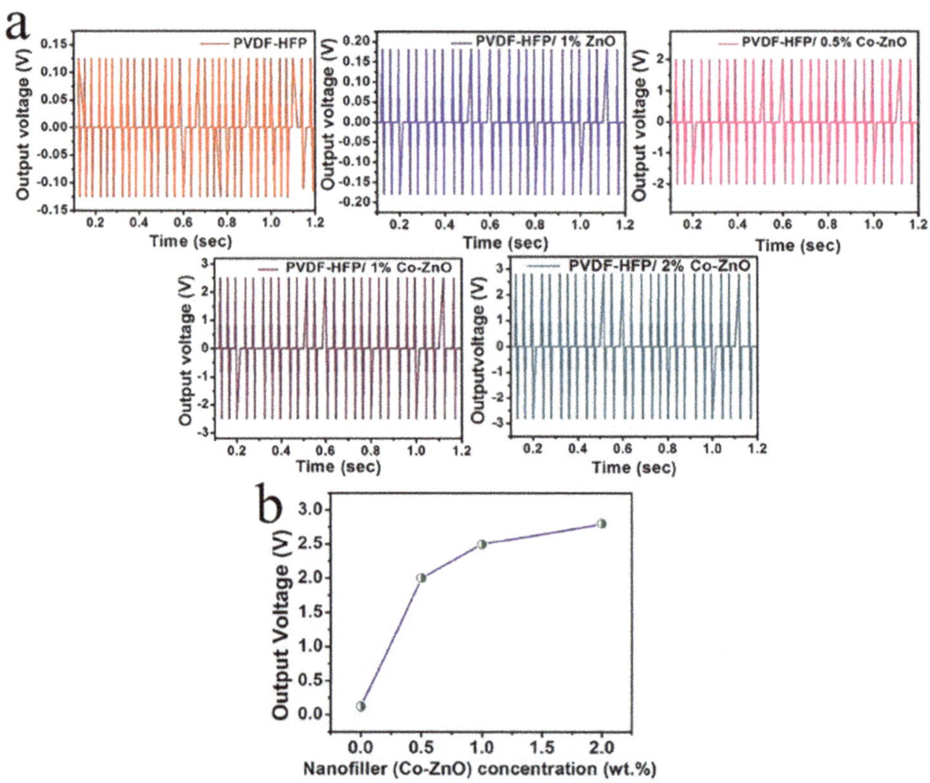

Figure 4.20. (a) Output voltage generation of neat PVDF–HFP and its composite devices; (b) voltage summary as a function of the Co–ZnO nanofiller concentration [104].

4.5 Piezoelectric–triboelectric hybrid energy generator devices

Hybrid energy generators with both piezoelectric and triboelectric mechanisms can be an effective way to enhance energy conversion performance. So far, there are two ways to fabricate hybrid energy generator devices. One uses piezoelectric materials as the triboelectric layers, and energy generators combining piezoelectricity and triboelectricity can be fabricated. The other method integrates the mechanical piezoelectric device unit and triboelectric device unit into one device.

Huang *et al* [105] developed a wearable TENG-based insole composed of electrospun PVDF nanofibers sandwiched between a pair of conducting fabric electrodes (figure 4.21). PVDF nanofibers produced by electrospinning formed a high β crystal phase content with strong piezoelectricity. The difference between the PENGs and TENGs is that for PENGs, the top electrode is firmly and directly attached to the PVDF nanofibers without any spacer so there is no relative displacement between the fabric electrode and PVDF nanofibers. The electrode in PENGs only collect piezoelectric charges. In this study, the authors prepared a PENG with a configuration of electrode/PVDF/electrode, and found that the direction of the generated voltage was reversed when the direction of dipoles was

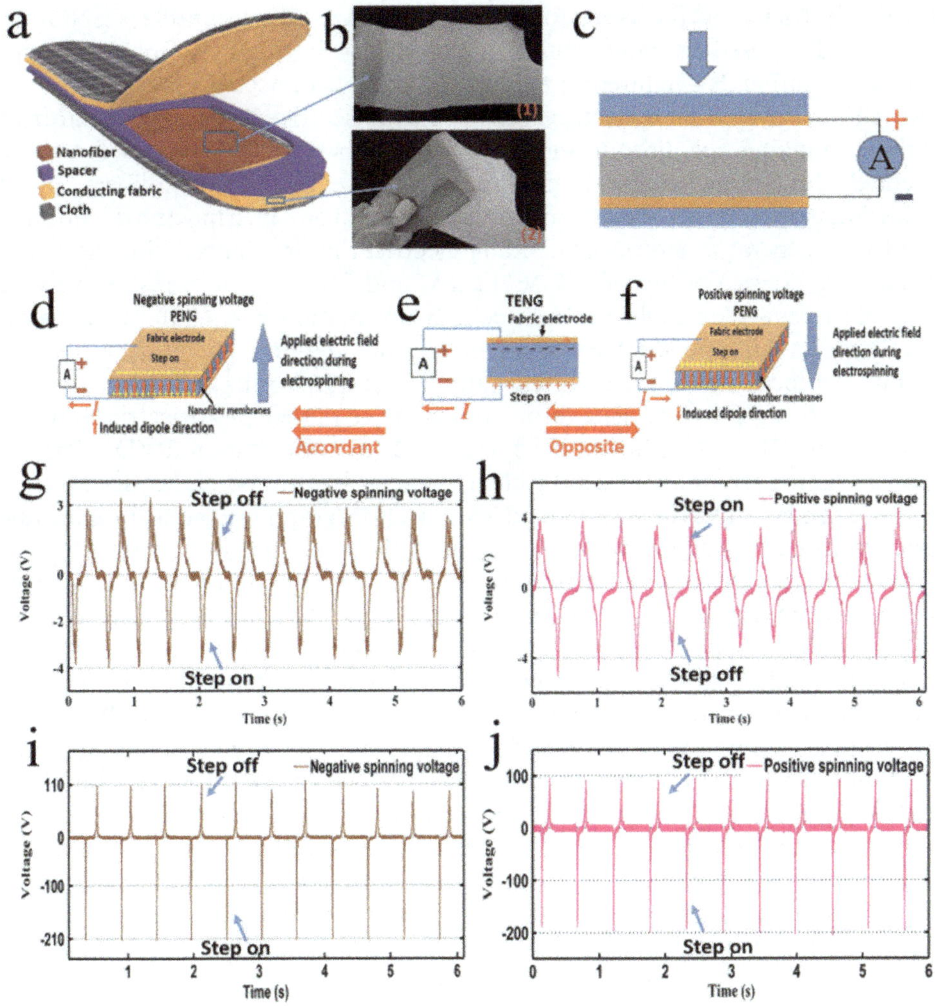

Figure 4.21. (a) Schematic structure illustration of the TENG-based insole; (b) digital photographs of the electrospun PVDF nanofibers on the conducting fabric (1) front side and (2) back side. (c) Schematic illustration of the TENG device. Schematic diagrams of (d) a PENG fabricated with negative spinning voltage; (e) a TENG fabricated by either negative or positive spinning voltage; and (f) a PENG fabricated by positive spinning voltage. Voltage generated by piezoelectricity of PVDF nanofibers fabricated under (g) negative and (h) positive spinning voltage. Voltage generated by triboelectricity of PVDF nanofibers fabricated under (i) negative and (j) positive spinning voltage [105]. Reproduced with permission. Copyright 2015, Elsevier.

reserved (figure 4.21). The PVDF PENG could produce a maximum voltage of 4 V. In comparison, the direction of voltage generated by triboelectricity would remain the same. Interestingly, the maximum voltage of the TENG was increased from 200 to 210 V when the voltage direction was the same as that of the voltage generated by piezoelectricity. These findings indicate that the overall voltage is the sum of the voltages generated by piezoelectricity and triboelectricity. A positive contribution achieved by piezoelectricity is tuning its voltage direction in accordance with that of the triboelectricity by reversing the dipole direction. The wearable TENG-based insole could be used to convert mechanical energy into light simply by stepping force, illuminating 214 light-emitting diodes (LEDs) connected in series.

An all-fiber hybrid piezoelectric-enhanced triboelectric nanogenerator fabricated by electrospinning silk fibroin and PVDF nanofibers on conductive fabrics was presented [106]. Due to the large specific surface area of nanofibers and the extraordinary ability of silk fibroin to donate electrons in triboelectrification, the hybrid nanogenerator shows outstanding electrical performance, with the voltage, current, and power density of 500 V, 12 μA, and 310 μW cm^{-2}, respectively. The ability of the smart textile device to detect the abrupt motion and falling of a person, as well as its high-power harvesting, make the hybrid generator a promising candidate in the area of wearable real-time health monitoring.

Another type of hybrid generator comprises cascaded piezoelectric and triboelectric units [107]. As shown in figure 4.22, an electrospun PVDF–TrFE/Ag nanowire nanofiber web was sandwiched between two fabric electrodes serving as a PENG, while a TENG was formed by a PDMS/graphite composite film and a fabric electrode. Ag nanowires could enhance the nanofiber web's piezoelectricity, electric conductivity, and dielectric constant. A typical hybrid generator showed a peak voltage output of 247.87 V by its piezoelectric unit and a voltage output of 84.43 V by its triboelectric unit, under a peak compression force of 1500 N and frequency of 3 Hz. At the same compression condition, the peak voltage reached around 300 V when the hybrid generator units were connected in parallel.

A flexible textile-based hybrid nanogenerator was prepared containing an electrospun PVDF/CNT/$BaTiO_3$ nanofiber nonwoven piezoelectric unit and a triboelectric unit with a microstructured surface configuration [108]. CNTs and $BaTiO_3$ were embedded into PVDF to modify the dielectric constant and improve the piezoelectric performance. The optimized electrospun piezoelectric web in their study consisted of 1 wt% CNT and 18 wt% $BaTiO_3$. The piezoelectric NG could generate a 21.31 V voltage while a voltage output of 157.39 V was achieved for the triboelectric NG. This hybrid NG has a low cost and can be suitable for industrial applications in the wearable device fields.

4.6 Summary

This chapter has provided a detailed introduction of significant progress in piezoelectric electrospun nanofibers, their fabrication, and their applications in mechanical-to-electrical energy conversion. Electrospun nanofibers exhibit a large surface area, high porosity, and unique web structure properties. PVDF is a widely studied piezoelectric polymer because of its high piezoelectric coefficient and easy electrospinning preparation. Nanofiber alignment, nanofiller doping, and device structure have been employed to enhance electrospun PVDF nanofiber device mechanical-to-electrical energy conversion. Besides PVDF, other piezoelectric polymers, including PVDF–TrFE, PVDF–HFP, PBLG, PLLA, and PAN, were also fabricated into nanofibers by electrospinning. These materials have shown comparable energy harvesting performance with PVDF, or even much better in some cases. The

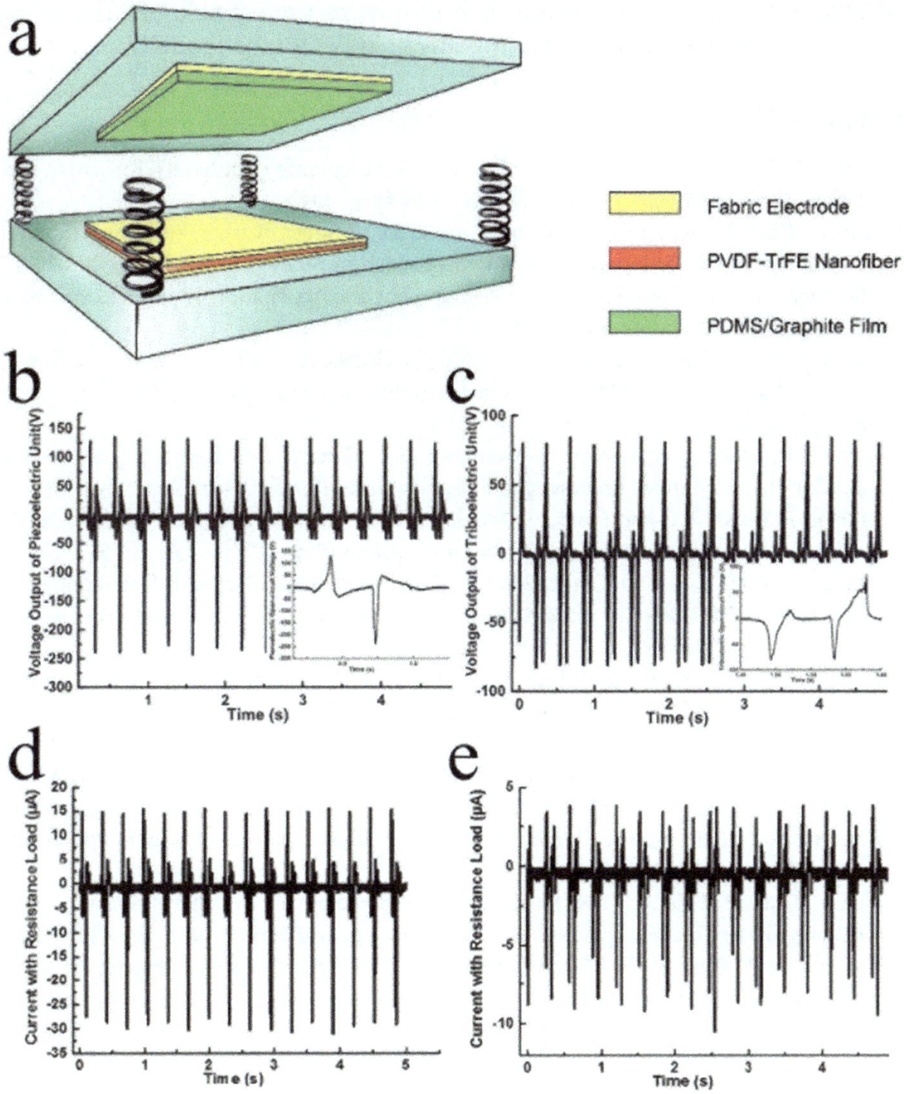

Figure 4.22. (a) Schematic illustration of hybrid generator. (b) Voltage output of the piezoelectric unit. (Inset shows an enlarged voltage curve in one cycle.) (c) Voltage output of the triboelectric unit. (Inset shows an enlarged voltage curve in one cycle.) (d) Current output of the piezoelectric unit with 5.1 MΩ external resistance. (e) Current output of the triboelectric unit with 5.1 MΩ external resistance. (Force: 1500 N, Frequency: 3 Hz) [107]. Reproduced with permission. Copyright 2016, Wiley.

ceramic piezoelectric nanofibrous materials are mainly studied in the direction of lead-free materials. They normally have much higher piezoelectric coefficients than the piezoelectric polymers. However, their rigidity and fragility can limit their wide application in flexible and stretchable mechanical-to-electrical energy conversion devices. Electrospun piezoelectric devices can be integrated with triboelectric devices to enhance electrical outputs. Many hybrid devices are capable of driving micro-electronic devices directly without using any energy storage modules.

References

[1] Dong Z, Kennedy S J and Wu Y 2011 Electrospinning materials for energy-related applications and devices *J. Power Sources* **196** 4886–904

[2] Huo L, Chen D, Kong Q, Li H and Song G 2017 Smart washer—a piezoceramic-based transducer to monitor looseness of bolted connection *Smart Mater. Struct.* **26** 025033

[3] Manbachi A and Cobbold R S 2011 Development and application of piezoelectric materials for ultrasound generation and detection *Ultrasound* **19** 187–96

[4] Kawai H 1969 The piezoelectricity of poly (vinylidene fluoride) *Jpn. J. Appl. Phys.* **8** 975

[5] Harrison J and Ounaies Z 2002 Piezoelectric polymers *Encyclopedia of Polymer Science and Technology* 3

[6] Martins P, Lopes A and Lanceros-Mendez S 2014 Electroactive phases of poly (vinylidene fluoride): determination, processing and applications *Prog. Polym. Sci.* **39** 683–706

[7] Martins P, Lopes A C and Lanceros-Mendez S 2014 Electroactive phases of poly(vinylidene fluoride): determination, processing and applications *Prog. Polym. Sci.* **39** 683–706

[8] Chang C, Fuh Y K and Lin L 2009 A direct-write piezoelectric PVDF nanogenerator *Transducers 2009–2009 Int. Solid-State Sensors, Actuators and Microsystems Conf. (21–25 June 2009)* pp 1485–8

[9] Chang C, Tran V H, Wang J, Fuh Y and Lin L 2010 Direct-write piezoelectric polymeric nanogenerator with high energy conversion efficiency *Nano Lett.* **10** 726–31

[10] Fang J, Wang X and Lin T 2011 Electrical power generator from randomly oriented electrospun poly(vinylidene fluoride) nanofiber membranes *J. Mater. Chem.* **21** 11088–91

[11] Fuh Y, Ye J, Chen P and Huang Z 2014 A highly flexible and substrate-independent self-powered deformation sensor based on massively aligned piezoelectric nano-/microfibers *J. Mater. Chem.* A **2** 16101–6

[12] Fuh Y, Chen P, Huang Z and Ho H 2015 Self-powered sensing elements based on direct-write, highly flexible piezoelectric polymeric nano/microfibers *Nano Energy* **11** 671–7

[13] Fuh Y, Ye J, Chen P, Ho H and Huang Z 2015 Hybrid energy harvester consisting of piezoelectric fibers with largely enhanced 20 V for wearable and muscle-driven applications *ACS Appl. Mater. Interfaces* **7** 16923–31

[14] Duan Y Q, Huang Y A, Yin Z P, Bu N B and Dong W T 2014 Non-wrinkled, highly stretchable piezoelectric devices by electrohydrodynamic direct-writing *Nanoscale* **6** 3289–95

[15] Ding Y, Duan Y and Huang Y 2015 Electrohydrodynamically printed, flexible energy harvester using *in situ* poled piezoelectric nanofibers *Energy Technol.* **3** 351–8

[16] Duan Y, Ding Y, Bian J, Xu Z, Yin Z and Huang Y 2017 Ultra-stretchable piezoelectric nanogenerators via large-scale aligned fractal inspired micro/nanofibers *Polymers* **9** 714

[17] Hansen B J, Liu Y, Yang R and Wang Z L 2010 Hybrid nanogenerator for concurrently harvesting biomechanical and biochemical energy *ACS Nano* **4** 3647–52

[18] Lei T, Yu L, Zheng G, Wang L, Wu D and Sun D 2015 Electrospinning-induced preferred dipole orientation in PVDF fibers *J. Mater. Sci.* **50** 4342–7

[19] Gheibi A, Bagherzadeh R, Merati A A and Latifi M 2014 Electrical power generation from piezoelectric electrospun nanofibers membranes: electrospinning parameters optimization and effect of membranes thickness on output electrical voltage *J. Polym. Res.* **21** 1–14

[20] Shao H, Fang J, Wang H and Lin T 2015 Effect of electrospinning parameters and polymer concentrations on mechanical-to-electrical energy conversion of randomly-oriented electrospun poly(vinylidene fluoride) nanofiber mats *RSC Adv.* **5** 14345–50

[21] Gheibi A, Latifi M, Merati A A and Bagherzadeh R 2014 Piezoelectric electrospun nanofibrous materials for self-powering wearable electronic textiles applications *J. Polym. Res.* **21** 1–7

[22] Pan X, Wang Z, Cao Z, Zhang S, He Y, Zhang Y, Chen K, Hu Y and Gu H 2016 A self-powered vibration sensor based on electrospun poly(vinylidene fluoride) nanofibres with enhanced piezoelectric response *Smart Mater. Struct.* **25** 105010

[23] Hu J, Gu Y, Zhang H, Zhu Y and Yang X 2018 Effect of electrospinning parameters on piezoelectric properties of electrospun PVDF nanofibrous mats under cyclic compression *J. Text. Inst.* **109** 843–50

[24] Hu P, Zheng D, Zhao C, Zhang Y and Niu J 2018 Linear dependence between content of effective piezo-phase and mechanical-to-electrical conversion in electrospun poly(vinylidene fluoride) fibrous membrane *Mater. Lett.* **218** 71–5

[25] Shao H, Fang J, Wang H and Lin T 2017 Effect of static charges on mechanical-to-electrical energy conversion of electrospun PVDF nanofiber mats *Adv. Mater. Lett.* **8** 418–22

[26] Fang J, Niu H, Wang H, Wang X and Lin T 2013 Enhanced mechanical energy harvesting using needleless electrospun poly(vinylidene fluoride) nanofiber webs *Energy Environ. Sci.* **6** 2196–202

[27] Wang Y R, Zheng J M, Ren G Y, Zhang P H and Xu C 2011 A flexible piezoelectric force sensor based on PVDF fabrics *Smart Mater. Struct.* **20** 045009

[28] Park B, Lee K, Park J, Kim J and Kim O 2013 Highly efficient hybrid energy generator: coupled organic photovoltaic device and randomly oriented electrospun poly(vinylidene fluoride) nanofiber *J. Nanosci. Nanotechnol.* **13** 2236–41

[29] Liu Z, Zhang S, Jin Y M, Ouyang H, Zou Y, Wang X X, Xie L X and Li Z 2017 Flexible piezoelectric nanogenerator in wearable self-powered active sensor for respiration and healthcare monitoring *Semicond. Sci. Technol.* **32** 064004

[30] Zandesh G, Gheibi A, Sorayani Bafqi M S, Bagherzadeh R, Ghoorchian M and Latifi M 2017 Piezoelectric electrospun nanofibrous energy harvesting devices: influence of the electrodes position and finite variation of dimensions *J. Ind. Text.* **47** 348–62

[31] Maity K, Mandal D and Mandal D 2018 All-organic high-performance piezoelectric nanogenerator with multilayer assembled electrospun nanofiber mats for self-powered multifunctional sensors *ACS Appl. Mater. Interfaces* **10** 18257–69

[32] Zaarour B, Zhu L, Huang C and Jin X 2019 Enhanced piezoelectric properties of randomly oriented and aligned electrospun PVDF fibers by regulating the surface morphology *J. Appl. Polym. Sci.* **136** 47049

[33] Kang S B, Won S H, Im M J, Kim C U, Park W I, Baik J M and Choi K J 2017 Enhanced piezoresponse of highly aligned electrospun poly(vinylidene fluoride) nanofibers *Nanotechnology* **28** 395402

[34] Liu Z H, Pan C T, Lin L W, Huang J C and Ou Z Y 2014 Direct-write PVDF nonwoven fiber fabric energy harvesters via the hollow cylindrical near-field electrospinning process *Smart Mater. Struct.* **23** 025003

[35] Pan C, Yen C, Wang S, Lai Y, Lin L, Huang J C and Kuo S 2015 Near-field electrospinning enhances the energy harvesting of hollow PVDF piezoelectric fibers *RSC Adv.* **5** 85073–81

[36] Fuh Y K and Wang B S 2016 Near field sequentially electrospun three-dimensional piezoelectric fibers arrays for self-powered sensors of human gesture recognition *Nano Energy* **30** 677–83
[37] Shao H, Fang J, Wang H, Lang C, Lin T and Lang C 2015 Robust mechanical-to-electrical energy conversion from short-distance electrospun poly(vinylidene fluoride) fiber webs *ACS Appl. Mater. Interfaces* **7** 22551–7
[38] Asadnia M, Kottapalli A G P, Karavitaki K D, Warkiani M E, Miao J, Corey D P and Triantafyllou M 2016 From biological cilia to artificial flow sensors: biomimetic soft polymer nanosensors with high sensing performance *Sci. Rep.* **6** 32955
[39] Jiang Y, Gong L, Hu X, Zhao Y, Chen H, Feng L and Zhang D 2018 Aligned P(VDF-TrFE) nanofibers for enhanced piezoelectric directional strain sensing *Polymers* **10** 364
[40] Wu Y, Hsu S L, Honeker C, Bravet D J and Williams D S 2012 The role of surface charge of nucleation agents on the crystallization behavior of poly(vinylidene fluoride) *J. Phys. Chem.* B **116** 7379–88
[41] Dhakras D, Borkar V, Ogale S and Jog J 2012 Enhanced piezoresponse of electrospun PVDF mats with a touch of nickel chloride hexahydrate salt *Nanoscale* **4** 752–6
[42] Mokhtari F, Shamshirsaz M and Latifi M 2016 Investigation of β phase formation in piezoelectric response of electrospun polyvinylidene fluoride nanofibers: LiCl additive and increasing fibers tension *Polym. Eng. Sci.* **56** 61–70
[43] Mokhtari F, Latifi M, Shamshirsaz M, Khelghatdoost M and Rahmani S 2017 Modeling of electrospun PVDF/LiCl nanogenerator by the energy approach method: determining piezoelectric constant *J. Text. Inst.* **108** 1917–25
[44] Yu B, Mao M, Yu H, Huang T, Zuo W, Wang H and Zhu M 2017 Enhanced piezoelectric performance of electrospun polyvinylidene fluoride doped with inorganic salts *Macromol. Mater. Eng.* **302** 1700214
[45] Wan C and Bowen C R 2017 Multiscale-structuring of polyvinylidene fluoride for energy harvesting: the impact of molecular-, micro- and macro-structure *J. Mater. Chem.* A **5** 3091–128
[46] Lee C, Wood D, Edmondson D, Yao D, Erickson A E, Tsao C T, Revia R A, Kim H and Zhang M 2016 Electrospun uniaxially-aligned composite nanofibers as highly-efficient piezoelectric material *Ceram. Int.* **42** 2734–40
[47] Lu X, Qu H and Skorobogatiy M 2017 Piezoelectric microstructured fibers via drawing of multimaterial preforms *Sci. Rep.* **7** 2907
[48] Kato M and Kakimoto K-I 2015 Processing and energy-harvesting ability of (Na,K)NbO3 particle-dispersed fibrous polyvinylidene fluoride multilayer composite *Mater. Lett.* **156** 183–6
[49] Ji S H, Cho J H, Jeong Y H, Paik J-H, Yun J D and Yun J S 2016 Flexible lead-free piezoelectric nanofiber composites based on BNT-ST and PVDF for frequency sensor applications *Sens. Actuators* A **247** 316–22
[50] Sorayani Bafqi M S, Bagherzadeh R and Latifi M 2015 Fabrication of composite PVDF-ZnO nanofiber mats by electrospinning for energy scavenging application with enhanced efficiency *J. Polym. Res.* **22** 1–9
[51] Khalifa M, Mahendran A and Anandhan S 2016 Probing the synergism of halloysite nanotubes and electrospinning on crystallinity, polymorphism and piezoelectric performance of poly(vinylidene fluoride) *RSC Adv.* **6** 114052–60

[52] Xin Y, Qi X, Tian H, Guo C, Li X, Lin J and Wang C 2016 Full-fiber piezoelectric sensor by straight PVDF/nanoclay nanofibers *Mater. Lett.* **164** 136–9
[53] Rahman W, Ghosh S K, Middya T R and Mandal D 1942 Enhanced mechanical energy harvesting ability of electrospun poly(vinylidene fluoride)/hectorite clay nanocomposites *AIP Conf. Proc.* **2018** 050081
[54] Liu Z-H, Pan C-T, Ou Z-Y and Wang W-C 2013 Piezoelectricity of well-aligned electrospun fiber composites *IEEE Sens. J.* **13** 4098–103
[55] Liu Z H, Pan C T, Su C Y, Lin L W, Chen Y J and Tsai J S 2014 A flexible sensing device based on a PVDF/MWCNT composite nanofiber array with an interdigital electrode *Sens. Actuators* A **211** 78–88
[56] Liu Z H, Pan C T, Yen C K, Lin L W, Huang J C and Ke C A 2015 Crystallization and mechanical behavior of the ferroelectric polymer nonwoven fiber fabrics for highly durable wearable sensor applications *Appl. Surf. Sci.* **346** 291–301
[57] Yu H, Huang T, Lu M, Mao M, Zhang Q and Wang H 2013 Enhanced power output of an electrospun PVDF/MWCNTs-based nanogenerator by tuning its conductivity *Nanotechnology* **24** 405401
[58] Abolhasani M M, Shirvanimoghaddam K and Naebe M 2017 PVDF/graphene composite nanofibers with enhanced piezoelectric performance for development of robust nanogenerators *Compos. Sci. Technol.* **138** 49–56
[59] Fashandi H, Abolhasani M M, Sandoghdar P, Zohdi N, Li Q and Naebe M 2016 Morphological changes towards enhancing piezoelectric properties of PVDF electrical generators using cellulose nanocrystals *Cellulose* **23** 3625–37
[60] Fu R, Chen S, Lin Y, Zhang S, Jiang J, Li Q and Gu Y 2017 Improved piezoelectric properties of electrospun poly(vinylidene fluoride) fibers blended with cellulose nanocrystals *Mater. Lett.* **187** 86–8
[61] Shao H, Fang J, Wang H, Lang C, Yan G and Lin T 2017 Mechanical energy-to-electricity conversion of electron/hole-transfer agent-doped poly(vinylidene fluoride) nanofiber webs *Macromol. Mater. Eng.* **302** 1600451
[62] Fuh Y K, Lee S C and Tsai C Y 2017 Application of highly flexible self-powered sensors via sequentially deposited piezoelectric fibers on printed circuit board for wearable electronics devices *Sens. Actuators* A **268** 148–54
[63] Fuh Y K, Huang Z M, Wang B S, Li S C and Fuh Y K 2017 Self-powered active sensor with concentric topography of piezoelectric fibers *Nanoscale Res. Lett.* **12** 44
[64] Dhakras D, Gawli Y, Chhatre S, Wadgaonkar P and Ogale S 2014 A high performance all-organic flexural piezo-FET and nanogenerator via nanoscale soft-interface strain modulation *Phys. Chem. Chem. Phys.* **16** 22874–81
[65] Mandal D, Henkel K and Schmeisser D 2014 Improved performance of a polymer nanogenerator based on silver nanoparticles doped electrospun P(VDF-HFP) nanofibers *Phys. Chem. Chem. Phys.* **16** 10403–7
[66] Pan C, Yen C, Lin L, Lu Y, Li H, Huang J C and Kuo S 2014 Energy harvesting with piezoelectric poly(γ-benzyl-L-glutamate) fibers prepared through cylindrical near-field electrospinning *RSC Adv.* **4** 21563–70
[67] Sultana A, Ghosh S K, Sencadas V, Zheng T, Higgins M J, Middya T R and Mandal D 2017 Human skin interactive self-powered wearable piezoelectric bio-e-skin by electrospun poly-L-lactic acid nanofibers for non-invasive physiological signal monitoring *J. Mater. Chem.* B **5** 7352–9

[68] Wang W, Zheng Y, Jin X, Sun Y, Lu B, Wang H, Fang J, Shao H and Lin T 2019 Unexpectedly high piezoelectricity of electrospun polyacrylonitrile nanofiber membranes *Nano Energy* **56** 588–94

[69] Street R M, Huseynova T, Xu X, Chandrasekaran P, Han L, Shih W Y, Shih W-H and Schauer C L 2018 Variable piezoelectricity of electrospun chitin *Carbohydr. Polym.* **195** 218–24

[70] Baniasadi M, Xu Z, Hong S, Naraghi M and Minary-Jolandan M 2016 Thermo-electro-mechanical behavior of piezoelectric nanofibers *ACS Appl. Mater. Interfaces* **8** 2540–51

[71] Neese B, Wang Y, Chu B, Ren K, Liu S, Zhang Q M, Huang C and West J 2007 Piezoelectric responses in poly(vinylidene fluoride/hexafluoropropylene) copolymers *Appl. Phys. Lett.* **90** 242917

[72] Adhikary P, Biswas A and Mandal D 2016 Improved sensitivity of wearable nanogenerators made of electrospun Eu^{3+} doped P(VDF-HFP)/graphene composite nanofibers for self-powered voice recognition *Nanotechnology* **27** 495501

[73] Beringer L T, Xu X, Shih W, Shih W-H, Habas R and Schauer C L 2015 An electrospun PVDF-TrFe fiber sensor platform for biological applications *Sens. Actuators* A **222** 293–300

[74] Park S, Lee H B, Yeon S M, Park J, Lee N K and Lee H B 2016 Flexible and stretchable piezoelectric sensor with thickness-tunable configuration of electrospun nanofiber mat and elastomeric substrates *ACS Appl. Mater. Interfaces* **8** 24773–81

[75] Yang E, Yang E, Xu Z, Chur L K, Behroozfar A, Baniasadi M, Moreno S, Huang J, Gilligan J and Minary-Jolandan M 2017 Nanofibrous smart fabrics from twisted yarns of electrospun piezopolymer *ACS Appl. Mater. Interfaces* **9** 24220–9

[76] Kim Y W, Lee H B, Yeon S M, Park J, Lee H J, Park S H, Kim Y W, Lee H B and Yoon J 2018 Enhanced piezoelectricity in a robust and harmonious multilayer assembly of electrospun nanofiber mats and microbead-based electrodes *ACS Appl. Mater. Interfaces* **10** 5723–30

[77] Guangyi R, Fanyi C, Baozhang L, Jianming Z and Chunye X 2013 Flexible pressure sensor based on a poly(VDF-TrFE) nanofiber web *Macromol. Mater. Eng.* **298** 541–6

[78] Ico G, Showalter A, Bosze W, Gott S C, Kim B S, Rao M P, Myung N V and Nam J 2016 Size-dependent piezoelectric and mechanical properties of electrospun P(VDF-TrFE) nanofibers for enhanced energy harvesting *J. Mater. Chem.* A **4** 2293–304

[79] Ma S, Ye T, Zhang T, Wang Z, Li K, Chen M, Zhang J, Wang Z, Ramakrishna S and Wei L 2018 Highly oriented electrospun P(VDF-TrFE) fibers via mechanical stretching for wearable motion sensing *Adv. Mater. Technol.* **3** 1800033

[80] Farrar D, Ren K, Cheng D, Kim S, Moon W, Wilson W L, West J E and Yu S M 2011 Permanent polarity and piezoelectricity of electrospun α-helical poly(α-amino acid) fibers *Adv. Mater.* **23** 3954–8

[81] Zhao G, Huang B, Zhang J, Wang A, Ren K and Wang Z L 2017 Electrospun poly(L-lactic acid) nanofibers for nanogenerator and diagnostic sensor applications *Macromol. Mater. Eng.* **302** 1600476

[82] Masamichi A, Hideki K, Keisuke K and Yoshiro T 2012 Film sensor device fabricated by a piezoelectric poly(L-lactic acid) film *Jpn. J. Appl. Phys.* **51** 09LD14

[83] Lee S J, Arun A P and Kim K J 2015 Piezoelectric properties of electrospun poly(L-lactic acid) nanofiber web *Mater. Lett.* **148** 58–62

[84] Zhu J, Jia L and Huang R 2017 Electrospinning poly(L-lactic acid) piezoelectric ordered porous nanofibers for strain sensing and energy harvesting *J. Mater. Sci. Mater. Electron.* **28** 12080–5
[85] Ueda H and Carr S H 1984 Piezoelectricity in polyacrylonitrile *Polym. J.* **16** 661
[86] Stupp S and Carr S 1978 Spectroscopic analysis of electrically polarized polyacrylonitrile *J. Polym. Sci., Part B: Polym. Phys.* **16** 13–28
[87] Fukada E and Yasuda I 1957 On the piezoelectric effect of bone *J. Phys. Soc. Jpn.* **12** 1158–62
[88] Ghosh S K, Adhikary P, Jana S, Biswas A, Sencadas V, Gupta S D, Tudu B and Mandal D 2017 Electrospun gelatin nanofiber based self-powered bio-e-skin for health care monitoring *Nano Energy* **36** 166–75
[89] Chen X, Xu S, Yao N and Shi Y 2010 1.6 V nanogenerator for mechanical energy harvesting using PZT nanofibers *Nano Lett.* **10** 2133–7
[90] Chen X, Guo S, Li J, Zhang G, Lu M and Shi Y 2013 Flexible piezoelectric nanofiber composite membranes as high performance acoustic emission sensors *Sens. Actuators* A **199** 372–8
[91] Wu W, Bai S, Yuan M, Qin Y, Wang Z L and Jing T 2012 Lead zirconate titanate nanowire textile nanogenerator for wearable energy-harvesting and self-powered devices *ACS Nano* **6** 6231–5
[92] Gu L, Cui N, Cheng L, Xu Q, Bai S, Yuan M, Wu W, Liu J, Zhao Y and Ma F 2012 Flexible fiber nanogenerator with 209 V output voltage directly powers a light-emitting diode *Nano Lett.* **13** 91–4
[93] Sutka A, Timusk M, Kisand V, Sutka A and Dauksta E 2015 Fabrication of lead titanate $PbTiO_3$ nanofiber mats via electrospinning *Int. J. Appl. Ceram. Technol.* **12** E117–21
[94] Suyitno S, Purwanto A, Hidayat R L L G, Sholahudin I, Yusuf M, Huda S and Arifin Z 2014 Fabrication and characterization of zinc oxide-based electrospun nanofibers for mechanical energy harvesting *J. Nanotechnol. Eng. Med.* **5** 011002
[95] Yan J and Jeong Y G 2016 High performance flexible piezoelectric nanogenerators based on $BaTiO_3$ nanofibers in different alignment modes *ACS Appl. Mater. Interfaces* **8** 15700–9
[96] Shirazi P, Ico G, Anderson C S, Ma M C, Kim B S, Nam J and Myung N V 2017 Size-dependent piezoelectric properties of electrospun $BaTiO_3$ for enhanced energy harvesting *Adv. Sustainable Syst.* **1** 1–8
[97] Xiao X, Hu P, Xiong Z, Xue H, Lu Q, Ye H and Zhang G 2012 Preparation and mechanical energy harvesting of BaTi0.9Zr0.1O3 ceramic nanofibers *Key Eng. Mater.* **512–515** 1359–62
[98] Wu W, Cheng L, Bai S, Dou W, Xu Q, Wei Z and Qin Y 2013 Electrospinning lead-free 0.5Ba(Zr0.2Ti0.8)O3-0.5 (Ba0.7Ca0.3)TiO3 nanowires and their application in energy harvesting *J. Mater. Chem.* A **1** 7332–8
[99] Wu Y, Ma F, Qu J, Luo Y, Lv C, Guo Q and Qi T 2019 Vertically-aligned lead-free BCTZY nanofibers with enhanced electrical properties for flexible piezoelectric nanogenerators *Appl. Surf. Sci.* **469** 283–91
[100] Wang Z, Zhang Y, Yang S, Hu Y, Wang S, Gu H, Wang Y, Chan H L W and Wang J 2015 $(K,Na)NbO_3$ nanofiber-based self-powered sensors for accurate detection of dynamic strain *ACS Appl. Mater. Interfaces* **7** 4921–7
[101] Gu L, Zhou D and Cao J C 2016 Piezoelectric active humidity sensors based on lead-free $NaNbO_3$ piezoelectric nanofibers *Sensors* **16** 833

[102] Zhu R, Jiang J, Wang Z, Cheng Z and Kimura H 2016 High output power density nanogenerator based on lead-free $0.96(K_{0.48}Na_{0.52})(Nb_{0.95}Sb_{0.05})O_3$-$0.04Bi_{0.5}(Na_{0.82}K_{0.18})0.5ZrO_3$ piezoelectric nanofibers *RSC Adv.* **6** 66451–6

[103] Kang H B, Han C S, Pyun J C, Ryu W H, Kang C-Y and Cho Y S 2015 $(Na,K)NbO_3$ nanoparticle-embedded piezoelectric nanofiber composites for flexible nanogenerators *Compos. Sci. Technol.* **111** 1–8

[104] Parangusan H, Ponnamma D and Al-Maadeed M A A 2018 Stretchable electrospun PVDF-HFP/Co-ZnO nanofibers as piezoelectric nanogenerators *Sci. Rep.* **8** 754

[105] Huang T, Wang C, Yu H, Wang H, Zhang Q and Zhu M 2015 Human walking-driven wearable all-fiber triboelectric nanogenerator containing electrospun polyvinylidene fluoride piezoelectric nanofibers *Nano Energy* **14** 226–35

[106] Guo Y, Zhang X, Wang Y, Gong W, Zhang Q, Wang H and Brugger J 2018 All-fiber hybrid piezoelectric-enhanced triboelectric nanogenerator for wearable gesture monitoring *Nano Energy* **48** 152–60

[107] Chen S, Tao X, Zeng W, Yang B and Shang S 2017 Quantifying energy harvested from contact-mode hybrid nanogenerators with cascaded piezoelectric and triboelectric units *Adv. Energy Mater.* **7** 1601569

[108] Song J, Yang B, Zeng W, Peng Z, Lin S, Li J and Tao X 2018 Highly flexible, large-area, and facile textile-based hybrid nanogenerator with cascaded piezoelectric and triboelectric units for mechanical energy harvesting *Adv. Mater. Technol.* **3** 1800016

Chapter 5

Enhancing β crystal phase content in electrospun PVDF nanofibers

Fatemeh Mokhtari, Javad Foroughi and Masoud Latifi

Recent developments in advanced materials are poised to create significant new opportunities for the development of smart materials. Various piezoelectric energy harvesting materials including PVDF have been developed to capture energy from human movement. To achieve high-performance PVDF piezofibers, a high content of the β phase in the PVDF material is desired. The electrospinning capabilities have received much attention in the last decade, with the aim to produce fibers in the submicron range. The β phase of the PVDF fiber can be enhanced due to simultaneous electric field poling and mechanical stretching during the spinning process.

Herein we discuss the advantages of using the electrospinning process to enhance the piezoelectric properties of electrospun PVDF fibers. The ability to control the physical and piezoelectric properties of a PVDF polymer holds great promise in the development of smart polymeric structures as energy harvesting materials, including films, fibers, and textiles. This comprehensive chapter discusses the effect of the electrospinning process on the piezoelectric properties of electrospun PVDF fibers. This chapter covers the various electrospinning conditions on the crystallinity structure, the β phase content, and the morphology of electrospun PVDF fibers.

5.1 Introduction

PVDF is known to be a semicrystalline polymer with complex molecular confirmations corresponding to five different crystalline phases (α, β, γ, δ, and ε) in the solid state. The α phase is dominant in raw PVDF materials because it has the most thermodynamically stable trans-gauche–trans-gauche form [1, 2]. The β phase has the highest dipolar moment per unit cell, accounting for the piezoelectricity, pyroelectricity, and ferroelectricity [3]. The unique β phase-derived properties of PVDF make it useful in a wide range of applications, including actuators,

doi:10.1088/978-0-7503-2005-4ch5

biosensors, energy harvesting materials, audio devices, transducers, and nonvolatile memories [4–8]. As well as its other useful properties, PVDF has flexibility, light weight, and long-term stability under high electric fields [9].

The biocompatible nature of PVDF creates significant opportunities for it to be used in wearable and implantable devices. However, it is known that PVDF exhibits relatively weak electromechanical coupling compared with other piezoelectric materials. This limitation is due to the fact that untreated PVDF can have α, β, and γ phases, and must be mechanically stretched and electrically poled to obtain the piezoelectric β phase for electricity generation [10]. The electrospinning process makes stretching and poling possible just in one step.

The β phase of the PVDF polymer has a planer all-trans (TTTT) conformation, and the H and F atoms are attached in the chain in such a way that the dipole moments associated with the two C–H and two C–F bonds add up and align in the direction perpendicular to the carbon backbone to give higher dipole moments ($\sim 8 \times 10^{-30}$ C m) per unit cell [11, 12]. An intermediate conformation is T3GT3G, which favors a parallel packing to generate the γ phase. The parallel version of the α phase is known as the δ phase. The δ and γ phases are polar, but their dipole moments are significantly smaller compared to the β phase [2, 13, 14].

The β phase can be obtained directly by melt crystallization at high pressure or by α phase film transformation through some special post-treatments, such as drawing at low temperature [15] (figure 5.1), ultra drawing at high temperature [16], annealing at very high pressure [17], and poling at a very high electrical field [18]. Furthermore, the β phase has been observed in solvent cast film using highly polar but toxic solvents, by blending the polymer with carbon nanotubes and organically modified nanoclay, and also by electrospinning through drawing and applying an electrical field at the same time [2, 19–21]. The nanoscale fibers of α, β, or γ phase PVDF with an increasing β phase content can be fabricated directly by electrospinning without any post treatment [22].

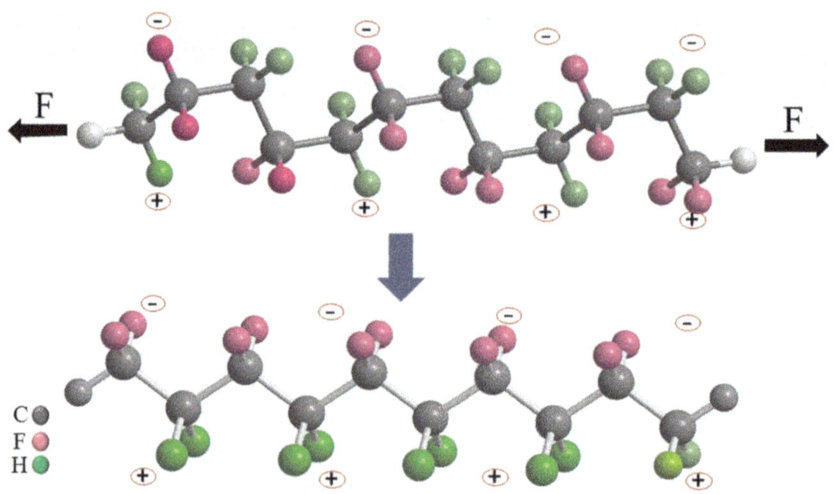

Figure 5.1. Conversion of α-crystalline phase into β-crystalline phase during stretching.

It is not easy to obtain PVDF consisting of entirely β phase crystals. This is because in all-trans PVDF neighboring fluorine atoms overlap since the diameter of the fluorine atom (0.270 nm) is slightly larger than the space provided by an all-trans carbon chain (0.256 nm) [23]. To diminish this overlap, CF_2 groups are tilted to the right and left, relative to their original conformation [24]. This deflection of CF_2 groups converts the all-trans form into TGTG (α form) or TTTGTTTG (γ form). Hence, the α phase is more easily formed than the β phase in normal circumstances. However, it was declared by Kabir *et al* that a completely pure (100%) β crystalline phase has been achieved in a nanocomposite of PVDF with CNTs with proper functionalization [2].

Phase transitions to obtain β phase PVDF can be divided into three methods depending on the processing conditions, as shown in figure 5.2.

Emerging techniques in nanofabrication have the potential to optimize piezoelectric responses, expand the range of device structures that can be considered, and simplify processing. For example, near- and far-field electrospinning methods can produce piezoelectric nanofibers in which the associated extensional forces and electric fields naturally cause local poling and, by consequence, enhanced properties [25].

As a low-cost, convenient, and environmentally friendly method, electrospinning or its improved technique is widely used to fabricate numerous nanofibers in a wide range of organic, inorganic, and inorganic/organic materials. Generally, almost all of the soluble polymers can be electrospun into nanofibers if the molecular weight is high enough, including water-soluble polymers and non-water-soluble polymers, such as PVDF [26].

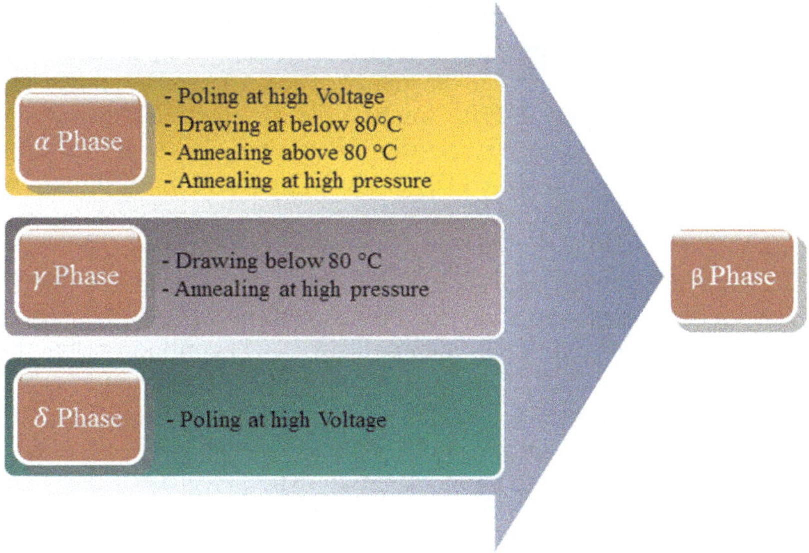

Figure 5.2. PVDF β phase formation from different phase conformations.

The main reason for getting the β phase from the PVDF nanofibers goes back to the physical and electrostatic forces applied to the polymer droplet due to the electrospinning process. When a droplet of the solution is subjected to an electric field, the high applied voltage turns the drop into an elongated cone. This process makes use of applied charges to provide a stretching force to a collector, which is a potential gradient. This mechanism changes the crystalline structure and leads to polarization of the fiber and formation of a β phase along the fiber axes [8]. During electrospinning, several factors contribute to the control of the crystal phase of PVDF, including high voltage, a high stretching ratio of the PVDF jet, rapid evaporation of the solvent, and low environmental temperature [27].

In this chapter the most common characterization methods, such as Fourier transform infrared (FTIR), x-ray diffraction (XRD), and differential scanning calorimetry (DSC), for measuring different phases of PVDF are presented and then discussed regarding three methods of β phase enhancing, which include processing condition optimization, the use of different additives and post-processing techniques.

5.2 Characterization techniques

FTIR, XRD, and DSC are the most common analysis techniques that prove the formation of the oriented β phase within PVDF materials [28]. The XRD and FTIR can implement the qualitative identification and quantitative analysis of the β phase. DSC is a thermoanalytical technique that has been complementary to the other identification techniques to identify the crystal phase of PVDF [29].

5.2.1 FTIR

The vibrational spectrum of PVDF can be used to gain information regarding the crystallinity and structure of PVDF. Although many of the peaks in the IR spectrum of PVDF result from overlapping absorptions of the α, β, and γ phases, several peaks are unique to one phase, and can be used to identify and quantify the relative amounts of the phases [30]. FTIR can be used to characterize both the dipole orientation and crystallographic structure of nanofibers based on the sensitivity of CF_2 orientation changes [31]. FTIR analysis provides information about the chemical bonding or molecular structure of materials, whether organic or inorganic. It is used to identify unknown materials and material phases present in a specimen. Many literature sources provide the locations of both α phase and β phase PVDF peaks. The FTIR absorption peaks for the three main α, β, and γ polymorphs of PVDF can be classified into three major categories: (1) common peaks that appear in all three phases; (2) exclusive peaks that only appear in one of the three phases; and (3) dual peaks that could come from two different phases [32]. The FTIR spectrum of PVDF pellets exhibited α phase peaks located at 531, 614, 763, 796, 870, and 970 cm^{-1} [33]. The β phase PVDF characteristic peaks are located at 440, 470, 510, 840, and 1280 cm^{-1} [34–36]. Webs produced by electrospinning display a higher degree of crystallites when compared to films obtained by electrospray. This result probably occurs due to the preferential orientation of the chains in the direction of the nanofibers, caused by the drawing experienced by the polymer during the electrospinning; this facilitates the crystallization, but this drawing depends on the production speed and collector speed. Sometimes it leads to fiber alignment and sometimes it only leads to the formation of the β phase along fiber axes without any improvement in crystallites of fibers. In this process, the addition of acetone promotes a small increase in the degree of the film crystallites for the PVDF α and β phases [37, 38].

The band at 880 cm^{-1} is assigned to the amorphous phase of CF_2–CH_2 bending vibration. Thus, one can notice clear changes when the polymer is processed as fibers, showing that the conformation of molecules in the amorphous region of PVDF is also influenced by the electrospinning process [39]. Lins *et al* [40] analyzed the effect of rotation speed of the drum collector on β phase formation of electrospun PVDF fibers. Their results showed that a higher speed of the drum leads to more β phase formation [40]. During electrospinning, the PVDF molecular chains are uniaxially stretched under high electrical forces (figure 5.3), resulting in an orientation of the entire chain including dipoles (CF_2 or CH_2 dipoles) [41].

To determine the percentage of β crystalline phase in each sample, the absorption peak of α and β phases at wavelengths 840 cm^{-1} and 764 cm^{-1}, respectively, are evaluated. The percentage of β crystalline phase is calculated using equation (5.1), where A_α and A_β are the absorptions at 764 and 840 cm^{-1} and X_α and X_β are crystalline mass fractions of the α and β phases [42]:

$$f(\beta) = \frac{X_\beta}{X_\alpha + X_\beta} \times 100 = \frac{A_\beta}{A_\beta + 1.26A_\alpha} \times 100. \tag{5.1}$$

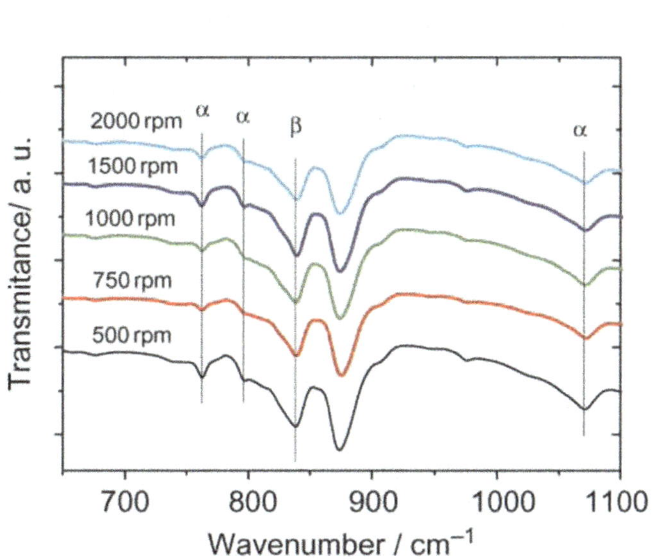

Figure 5.3. Detail of the FTIR spectra for the PVDF electrospun mat obtained at different rotations (lines indicates the absorption bands characteristics of α and β-phase). Reprinted with permission of the publisher (Taylor & Francis Ltd) [41], copyright 2010.

Sengupta *et al* [43] showed that for increasing the piezoelectric coefficient (d_{33}) of the PVDF nanofiber, the fraction of β phase (F(β)) should be maximum. The piezoelectric coefficient depends on various electrospinning parameters such as the electrospinning voltage, gap between the needle and collecting electrode, feed rate, and size of the needle. The electric field applied during the electrospinning predominantly controls the piezoelectric performance. Therefore, it is essential to know the piezoelectric ratio or the fraction of β phase in the electrospun nanofibers [43].

5.2.2 XRD

XRD is one of the most important nondestructive tools used to analyze all kinds of materials ranging from fluids to powders and crystals. Studies in production and engineering have shown that XRD is an essential method for materials characterization and quality control [44]. X-ray crystallography, the study of crystal structures through XRD techniques, generally leads to an understanding of the material and molecular structure of a substance. When an x-ray beam bombards a crystalline lattice in a given orientation, the beam is scattered in a definite manner characterized by the atomic structure of the web [45]. There are three weak α phase crystal peaks at $2\theta = 18.4°$, $2\theta = 26.6°$, and $2\theta = 35.7°$, which are assigned to (020), (022), and (200) planes, respectively. These suggests that the PVDF nanofibers mainly contain α and β phases, and less α phases and more β phases are formed during the electrospinning [46].

In figure 5.4, there is a typical β phase crystal peak at around $2\theta = 20.8°$, which is assigned to the total diffraction at (110) and (200) crystal planes; the β phase is represented by a broad peak with growing intensity as applied voltage increases up to 20 kV (table 5.1). The broadness of the β phase peak may be attributed to high elongation and rapid vitrification of the polymer solution during the electrospinning process [46].

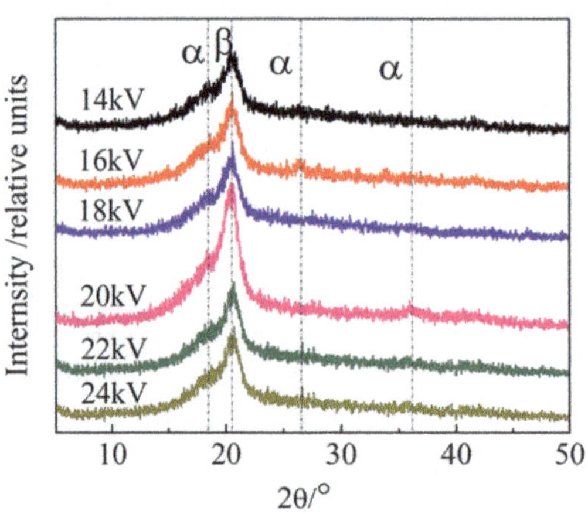

Figure 5.4. XRD patterns of PVDF nanofiber mats under different electrospinning voltages [46].

Table 5.1. Crystal planes for β phase of PVDF and their corresponding 2θ peaks. Reproduced with permission [44]. Copyright 2003, Wiley.

Crystal planes	2θ (°)
(110)	20.8
(200)	20.7
(020)	36.6
(101)	36.6
(221)	56.1

5.2.3 DSC

DSC is a thermal analysis technique used to study the thermal transitions in polymers, such as the melting point temperature T_m and the glass transition temperature T_g. DSC is an important tool for studying the thermal transitions in PVDF. The main use of DSC in PVDF studies is to determine the melting point, percentage of crystallinity in the sample, solvent evaporation, and annealing [45]. Depending on the crystalline phase of PVDF, different melting peaks appear on the DSC thermogram. The DSC characteristics not only depend on the crystalline phase, but also on the characteristics of the morphology, such as defects and crystalline size, as well as the processing history of the sample. Because of this, the existing literature does not define a temperature range for the different phases [47]. According to Prest and Luca [48], the expected melting temperature, represented by the endothermal peak, of the α phase of PVDF is 172 °C while Gregorio and Cestarini state that it occurs at 167 °C [49]. Meanwhile, β-crystallites present a melting temperature similar to the one for α-PVDF, and therefore, DSC is not used to distinguish these two phases but just to calculate the crystalline percentage of the sample. The change in thermal properties of PVDF nanofibers are directly reflected by the crystalline melting temperature and heat of fusion, which are determined by thermal analyses. The crystalline structure of β phase formation has a lower temperature region while α phase formation leads to a higher one. The melting point of the β phase is less than the melting point of the α phase [8].

The degree of crystallinity (X_c) of the samples was calculated from the DSC thermograms according to

$$X_c = \frac{\Delta H_f}{x\Delta H_\alpha + y\Delta H_\beta} \tag{5.2}$$

where ΔH_f is the melting enthalpy of the sample, x and y represent the α and β phase contents present in the sample from FTIR, respectively, and ΔH_α and ΔH_β are the melting enthalpies for a 100% α-PVDF (93.04 J g^{-1}) and β-PVDF (104.4 J g^{-1}) crystalline sample, respectively [50].

Ribeiro *et al* [41] made DSC measurements on PVDF nanofibers prepared at different electrospinning voltages, and found that all samples showed similar endothermic peaks. It is worth noticing that the DSC curves show two melting

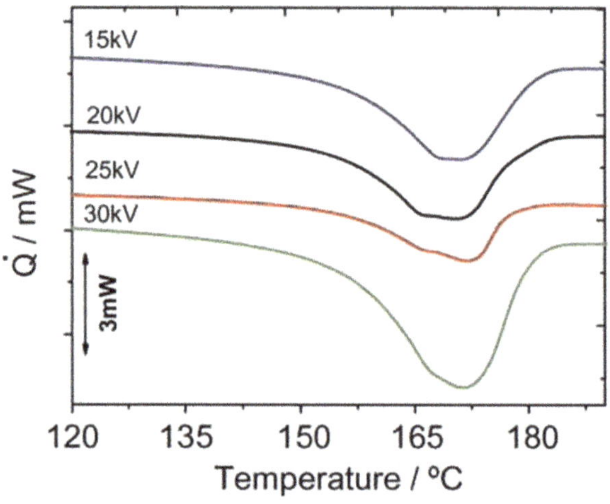

Figure 5.5. DSC thermograms for the electrospun PVDF membranes processed at different applied voltages. Reprinted with permission of the publisher (Taylor & Francis Ltd) [41], copyright 2010.

peaks. The melting point of the α phase of PVDF is several degrees lower than that of the β phase. Additionally, since FTIR results probe the presence of both phases in the sample, the two melting peaks that are superposed in the DSC thermogram could be ascribed to the melting of both phases. Nevertheless, this kind of result has to be considered with care since the DSC heating thermograms recorded at low heating rates can present more than one endotherm due to recrystallization taking place during the scan itself after the first melting. A distribution of crystal sizes could produce the same effect. Therefore, a PVDF sample consisting of only the α phase could present a thermogram similar to the ones shown in figure 5.5. In the present case, contributions of the two existing phases (proven by FTIR) and a distribution of crystal sizes must be the main reasons for the observed thermograms [41].

5.3 Enhancing β phase formation in electrospun PVDF nanofibers

The electroactive properties of PVDF heavily depend on the β phase content, microstructure, and degree of crystallinity of the samples, which in turn depend on the processing conditions. Interestingly enough, it has been shown that the electrospinning conditions, namely, solution parameters, processing, and environmental conditions, strongly affect not only fiber morphology but also the phase in which PVDF crystallizes [41]. The PVDF nanofiber membrane produced by electrospinning can be directly used for energy harvesting without any additional treatment. This is because during the electrospinning process, the applied high voltage, rapid evaporation of the solvent, and high stretching ratio of the PVDF solution jet benefit crystallization in the β phase, which possesses the most favorable piezoelectric and pyroelectric properties [51]. Different approaches can be combined with the electrospinning process to increase PVDF ability for more output voltage, which can be benefited by enhancing the β phase. These approaches can be divided into three main groups: adjusting processing conditions, adding an additive to polymer solutions (NaCl, LiCl, KCl, BTEAC, CNTs, ZnO, BTO, etc), and post treatment (drawing, annealing, poling, etc) [8]. In this section, the effects of these approaches on the formation of the β phase will be discussed.

5.3.1 Effect of processing conditions on polymorphism of PVDF nanofibers (Influence of preparation conditions on crystalline-phase behavior)

5.3.1.1 Applied voltage

The fundamental principle of electrospinning is based on a high voltage applied onto the polymer solution. In this process, the electrostatic force overcomes the surface tension of the polymer solution, causing the initiation of the jet. Increasing the applied voltage results in a higher electrostatic force to stretch the jet and filament during electrospinning, therefore decreasing fiber diameter. Shao *et al* showed that increasing the applied voltage beyond 15 kV leads to an increase in fiber diameter; this is attributed to the intensive bending instability, which increases the diameter distribution. The formation of more coarse fibers when the applied voltage is above 15 kV could be the reason for the large diameter distribution [1]. Electrospun nanofibers with smaller diameters were found to have higher contents of β phase, probably due to the strong stretching effect [52]. A high voltage or high stretching ratio of the jets also benefits crystallization in the β phase, and the alignment of the electrospun fibers by using a rotating collector also favors formation of the β phase. A direct-write electrospinning technique using near-field electrospinning (NFES) was developed to achieve controllable fiber deposition for various materials. Unlike the conventional electrospinning process, NFES only needs a low applied voltage to produce continuous fibers with fine diameters. This approach can be used to form PVDF fibers with *in situ* electrical poling with dipoles that are supposed to be aligned along the fibers (figure 5.6). Therefore, electrospinning of a PVDF solution can transform the nonpolar α phase into the polar β phase [53].

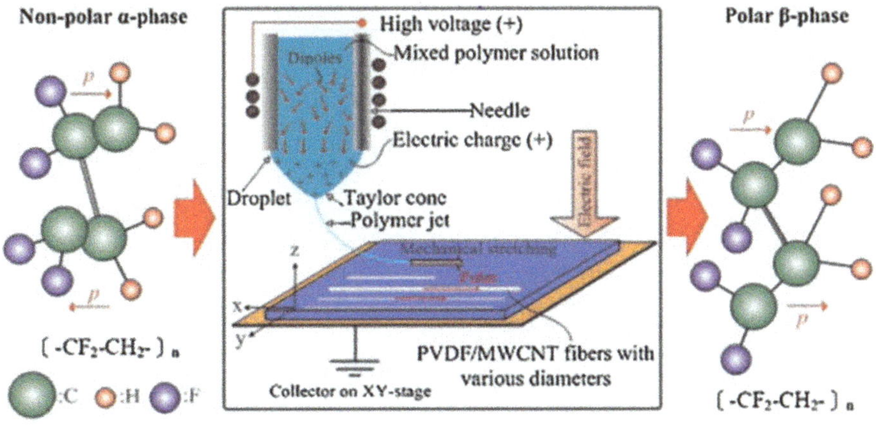

Figure 5.6. Direct-write PVDF fiber processed with *in situ* poling: the dipoles in the nonpolar, α phase PVDF could be stretched and oriented by a high electrical field with *in situ* strong mechanical stretching to transition into the polar β phase. Reproduced with permission [53]. Copyright 2013, Elsevier.

Jiyong *et al* [46] reported a strong correlation between PVDF nanofiber crystallinity and the applied voltage. They showed that crystallinity increased gradually until the applied voltage of electrospinning reached about 20 kV, and the effect of applied voltage on the crystallinity assisted majority formation of the β phase (figure 5.7).

The fraction of β phase is commonly over 70%, and it increases with the applied voltage up to 20 kV or so (figure 5.8). That is to say, the applied voltage in the range of 14–24 kV has an optimal value, which, in turn, leads to the further formation of the β phase in the electrospinning of the PVDF nanofiber [46].

Damaraju *et al* showed that the highest β phase fraction was obtained at 25 kV, where the fraction of β phase was calculated by FTIR to be 72.4% ± 1.0% [54]. With increasing *F*(β), the output voltage increases. It has been reported that poled samples

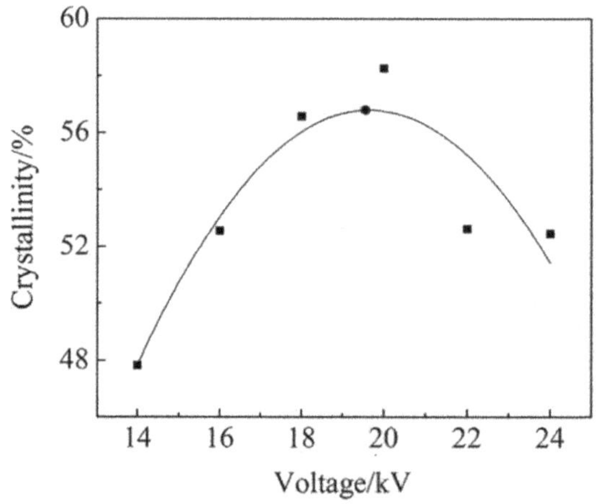

Figure 5.7. XRD patterns of PVDF nanofiber mats prepared at different electrospinning voltages [46].

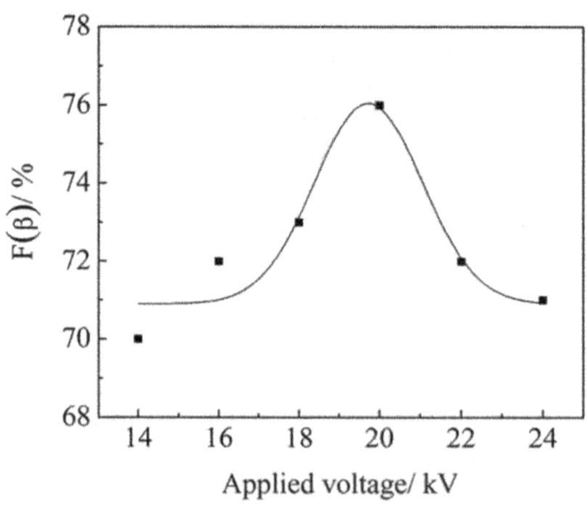

Figure 5.8. The relationship between the β crystalline phase in PVDF nanofibers and electrospinning voltage [46].

with higher fractions of the β phase show higher d_{33} piezoelectric coefficients, and therefore probably exhibit higher voltage outputs during energy harvesting [18]. However, it should be noted that the piezoelectric performance of PVDF fibers is affected by both the fraction of the β phase and the total dipole moment (closely related to the poling effect). Hence, figure 5.9 shows the effect of applied electric field on the piezoelectricity of electrospun fibers with a thickness around 20 μm under a continuous input pressure of 0.2 MPa at a constant frequency of 3 Hz [55]. Poling aligns electret dipoles of electrospun nanofibers, and researchers use this method to maximize the electrical energy output performance of piezoelectric materials by applying an electrical field under the temperature.

Andrew *et al* [28] showed that the fraction of the β phase decreased with increasing fiber diameter for fibers spun at lower voltages, i.e. ⩽10 kV, whereas for samples spun at 15 and 20 kV the fraction of the β phase increased with increasing fiber diameter [28] (figure 5.10).

5.3.1.2 Feeding rate

Feeding rate is another parameter that significantly influences the polymorphous behavior of electrospun PVDF fibrous membranes. It is clear that the feeding rate determines the amount of polymer solution available for electrospinning. As a result, a lower feeding rate produces fibers with smaller diameters [56]. The fraction of the β phase generally decreases with an increase in feeding flow rate. A lower feeding rate leads to higher stretching of the electrospinning jet. However, a high flow rate will decrease the traveling time of the jet in the electric field, and the solvent does not evaporate completely; thus, the stretching poling is weakened and the β phase decreases. It is important to note that the extremely low flow rate can cause a discontinuous electrospinning jet, and this discontinuity will hamper stretching of the PVDF solution during poling [46].

In one study, PVDF electrospinning at a feeding rate of 75 μl min^{-1} resulted in a complete predominance of α phase content. With the decrease in the feeding rate

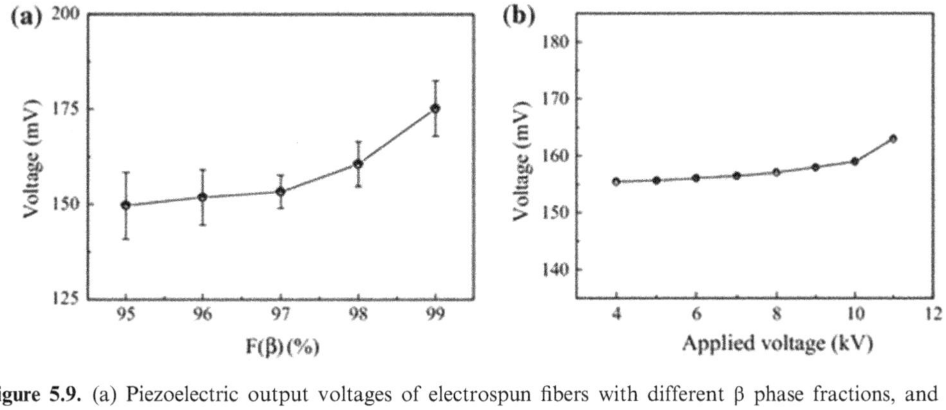

Figure 5.9. (a) Piezoelectric output voltages of electrospun fibers with different β phase fractions, and (b) piezoelectric output voltage of electrospun fibers by varying the applied voltage from 4 to 11 kV. Reprinted with permission from Springer [55], copyright 2015.

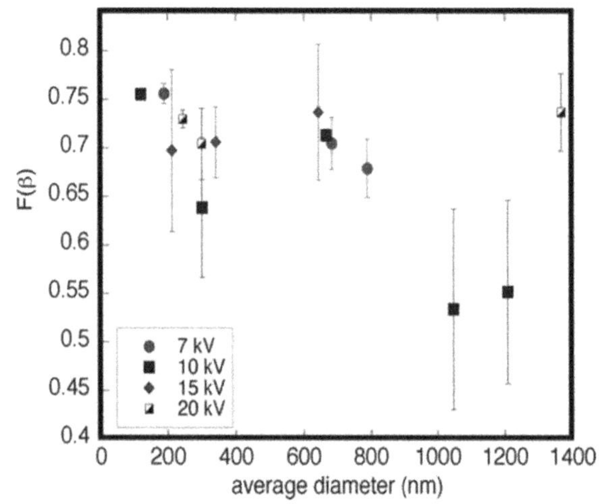

Figure 5.10. Fraction of β phase as a function of the average diameter for electrospun PVDF fibers. Reprinted with permission from the American Chemical Society [28], copyright 2008.

from 75 to 5 μl min^{-1}, a considerable decrease in α phase absorption at 613, 763, 795, and 975 cm^{-1} was observed. Meanwhile, there was an increase in β phase absorption at 470 and 1276 cm^{-1} [27].

The fraction of β phase within PVDF nanofibers was found to increase with the feeding flow rate (0.5–2.4 ml h^{-1}) up to a certain critical value, and then decrease [57]. Other studies reported a negative effect of the feeding flow rate [27], where the fraction of the β phase even rapidly decreased with an increasing flow rate of 0.3–4.5 ml h^{-1} [41]. Intrinsically, a low feeding rate leads to a higher stretching of the electrospinning jet, and this stretching benefits the formation of β phase nuclei. The effect of feeding flow rate is probably related to other electrospinning parameters.

Figure 5.11 shows that there is little change in the β phase fraction with the feeding flow rate in the range of 0.5–2.0 ml h^{-1}. Generally, the β phase fraction decreases with an increase in feeding flow rate [46]. Meanwhile, the XRD curves of electrospinning PVDF nanofibers show a strong β phase peak and weak α phase peak in the feeding flow range of 0.5–3.0 ml h^{-1}, and the peak intensity of β phase is strongest at 1.0 ml h^{-1} [46].

Ghafari *et al* [58] developed a reliable model to design and engineer PVDF nanofibers in terms of both morphology and the fraction of β phase content based on response surface methodology. As presented in the trace plot (figure 5.12), the flow rate does not have any significant effect on the β phase fraction [58].

5.3.1.3 Needle tip diameter and tip-to-collector distance

The local FTIR spectra at 841 cm^{-1} is shown in figure 5.13(a), and the absorbance of the β phase generally decreases with increasing needle tip diameter. However, there is no observable difference of FTIR absorbance between the needle tip diameters of

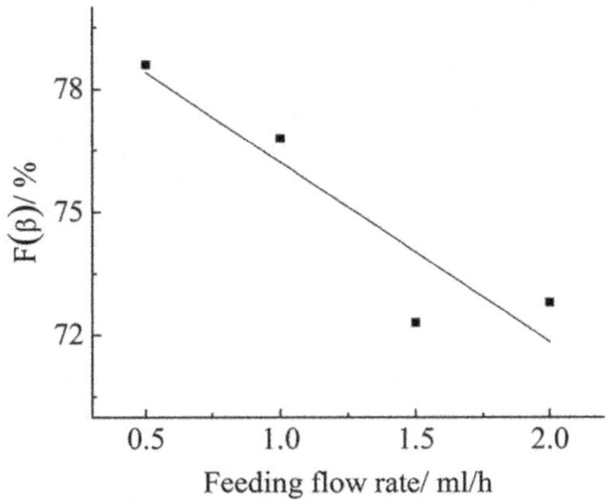

Figure 5.11. Crystalline structure characterization of PVDF nanofibers: β phase fraction versus feeding flow rate [46].

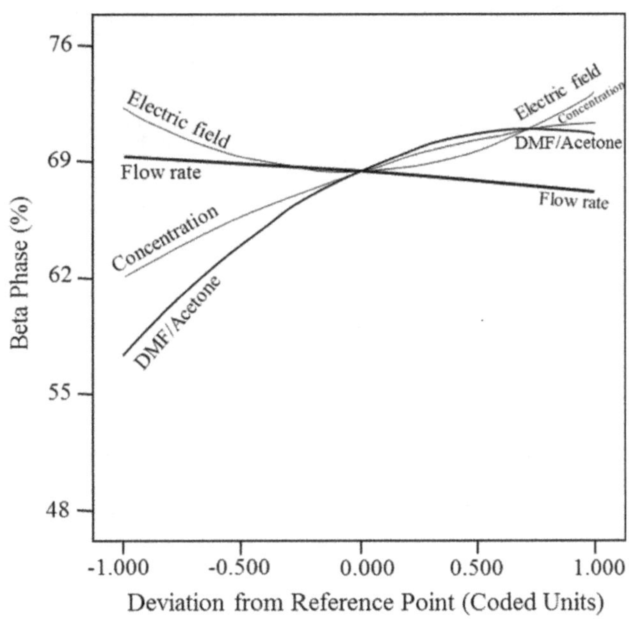

Figure 5.12. Trace plot for the fraction of β phase in PVDF nanofibers. Reprinted with permission from Springer [58], copyright 2018.

0.3 and 0.4 mm (figure 5.13(b)). Additionally, the fraction of β phase is plotted against the needle tip diameter in figure 5.13(c), and it decreases with increasing needle tip diameter. Wang *et al* [59] have numerically concluded that the increasing needle tip diameter will decrease the electrical field intensity. In this sense, the increase in needle tip diameter will weaken the stretching poling of the electrospinning jet so that the fraction of β phase decreases. Both the crystallinity and the β phase fraction results support that a large needle tip diameter is not helpful to

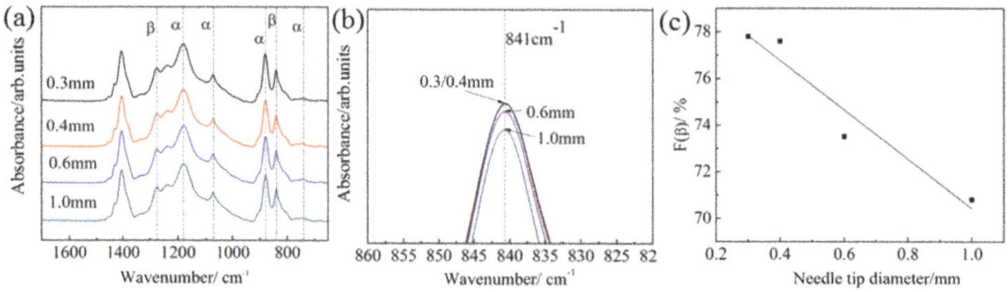

Figure 5.13. Crystalline structure characterization of PVDF nanofibers. (a) FTIR spectrum, (b) local magnification of typical vibration peaks near the β phase, and (c) β phase fraction versus needle tip diameter [46].

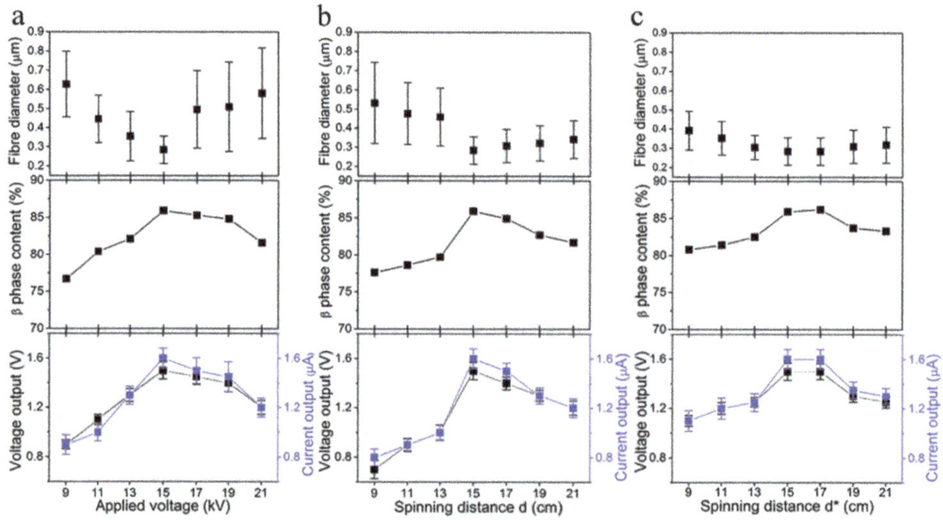

Figure 5.14. Effect of (a) applied voltages, (b) spinning distances, and (c) electric field intensity on PVDF fiber diameters, β crystal phase contents, and electrical outputs of PVDF nanofiber mats (PVDF concentration 20%; nanofiber mat thickness 100 μm). Reprinted with permission from the Royal Society of Chemistry [1], copyright 2015.

the formation of a β phase crystalline structure within electrospinning PVDF nanofibers [46].

The spinning distance (nozzle tip-to-collector distance) in electrospinning affects fiber stretching and deposition [60]. When the electric field force is sufficient to maintain the electrospinning process, increasing the spinning distance provides a larger space for jet stretching and a longer time for solvent evaporation from the fibers. A spinning distance shorter than a critical length often leads to insufficient fiber stretching and solvent evaporation. As a result, wet and coarse fibers, and even porous structures, will be formed [1] (figure 5.14).

Shao *et al* showed that at a constant applied voltage, the electric field intensity reduced with increasing spinning distance. Increasing the spinning distance initially provides more time and space for the jets to be stretched. The highest β crystal phase content of 85.9% was obtained at a spinning distance of 15 cm [1].

5.3.1.4 Collection method

In the aligned electrospun PVDF fibers, the *c*-axis of the β phase crystallites had preferred orientation along the fiber axis. The degree of orientation and the polymorphism behavior of the fibers change significantly with either the rotating collector speed or the size of the spinneret used. This implies that the formation of the β phase is likely to be caused by the Columbic force imposed by the electric field rather than the mechanical and shear force exerted by the rotation disc collector and spinnerets. The Columbic force may cause conformational changes to a straighter TTTT conformation, and hence promote the β phase [61]. Comparison of laser-melt electrospinning and electrospinning by Asai *et al* [62] shows that the rotating speed of a collector and applied voltage only affect the fiber structure and do not strongly influence the crystal structure of the fibers obtained by laser-melt electrospinning. XRD and attenuated total reflectance (ATR)-FTIR measurements revealed that the crystal structure of laser-melt electrospun PVDF fibers was mainly α phase, whereas the solution electrospun fibers showed a higher β phase fraction. From DSC measurements, it was found that the crystallinity of laser-melt electrospun fibers was comparable with that of the solution electrospun fibers [62].

The influence of rotation was investigated at a constant voltage of 20 kV, a flow rate of 4 ml h^{-1}, and a needle diameter of 0.5 mm. Electrospun fibers obtained from a 0.5 mm diameter needle contained around 50% of the PVDF crystalline fraction in the β phase, and a similar result was obtained in the case of the aligned fibers collected at a low rotating speed. Interestingly, higher rotation speeds yielded fibers with the same β to α phase ratios that can be obtained with narrower needles (figure 5.15) [41].

The aligned nanofibers of PVDF are fabricated by two different methods. In the first method, oriented nanofibers were fabricated by a rotational collector at

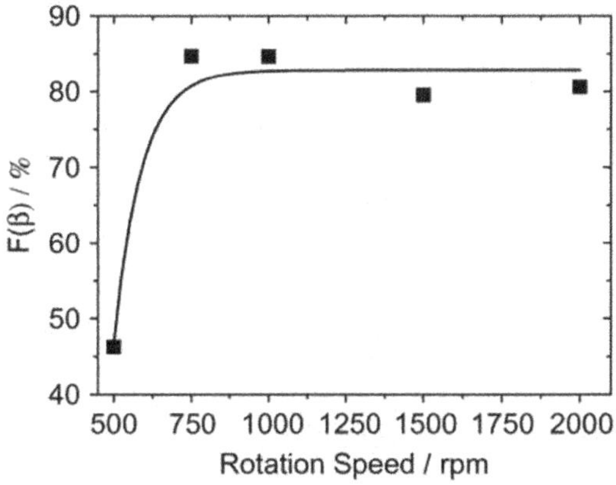

Figure 5.15. Crystalline phase content for the PVDF electrospun membranes obtained at different collector rotation speeds. Reprinted with permission of the publisher (Taylor & Francis Ltd) [41], copyright 2010.

different rotating speeds (0, 200, and 400 r min^{-1}) (figure 5.16). The result showed that by increasing the rotation speed, fiber morphology and arrangement improved significantly; moreover, the percentage of β phase increased [8]. For the second method, a pair of magnets were placed 5 cm away from each other and a flat plate was placed between them to collect the fibers. By applying a magnetic field, electromagnetic force applied tension on the nanofibers, which leads to an increment in the percentage of β phase formation [63].

Wu *et al* [64] also compared the electrospun PVDF fibrous membranes obtained under different drum rotation speeds and reported that the alignment of the PVDF nanofibers was significantly affected by the rotation speed of the drum collector. High rotation speeds provided enhanced mechanical drawing on the PVDF fibers, thus increasing the β phase content [64].

Figure 5.16. SEMs of aligned electrospun PVDF fibers at different drum speeds: (a) 800, (b) 1000, (c) 1500, and (d) 2000 rpm. Reproduced with permission [8]. Copyright 2015, Wiley-VCH.

5.3.1.5 Solution viscosity and solvent

The electrospun fiber diameter is controlled by the viscosity of the spinning solution, with higher viscosity solutions producing fibers with larger diameters. For the lower viscosity spinning solution, which produces finer fibers, the polymer can be elongated more easily during the electrospinning process, resulting in the formation of the β phase [28]. The polymer chains from the solution get oriented along the applied electric field and hence polarization takes place concurrently. During electrospinning, the favored arrangement of the polymer chains through the stretching of the solution droplet also creates a certain preferential orientation of the dipolar moments. The simultaneous elongation forces, stretching, and poling during the process promote the formation of the polar β phase. The presence of the β phase in PVDF can thus be attributed to the solidification during electrospinning [65]. The maximum β phase content fibers are obtained by electrospinning from low viscosity solutions or under a higher applied voltage for high-viscosity solutions. The increase in the formation of the β phase for a high-viscosity solution spun at increased voltages can also be attributed to electrical poling by the increased electric field that forms between the needle tip and the collector [28].

The formation of α and β phases is related primarily to the evaporation rate of the solvent. High rates favor the formation of the α phase, whereas low rates favor the β phase. Under the same conditions of applied voltage and needle–collector distance, the boundary concentration between electrospinning depends on the solvent used. Moreover, addition of acetone to the solution reduced average fiber diameter. Nanofiber mats produced by electrospinning present a higher degree of crystallinity than the films obtained by electrospraying, for both solvents used [66].

It was found that an increase in β phase content was realized when a solvent with a low boiling point was added into the electrospinning solutions. Dimethylformamide (DMF) has a relatively high boiling point of 153 °C, while acetone has a low boiling point of 56 °C. When the electrospinning temperature was constant, for example, at 40 °C, the use of pure DMF as the solvent could lead to the coexistence of both α and β phase in PVDF fibers, and strong absorptions at 613, 763, 795, and 975 cm^{-1} attributed to the α phase were observed. However, when mixed DMF/acetone was used as the solvent, the β phase became dominant [27].

The increase in the DMF/acetone ratio leads to fiber mats with wider fiber diameter distributions and numerous beads. This is probably due to the incomplete solvent evaporation. These results can be explained by taking into account acetone's vapor pressure (v_p) being higher than that of DMF [67]. In these conditions, acetone weakens the interactions between DMF and PVDF, and increases the vapor pressure of the polymer solution. Therefore, by increasing the DMF/acetone ratio, the solution's vapor pressure decreases and the interactions between the polymer and the solvent mixture increase, thus limiting the solvent evaporation and consequently the stretching of the polymer during nanofiber formation. This effect leads to nanofibers with defective morphologies and incomplete solvent evaporation. Moreover, as a nonsolvent of PVDF, acetone reduces the hydrodynamic radius of the polymer in solution, thus leading to a lower viscosity and fewer molecular entanglements between PVDF chains, and promoting beaded fibers [19].

Costa *et al* [66] showed that the formation of the β crystalline phase of PVDF in electrospun mats is not related to drawing experienced by the polymer during the process, as suggested by some authors. Formation of the α and β phase is related primarily to the evaporation rate of the solvent [66].

DMF and *N*-methyl-2-pyrrolidinone (NMP) are commonly used in mat fabrication processes despite being classified as hazardous and having possible toxic health effects. Dimethyl sulfoxide (DMSO) is less hazardous both to human health and the environment, making it an appealing substitute. Gee *et al* show that mats produced with DMF as the major solvent had the highest average β phase fraction, followed closely by DMSO and lastly by NMP [68].

In another study, NMP and acetone with different volume ratios ($V_{NMP}/V_{acetone}$) were used for producing randomly distributed electrospun fibers. The results show that 16 wt% solutions of $V_{NMP}/V_{acetone} = 6/4$ has a higher amount of β phase (99%) and that under a continuous input pressure of 0.2 MPa at a constant frequency of 3 Hz they can generate a voltage around 175 mV [55].

5.3.1.6 Temperature and humidity

Jianfen *et al* [27] electrospun PVDF solution at different temperatures varying from 20 °C to 60 °C. It was found that at a low electrospinning temperature, which indicated a low crystallization temperature, trans-state (β and γ phase) PVDF fibers were more likely to be formed. Compared to the samples prepared from the mixed solvent of DMF/acetone, the samples from pure DMF solution were much more sensitive to the changes in temperature; this may be caused by the temperature sensitivity of intermolecular interactions between PVDF and DMF [27]. Adding a solvent with a low boiling point, decreasing the electrospinning temperature, and decreasing the feeding rate and tip-to-collector distance can all bring about an increase in β phase content. Furthermore, these electrospun PVDF fibrous membranes with high β phase content are expected to show excellent piezoelectricity and have potential for use in nanosensors and electroacoustic transducers [27].

Another electrospinning parameter, generally not comprehensively studied, is the relative humidity (RH), namely, the amount of water vapor that is contained in the gaseous environment during electrospinning. Cozza *et al* [19] showed that increasing RH (figures 5.17(a) and (d)) caused the fraction of fibers having diameters under 100 nm to decrease. In the case of nanofibers prepared by using 40% RH (figure 5.17(e)), the diameters are predominantly within the range of 180–200 nm. At an RH of 50% (figure 5.17(f)), nanofibers with fine diameters (around 10 nm) completely disappeared and most of the electrospun filaments tended to have diameters in the range of 260–280 nm. It is worth underlining that, using the described system, it was not possible to apply an RH greater than 50% because of polymer solution solidification on the needle. With increasing RH, the uniformity of the nanofiber diameters tends to increase, and as a consequence the β phase formation is increased [19].

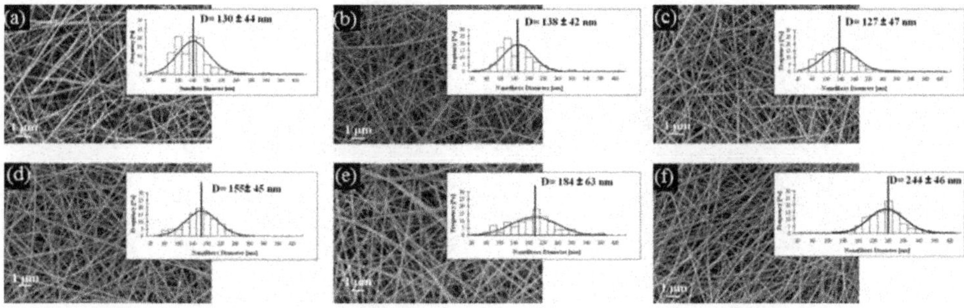

Figure 5.17. SEM micrographs of PVDF electrospun nanofibers prepared by applying different humidity conditions: (a) PVDF (RH 0%), (b) PVDF (RH 10%), (c) PVDF (RH 20%), (d) PVDF (RH 30%), (e) PVDF (RH 40%), and (f) PVDF (RH 50%). Reproduced with permission [19]. Copyright 2013, Wiley.

5.3.2 Effect of additives on polymorphism of PVDF nanofibers

In recent years, many studies have indicated that mixing PVDF with various nano-fillers facilitates the formation of the β phase in PVDF and improves its piezoelectricity. Moreover, particulate materials such as pigments and carbon black particles are easily incorporated into the nanofibers via the electrospinning process [61, 69]. The β phase in PVDF has been obtained by the addition of $BaTiO_3$ [70, 71], clay [72], hydrated ionic salt [73], PMMA [74], TiO_2 [75], $CoFe_2O_4$ [76], and nanoparticles such as palladium, ferrite, and gold [77] as nucleating fillers. It is also reported that β phase PVDF nanocomposites can be fabricated with the assistance of ionic liquid [78]. Sorayani *et al* introduced ZnO particles in preparing electrospun PVDF nanofibers and found that blending 15% ZnO particles promoted the formation of the β phase to 87% [42].

Adding soluble LiCl salt to the polymer solution, which separates into equal numbers of positive and negative ions, increases the electrical conductivity of the solution by increasing the number of ions per unit volume. During electrospinning, when an external electrical field is applied to the solution, the positive and negative ions in the polymer fluid tend to move in opposite directions. Negative ions are forced toward the positive electrode, and positive ions are forced toward the negative electrode; stretching and elongation result from this phenomenon, as mentioned above. An inorganic LiCl salt added into the spinning solutions helps increase the conductivity of the solutions considerably [79, 80].

Adding just 0.00133% LiCl leads to increasing the β phase content to 93.6%. The salt-containing webs have stronger β-phase crystalline peaks than salt-free electrospun webs [8]. Investigation of the crystal polymorphism of PVDF/PCDA (the self-assembly of diacetylene monomers) nanofibers showed that PCDA inhibited the growth of nonpolar α phase crystals while promoting the growth of the polar β phase. The highly ordered assembly of PCDA monomers and the interaction between OH groups of PCDA and fluorine atoms of PVDF are possible reasons to align the PVDF polymer chains to form more trans-conformation and, consequently, more β phase crystals. Moreover, the β phase fraction of electrospun PVDF/PCDA nanofibers increased after UV irradiation. After irradiation, the

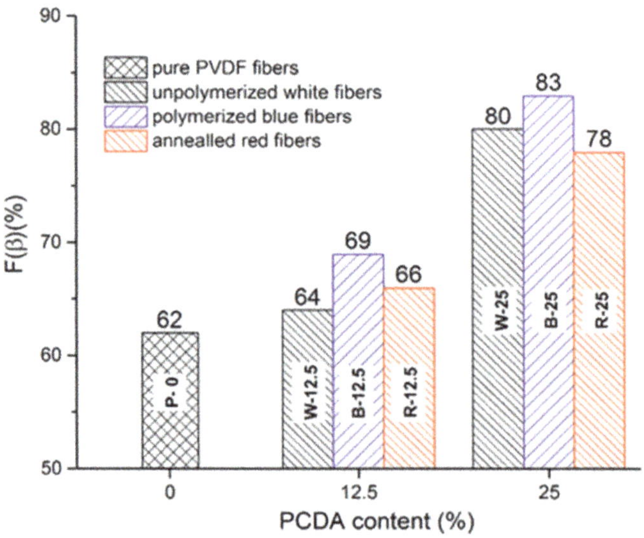

Figure 5.18. The calculated *F*(β) for different electrospun fiber samples with different PCDA contents. Reprinted with permission from Springer [81], copyright 2017.

assembled PCDA monomers in the fibers were polymerized to form PDA with alternating ene–yne main chains. This polymerization can help stabilize the monomer position where there is interaction with fluorine atoms of PVDF. This might lead to less freedom of movement of PVDF chains and help preserve all-trans configurations of the β phase of PVDF [81] (figure 5.18).

Apart from inorganic piezoelectric particles, conductive particles are worthy of consideration for the fine-tuning of fiber composition and structure. Research on this is mainly focused on graphene and CNTs. There are fluorine atoms with highly electron-dense characteristics in PVDF, and they will interact with the large π bond in graphene if the graphene is mixed with PVDF. Firstly, the interactions lead to β phase formation in PVDF/graphene fibers. In addition, the dielectric constant of PVDF/graphene fibers is higher than that of neat PVDF fibers because of the high electric conductivity of graphene [82].

The FTIR spectra of aligned electrospun PVDF/MWCNT fibers (infrared bands at 837 and 1273 cm^{-1}) show the enhancement of the β phase with increasing MWCNT content. This is because the higher electric field during electrospinning produces a completely different arrangement of the polarity direction of the PVDF fibers, facilitating the growth of the crystalline structure [83].

When the mass fraction of MWCNTs was increased to 3% and 5%, a stronger β peak at $2\theta = 20.6°$ became dominant compared with the bare PVDF fiber mats. It was also found that both the crystallinity and β phase proportion decreased upon further increase of the MWCNT concentration to 7% and 10%, as shown in table 5.2 [84].

Dhakras *et al* [65] examined the effect of a hydrated salt (NC), nickel chloride hexahydrate ($NiCl_2{\cdot}6H_2O$), on the phase formation of electrospun PVDF

Table 5.2. The crystallinity and proportion of the β phase of PVDF nanofibers with different weight ratios of MWCNTs [84].

Sample name	Crystallinity (%)	β phase proportion (%)
PVDF powder	47.1	0
PVDF fiber mats	31.0	54.8
PVDF-3% CNTs	36.9	56.8
PVDF-5% CNTs	38.1	68.4
PVDF-7% CNTs	33.4	48.5
PVDF-10% CNTs	25.0	48.0

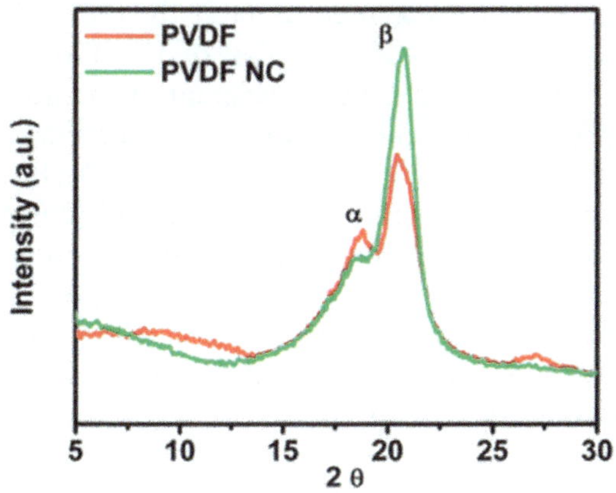

Figure 5.19. XRD patterns for PVDF and PVDF NC Reprinted with permission from RSC Publishing [65], copyright 2012.

nanofibers. The addition of this hydrated salt was found to enhance the polar β phase by about 30%. The XRD patterns revealed the coexistence of α and β phases in both samples. The intensity ratio I_β/I_α corresponding to β and α phases was dramatically enhanced from 1.45 for PVDF to 2.53 for PVDF NC (figure 5.19). The addition of the salt results in a considerable enhancement of the β phase in PVDF NC. In the electrospinning process when the droplet of the viscous polymer solution is subjected to an electric field, its elongation takes place and it undergoes solidification. Addition of NC can further enhance β phase formation due to ionic interactions of the polymer with the hydrated salt and the polar solvent. The hygroscopic nature of the salt helps in the hydrogen bonding between the water of crystallization and PVDF chains [65].

5.3.3 Post treatment processing

In the electrospinning process, mechanical stretching and electrical poling occur simultaneously, which contributes toward the orientation of the dipoles in the preferential direction, leading to a slightly higher β phase. Therefore, many researchers believe that PVDF nanofibers do not need any post treatment. Normally, a piezoelectric material needs a poling process to align all the dipoles along one direction in the material. In one study, a DC poling method was used to obtain stable piezoelectricity in electrospun PVDF–TrFE nanofibers in silicone oil at 100 °C under 2 kV for 30 min [85].

Post processes such as thermal annealing and drawing affect the morphology and structure (crystallinity, orientation and size of the crystallites, and alignment of the amorphous chains, etc) of polymers. Such changes in morphology can influence the material and mechanical properties of the polymer. In the case of piezoelectric polymers, post processing also affects the electromechanical coupling behavior [86].

This was confirmed by Hansen *et al* [87] in their study: when aligned electrospun PVDF nanofibers were used to make power generation devices, a post electrospinning poling treatment had to be used to make the molecular dipoles align along the fiber length direction [87]. This is also supported by Farrar *et al* [88], who stated that electrospinning fails to induce dipole orientation because the stretching force and electric field during this process are in the same direction.

It is also known that the piezoelectric properties of PVDF can be lost over time and with temperature, which impedes wide device and sensor applications of this material. A recent study of a piezoelectric PVDF thin film with nondestructive second-harmonic generation laser spectroscopy showed that the piezoelectric properties are lost near the temperature of 90 °C–100 °C for the 'clamped' film; the free-standing film, on the other hand, will lose its piezoelectric properties as the temperature increases and the film shrinks [89]. Thus, the study of heat treatment of PVDF nanofibers is an important first step toward sensing phenomena such as potentially dangerous temperature changes in lithium ion battery cells [90].

Parker *et al* [90] showed that the thermal annealing of the deposited nanofibers are stable up to 100 °C and results in even more β phase. Thus, thermal annealed nanofibers may be even more piezoelectric and pyroelectric for sensory applications [90].

Baniasadi *et al* [86] investigated the effects of thermal annealing and stretching (drawing) as two post processing treatments on mechanical and piezoelectric properties of PVDF–TrFE (figure 5.20). The experiments included an as-spun fiber mat, a fiber mat annealed at 135 °C for 2 h, and a fiber mat stretched to 50% strain and then annealed at 135 °C for 2 h. The FTIR results show that the intensity of peaks for β phase increased. Polarized-FTIR also explained the orientation of electric dipoles and enhancement of this orientation because of annealing and stretching, which was quantified by local piezo response measurement using the piezoresponse force microscopy (PFM) technique. The mechanical strength in the stretched-then-annealed fiber was enhanced [86].

5.4 Conclusion

Among the five polymorphs of PVDF, the β phase is polar and the dipole moments are favorably oriented in the same direction, thus providing the piezoelectric, pyroelectric, and ferroelectric properties of PVDF and exhibiting the highest

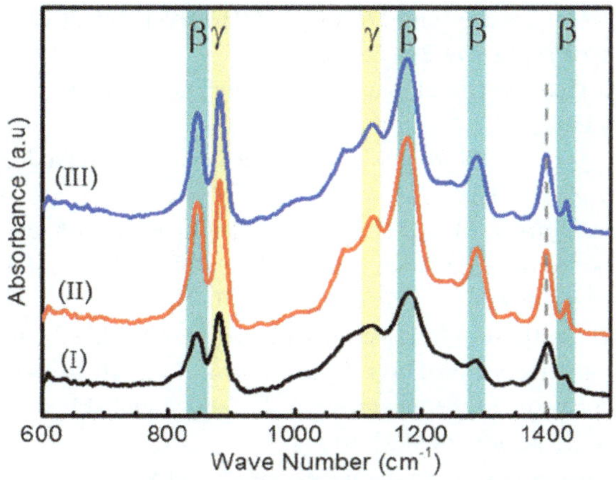

Figure 5.20. FTIR-attenuated total reflectance (ATR) spectra of (I) as-spun, (II) annealed, and (III) stretched-then-annealed PVDF–TrFE nanofiber mats. The β phase peaks are shown as green bands. Reprinted with permission from Springer [86], copyright 2017.

electroactive properties. Electrospinning has been the technique most widely adopted to produce electroactive β-PVDF nanofibers. It provides high orientation of the polymer chains under the high shear field. However, the formation of the β phase in the spinning process is affected by many factors, and other co-existing crystalline phases, such as α and γ phases, are often produced. Hence, the enhancement of the β phase during electrospinning has been examined in recent years. The polar β phase can be obtained through various post treatments, such as mechanical stretching and high electric field poling. The advantage of electrospinning is the capability to do these post treatments simultaneously. The blending of PVDF with iron-based nanomaterials and conductive additives, piezo ceramic nanoparticles, and ionic salts is the most common way to obtain the β phase and improve the electrical properties. Apart from the predominant parameters of precursor solution preparation and the spinning condition, different collector types used in electrospinning can affect the morphologies of the electrospun PVDF mats.

References

[1] Shao H *et al* 2015 Effect of electrospinning parameters and polymer concentrations on mechanical-to-electrical energy conversion of randomly-oriented electrospun poly(vinylidene fluoride) nanofiber mats *RSC Adv.* **5** 14345–50

[2] Ekramul K *et al* 2017 Pure β-phase formation in polyvinylidene fluoride (PVDF)-carbon nanotube composites *J. Phys. D: Appl. Phys.* **50** 163002

[3] Hao S *et al* 2017 Mechanical energy-to-electricity conversion of electron/hole-transfer agent-doped poly(vinylidene fluoride) nanofiber webs *Macromol. Mater. Eng.* **302** 1600451

[4] Fujisaki S, Ishiwara H and Fujisaki Y 2007 Low-voltage operation of ferroelectric poly (vinylidene fluoride-trifluoroethylene) copolymer capacitors and metal-ferroelectric-insulator-semiconductor diodes *Appl. Phys. Lett.* **90** 162902

[5] In'acio P, Marat-Mendes J N and Dias C J 2003 Development of a biosensor based on a piezoelectric film *Ferroelectrics* **293** 351–6
[6] Xin Y *et al* 2006 Piezoelectric diaphragm as a high performance biosensor platform *Appl. Phys. Lett.* **89** 223508
[7] Niell G E, Alex A E and Myron S 2001 A self-powered mechanical strain energy sensor *Smart Mater. Struct.* **10** 293
[8] Mokhtari F, Shamshirsaz M and Latifi M 2016 Investigation of β phase formation in piezoelectric response of electrospun polyvinylidene fluoride nanofibers: LiCl additive and increasing fibers tension *Polym. Eng. Sci.* **56** 61–70
[9] Wang J *et al* 2003 On the α → β transition of carbon-coated highly oriented PVDF ultrathin film induced by melt recrystallization *J. Am. Chem. Soc.* **125** 1496–7
[10] Hao W *et al* 2016 Energy harvesters for wearable and stretchable electronics: from flexibility to stretchability *Adv. Mater.* **28** 9881–919
[11] Singh P *et al* 2014 Ferroelectric polymer-ceramic composite thick films for energy storage applications *AIP Adv.* **4** 087117
[12] Martins P *et al* 2012 On the origin of the electroactive poly(vinylidene fluoride) β-phase nucleation by ferrite nanoparticles via surface electrostatic interactions *Cryst. Eng. Comm.* **14** 2807–11
[13] Wang J L *et al* 2014 Transition of the polarization switching from extrinsic to intrinsic in the ultrathin polyvinylidene fluoride homopolymer films *Appl. Phys. Lett.* **104** 182907
[14] Li M *et al* 2013 Revisiting the δ-phase of poly(vinylidene fluoride) for solution-processed ferroelectric thin films *Nat. Mater.* **12** 433
[15] Yang L *et al* 2018 Effect of rolling temperature on the microstructure and electric properties of β-polyvinylidene fluoride films *J. Mater. Sci. Mater. Electron.* **29** 15957–65
[16] Lovinger A J 1982 Poly (vinylidene fluoride) *Developments in Crystalline Polymers—1* (Berlin: Springer), pp 195–273
[17] Ippili S *et al* 2019 An eco-friendly flexible piezoelectric energy harvester that delivers high output performance is based on lead-free $MASnI_3$ films and $MASnI_3$-PVDF composite films *Nano Energy* **57** 911–23
[18] Mokhtari F *et al* 2019 Triaxial braided piezo fibers energy harvesters for self-powered wearable technologies *J. Mater. Chem.* A **7** 8245–57
[19] Simona C E *et al* 2013 On the electrospinning of PVDF: influence of the experimental conditions on the nanofiber properties *Polym. Int.* **62** 41–8
[20] Ahn Y *et al* 2013 Enhanced piezoelectric properties of electrospun poly(vinylidene fluoride)/multiwalled carbon nanotube composites due to high β-phase formation in poly(vinylidene fluoride) *J. Phys. Chem.* C **117** 11791–9
[21] Jian M *et al* 2018 Piezoelectric and optoelectronic properties of electrospinning hybrid PVDF and ZnO nanofibers *Mater. Res. Express* **5** 035057
[22] Wang X *et al* 2018 Tactile-sensing based on flexible PVDF nanofibers via electrospinning: a review *Sensors* **18** 330
[23] Kim G H, Hong S M and Seo Y 2009 Piezoelectric properties of poly(vinylidene fluoride) and carbon nanotube blends: β-phase development *Phys. Chem. Chem. Phys.* **11** 10506–12
[24] Lovinger A J 1983 Ferroelectric polymers *Science* **220** 1115–21
[25] Persano L *et al* 2013 High performance piezoelectric devices based on aligned arrays of nanofibers of poly (vinylidenefluoride-co-trifluoroethylene) *Nat. Commun.* **4** 1633

[26] Liu Q *et al* 2018 Recent advances in energy materials by electrospinning *Renew. Sustain. Energy Rev.* **81** 1825–58

[27] Jianfen Z *et al* 2007 Polymorphism control of poly(vinylidene fluoride) through electrospinning *Macromol. Rapid Commun.* **28** 2159–62

[28] Andrew J S and Clarke D R 2008 Effect of electrospinning on the ferroelectric phase content of polyvinylidene difluoride fibers *Langmuir* **24** 670–2

[29] Ruan L *et al* 2018 Properties and applications of the β phase poly(vinylidene fluoride) *Polymers* **10** 228

[30] Ma W *et al* 2008 Crystalline phase formation of poly(vinylidene fluoride) from tetrahydrofuran/N,N-dimethylformamide mixed solutions *J. Macromol. Sci. Part B* **47** 434–49

[31] Sánchez J A G *et al* 2014 Piezoelectric effect in nanofibers deposited with magnetic field assisted electrospinning using solutions with PVDF and Fe_3O_4 nanoparticles *2014 29th Symp. on Microelectronics Technology and Devices (SBMicro)*

[32] Cai X *et al* 2017 A critical analysis of the α, β and γ phases in poly(vinylidene fluoride) using FTIR *RSC Adv.* **7** 15382–9

[33] Salimi A and Yousefi A A 2003 Analysis method: FTIR studies of β-phase crystal formation in stretched PVDF films *Polym. Test.* **22** 699–704

[34] Enomoto S, Kawai Y and Sugita M 1968 Infrared spectrum of poly (vinylidene fluoride) *J. Polym. Sci. Part A-2: Polym. Phys.* **6** 861–9

[35] Bormashenko Y *et al* 2004 Vibrational spectrum of PVDF and its interpretation *Polym. Test.* **23** 791–6

[36] Benz M and Euler W B 2003 Determination of the crystalline phases of poly (vinylidene fluoride) under different preparation conditions using differential scanning calorimetry and infrared spectroscopy *J. Appl. Polym. Sci.* **89** 1093–100

[37] Gregorio R Jr 2006 Determination of the α, β, and γ crystalline phases of poly (vinylidene fluoride) films prepared at different conditions *J. Appl. Polym. Sci.* **100** 3272–9

[38] Gregorio R Jr and Cestari M 1994 Effect of crystallization temperature on the crystalline phase content and morphology of poly (vinylidene fluoride) *J. Polym. Sci., Part B: Polym. Phys.* **32** 859–70

[39] Xing C *et al* 2014 Effect of a room-temperature ionic liquid on the structure and properties of electrospun poly (vinylidene fluoride) nanofibers *ACS Appl. Mater. Interfaces* **6** 4447–57

[40] Lins L C *et al* 2017 Effect of polyvinylidene fluoride electrospun fiber orientation on neural stem cell differentiation *J. Biomed. Mater. Res., Part B* **105** 2376–93

[41] Ribeiro C *et al* 2010 Influence of processing conditions on polymorphism and nanofiber morphology of electroactive poly (vinylidene fluoride) electrospun membranes *Soft Mater.* **8** 274–87

[42] Sorayani Bafqi M S, Bagherzadeh R and Latifi M 2015 Fabrication of composite PVDF-ZnO nanofiber mats by electrospinning for energy scavenging application with enhanced efficiency *J. Polym. Res.* **22** 130

[43] Sengupta D *et al* 2017 Characterization of single polyvinylidene fluoride (PVDF) nanofiber for flow sensing applications *AIP Adv.* **7** 105205

[44] Esterly D M and Love B J 2004 Phase transformation to β-poly(vinylidene fluoride) by milling *J. Polym. Sci., Part B: Polym. Phys.* **42** 91–7

[45] Mokhtari F, Latifi M and Shamshirsaz M 2016 Electrospinning/electrospray of polyvinylidene fluoride (PVDF): piezoelectric nanofibers *J. Text. Inst.* **107** 1037–55

[46] Hu J *et al* 2017 Mixed effect of main electrospinning parameters on the β-phase crystallinity of electrospun PVDF nanofibers *Smart Mater. Struct.* **26** 085019
[47] Martins P, Lopes A C and Lanceros-Mendez S 2014 Electroactive phases of poly(vinylidene fluoride): determination, processing and applications *Prog. Polym. Sci.* **39** 683–706
[48] Prest W M Jr and Luca D J 1978 The formation of the γ phase from the α and β polymorphs of polyvinylidene fluoride *J. Appl. Phys.* **49** 5042–7
[49] Gregorio R Jr and Cestari M 1994 Effect of crystallization temperature on the crystalline phase content and morphology of poly(vinylidene fluoride) *J. Polym. Sci., Part B: Polym. Phys.* **32** 859–70
[50] Maciel M M *et al* 2018 Relation between fiber orientation and mechanical properties of nano-engineered poly(vinylidene fluoride) electrospun composite fiber mats *Composites, Part B* **139** 146–54
[51] You M-H *et al* 2018 A self-powered flexible hybrid piezoelectric–pyroelectric nanogenerator based on non-woven nanofiber membranes *J. Mater. Chem.* A **6** 3500–9
[52] Chang J *et al* 2012 Piezoelectric nanofibers for energy scavenging applications *Nano Energy* **1** 356–71
[53] Liu Z H *et al* 2013 Piezoelectric properties of PVDF/MWCNT nanofiber using near-field electrospinning *Sens. Actuators,* A **193** 13–24
[54] Sita M D *et al* 2013 Structural changes in PVDF fibers due to electrospinning and its effect on biological function *Biomed. Mater.* **8** 045007
[55] Lei T *et al* 2015 Electrospinning-induced preferred dipole orientation in PVDF fibers *J. Mater. Sci.* **50** 4342–7
[56] Zong X *et al* 2002 Structure and process relationship of electrospun bioabsorbable nanofiber membranes *Polymer* **43** 4403–12
[57] Wang Y R *et al* 2011 A flexible piezoelectric force sensor based on PVDF fabrics *Smart Mater. Struct.* **20** 045009
[58] Ghafari E, Jiang X and Lu N 2018 Surface morphology and beta-phase formation of single polyvinylidene fluoride (PVDF) composite nanofibers *Adv. Compos. Hybrid Mater.* **1** 332–40
[59] Wang C *et al* 2007 Electrospinning of polyacrylonitrile solutions at elevated temperatures *Macromolecules* **40** 7973–83
[60] Subbiah T *et al* 2005 Electrospinning of nanofibers *J. Appl. Polym. Sci.* **96** 557–69
[61] Yee W A *et al* 2007 Morphology, polymorphism behavior and molecular orientation of electrospun poly(vinylidene fluoride) fibers *Polymer* **48** 512–21
[62] Asai H *et al* 2017 Effect of melt and solution electrospinning on the formation and structure of poly(vinylidene fluoride) fibres *RSC Adv.* **7** 17593–8
[63] Sorayani Bafqi M S, Bagherzadeh R and Latifi M 2017 Nanofiber alignment tuning: an engineering design tool in fabricating wearable power harvesting devices *J. Ind. Text.* **47** 535–50
[64] Wu C-M, Chou M-H and Zeng W-Y 2018 Piezoelectric response of aligned electrospun polyvinylidene fluoride/carbon nanotube nanofibrous membranes *Nanomaterials* **8** 420
[65] Dhakras D *et al* 2012 Enhanced piezoresponse of electrospun PVDF mats with a touch of nickel chloride hexahydrate salt *Nanoscale* **4** 752–6
[66] Costa L M M, Bretas R E S and Gregorio R 2010 Effect of solution concentration on the electrospray/electrospinning transition and on the crystalline phase of PVDF *Mater. Sci. Appl.* **1** 247

[67] Zhao Z Z *et al* 2005 Preparation and properties of electrospun poly(vinylidene fluoride) membranes *J. Appl. Polym. Sci.* **97** 466–74
[68] Gee S, Johnson B and Smith A L 2018 Optimizing electrospinning parameters for piezoelectric PVDF nanofiber membranes *J. Membr. Sci.* **563** 804–12
[69] Reneker D H *et al* 2000 Bending instability of electrically charged liquid jets of polymer solutions in electrospinning *J. Appl. Phys.* **87** 4531–47
[70] Ye H J, Shao W Z and Zhen L 2013 Crystallization kinetics and phase transformation of poly (vinylidene fluoride) films incorporated with functionalized $BaTiO_3$ nanoparticles *J. Appl. Polym. Sci.* **129** 2940–9
[71] Mendes S F *et al* 2012 Effect of filler size and concentration on the structure and properties of poly (vinylidene fluoride)/$BaTiO_3$ nanocomposites *J. Mater. Sci.* **47** 1378–88
[72] Priya L and Jog J P 2002 Poly(vinylidene fluoride)/clay nanocomposites prepared by melt intercalation: crystallization and dynamic mechanical behavior studies *J. Polym. Sci., Part B: Polym. Phys.* **40** 1682–9
[73] Benz M, Euler W B and Gregory O J 2002 The role of solution phase water on the deposition of thin films of poly(vinylidene fluoride) *Macromolecules* **35** 2682–8
[74] Ma W *et al* 2008 β-phase of poly(vinylidene fluoride) formation in poly(vinylidene fluoride)/poly(methyl methacrylate) blend from solutions *Appl. Surf. Sci.* **254** 5635–42
[75] An N *et al* 2011 Preparation and electroactive properties of a PVDF/nano-TiO_2 composite film *Appl. Surf. Sci.* **257** 3831–5
[76] Martins P *et al* 2012 Role of nanoparticle surface charge on the nucleation of the electroactive β-poly (vinylidene fluoride) nanocomposites for sensor and actuator applications *J. Phys. Chem.* C **116** 15790–4
[77] Wang W *et al* 2011 Gold-nanoparticle- and gold-nanoshell-induced polymorphism in poly (vinylidene fluoride) *Macromol. Mater. Eng.* **296** 178–84
[78] Xing C *et al* 2012 Impact of ionic liquid-modified multiwalled carbon nanotubes on the crystallization behavior of poly (vinylidene fluoride) *J. Phys. Chem.* B **116** 8312–20
[79] Mokhtari F *et al* 2017 Comparative evaluation of piezoelectric response of electrospun PVDF (polyvinilydine fluoride) nanofiber with various additives for energy scavenging application *J. Text. Inst.* **108** 906–14
[80] Mokhtari F *et al* 2017 Modeling of electrospun PVDF/LiCl nanogenerator by the energy approach method: determining piezoelectric constant *J. Text. Inst.* **108** 1917–25
[81] Moazeni N *et al* 2018 Fabrication and characterization of polydiacetylene supramolecules in electrospun polyvinylidene fluoride nanofibers with dual colorimetric and piezoelectric responses *Polymer* **134** 211–20
[82] Xin Y *et al* 2018 A brief review on piezoelectric PVDF nanofibers prepared by electrospinning *Ferroelectrics* **526** 140–51
[83] Wang S-H *et al* 2014 Mechanical and electrical properties of electrospun PVDF/MWCNT ultrafine fibers using rotating collector *Nanoscale Res. Lett.* **9** 522
[84] Hao Y *et al* 2013 Enhanced power output of an electrospun PVDF/MWCNTs-based nanogenerator by tuning its conductivity *Nanotechnology* **24** 405401
[85] Wang A *et al* 2018 Piezoelectric nanofibrous scaffolds as *in vivo* energy harvesters for modifying fibroblast alignment and proliferation in wound healing *Nano Energy* **43** 63–71
[86] Baniasadi M *et al* 2017 Effect of thermomechanical post-processing on chain orientation and crystallinity of electrospun P(VDF-TrFE) nanofibers *Polymer* **118** 223–35

[87] Hansen B J *et al* 2010 Hybrid nanogenerator for concurrently harvesting biomechanical and biochemical energy *ACS Nano* **4** 3647–52

[88] Farrar D *et al* 2011 Permanent polarity and piezoelectricity of electrospun α-helical poly (α-amino acid) fibers *Adv. Mater.* **23** 3954–8

[89] Jones J *et al* 2013 Investigation of ferroelectric properties and structural relaxation dynamics of polyvinylidene fluoride thin film via second harmonic generation *Appl. Phys. Lett.* **103** 072901

[90] Parker A *et al* 2018 Structural and thermal treatment evaluation of electrospun PVDF nanofibers for sensors *J. Polym. Sci. Appl.* **2018**

Chapter 6

Acoustoelectric energy conversion of nanofibrous materials

Chenhong Lang, Jian Fang and Tong Lin

Acoustic transducers, with the ability of acoustoelectric energy conversion, can be used for highly efficient sound detecting or energy harvesting devices. They also have the potential to be applied in various other fields such as industrial manufacturing, healthcare and medical, military defense, intelligence techniques, environmental protection, energy recycling, and stand-alone electric devices. Owing to high length-to-diameter ratio, specific surface area, and surface charge density, acoustic transducers (acoustoelectric sensors and energy harvesters) based on nanofibrous materials can be produced by the simple electrospinning process. More importantly, they show higher sensitivity and energy conversion efficiency than those based on conventional techniques and materials. Nanofibrous materials, especially electrospun nanofibers, are lightweight and have great flexibility, which make them particularly suitable for the emerging applications of wearable electronics. In this chapter, after introducing basic principles of acoustic transducing, common materials, and key evaluation parameters, we summarize recent developments in using nanofibrous materials for acoustic energy conversion, particularly for sensing and energy harvesting applications.

6.1 Introduction

Sound is a wave that is formed and propagated due to the vibration of an object in a medium, and it exists everywhere around us. In general, sound is specifically referred to those acoustic vibrations in air. It is well known that our human ears can only hear a limited section of sounds, in the frequency range of 20–20 000 Hz. Beyond this range, sound is unhearable by us; sounds below 20 Hz are termed as infrasound while those above 20 kHz as ultrasound. Within the hearable range, sounds are further classified according to the frequency, into low-frequency (<200 Hz), mid-frequency (200–6000 Hz), and high-frequency (6000–20 000 Hz) sound.

doi:10.1088/978-0-7503-2005-4ch6

Sound pressure level (SPL), the amplitude of acoustic vibrations, is an important parameter to indicate sound intensity and loudness. Human ears are sensitive to sound intensity, and when the SPL is greater than 90 dB, the sounds are usually too loud to be tolerable, and therefore considered as noise. It has been widely accepted now that noise is a type of environmental pollution causing health issues, and considerable attention has been devoted to monitor and control harmful noise in working environments and living spaces.

Traditional acoustic technologies mainly involved sound recording and reproduction. They have been gradually broadened to sound detection, cancellation, energy conversion, and harvesting, including various applications in information communication, entertainment, environment protection [1], industrial manufacturing [2], healthcare [3], medical settings [4], and defense [5]. Due to the considerable amount of energy carried by acoustic vibration of sounds, there have been growing efforts in recent years to convert acoustic energy into electric signals through energy harvesting and conversion techniques. The main applications of acoustoelectric energy conversion can be divided into two main categories: acoustic sensors functioning to detect sound sources, and acoustic energy generators that convert sounds into electrical power high enough to drive other microelectronic devices. For all these proposed applications, the most critical component in achieving efficient acoustoelectric energy conversion is the acoustic transducer. Acoustic transducers can convert acoustic vibrations into electric signals with different energy conversion mechanisms, such as electromagnetic induction, the piezoelectric effect, and the triboelectric effect. Electromagnetic induction-based transducers have been conventionally used in many applications, such as microphones and speakers. With the growing needs of miniatured device dimensions and flexible electronics in recent years, more research efforts have been devoted to the development of piezoelectric and triboelectric acoustic transducers and energy conversion systems. These recent developments have benefited from promising advances in functional piezoelectric and triboelectric materials, such as nanofibers. Nanofibrous materials, including electrospun polymeric nanofibers and many other inorganic nanowires, can be effectively prepared with different sizes, morphologies, and fibrous structures. They are lighter than their microsized counterparts, and much more sensitive to acoustic vibrations, and therefore can achieve a much higher acoustoelectric energy conversion efficiency.

In this chapter, after introducing the basic principles of acoustic transducing, common materials, and key evaluation parameters, we will summarize the recent developments in using nanofibrous materials for acoustic energy conversion, particularly for sensing and energy harvesting applications.

6.2 Conventional acoustic transducers

6.2.1 Basic principles

The basic function of an acoustic transducer is to convert sound vibrations to testable electric signals. This can be achieved by a few different principles (table 6.1), such as electromagnetic induction [6–10], piezo-resistance [11], piezo-capacitance [12],

Table 6.1. Comparison of conventional acoustic transducers.

Method	Principle	Sensing element materials	Devices	External power	Applications		
					Sensor	Energy harvester	Application example
Piezoresistive [11]	Change in electrical resistance under acoustic vibration	Metallic plates, carbon granules	A vibration diaphragm, two metallic plates separated by carbon granules, connected to an external power source	√	√	×	Button microphone, repeater, etc
Piezo-capacitive [12]	Change in capacitance under acoustic vibration	Metallic plates, a dielectric medium	A vibration diaphragm, two metallic plates separated by a dielectric medium, connected to an external power source	√	√	×	DC-biased microphones, radio frequency or high-frequency condenser microphones, electret microphone, etc
Piezo-optical [13]	Change in light intensity under acoustic vibration	optical fiber	A laser source, optical fibers, a vibration and reflecting diaphragm, a photo detector	√	√	×	Industrial and surveillance acoustic monitoring, audio calibration and measurement, high-fidelity recording and law enforcement, etc
Electromagnetic [6–10]	Electromagnetic induction under acoustic vibration	Diamagnetic, ferromagnetic, and paramagnetic materials	A vibration diaphragm, a magnet, a fixed coil placed in a coil holder, a resonator (acoustic energy harvester)	×	√	√	Stage microphone, acoustic energy harvester, etc

(*Continued*)

Table 6.1. (*Continued*)

Method	Principle	Sensing element materials	Devices	External power	Applications		
					Sensor	Energy harvester	Application example
Piezoelectric [14–16]	Piezoelectric charge generation under acoustic vibration	Piezoelectric materials	A vibration diaphragm, two electrodes, piezoelectric materials, a resonator (acoustic energy harvester)	×	√	√	Contact microphone, amplify sound from acoustic musical instruments, record sound in challenging environments, such as underwater under high pressure; acoustic energy harvester, etc
Triboelectric [17–19]	Triboelectric charge generation under acoustic vibration	Positively charged materials: metal, such as Cu; negatively charged materials: fluorinated polymers, such as PTFE and PVDF	A vibration diaphragm, two electrodes, triboelectric materials, a resonator (acoustic energy harvester)	×	√	√	Wearable and flexible electronics, military surveillance, jet engine noise reduction, low-cost implantable human hearing aids, and wireless technology applications, etc

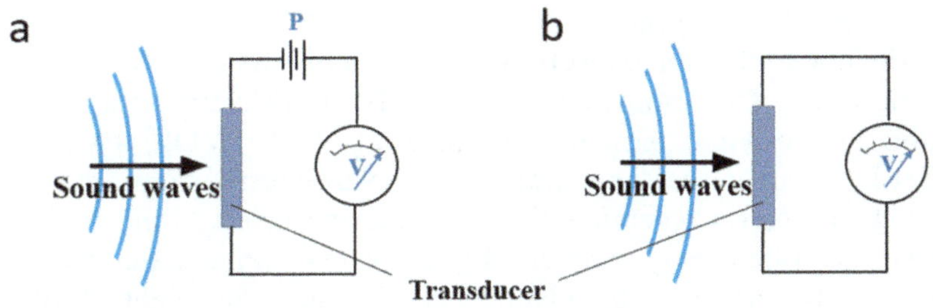

Figure 6.1. Acoustoelectric conversion (a) with and (b) without an external power source.

piezo-optics [13], the piezoelectric effect [14–16], and the triboelectric effect [17–19]. For example, a diaphragm vibrates when it receives air vibrations caused by sound, inducing changes in resistance, capacitance, or light intensity, which are further converted into voltage changes or digital signals. Consequently, some acousto-electric conversion processes need an external power to drive the conversion of diaphragm vibration into electric signals, as illustrated in figure 6.1(a). In some other cases, electric signals can be directly generated from sound vibrations through electromagnetic induction, piezoelectric effects, or triboelectric effects (figure 6.1(b)); therefore, no external power supply is required.

In an electromagnetic acoustoelectric energy conversion device, a magnet is normally attached to a flexible membrane fixed with an electrical coil. The membrane movement caused by an acoustic wave induces electrical current and voltage within the coil. For the piezoelectric effect, electric charges with opposite polarities are generated on piezoelectric material surfaces under a mechanical strain of the material. Piezoelectric membranes are typically used for acoustoelectric conversion, and the vibration of the piezoelectric membrane leads to an alternating voltage signal. The triboelectric effect generates opposite electric charges on material surfaces when two materials with different electric properties are brought together and then separated; an electric potential is therefore formed across these two surfaces. In comparison with electromagnetic systems that are normally with a bulky device structure, piezoelectric and triboelectric energy converters have a simpler device structure and are more dependent on material performance.

6.2.2 Materials

In this section, the material selection for the above-mentioned three types of acoustic transducers without an external power supply are briefly introduced.

6.2.2.1 Electromagnetic transducer

A permanent magnet and a fixed coil are the key components in an electromagnetic acoustoelectric conversion device. A neodymium iron boron magnet is usually used for its high residual magnetic flux density. Diamagnetic, paramagnetic, and ferromagnetic materials can be used for making the coil, and diamagnetic materials are preferred over others because they result in a very weak repletion with the magnet [20].

6.2.2.2 Piezoelectric transducer

Piezoelectric materials can convert acoustic energy to electrical signals by the piezoelectric effect. Common materials include piezoelectric ceramics (e.g. ZnO, PZT, AlN) and organic piezoelectric materials (e.g. PVDF, PVDF–HFP and PVDF–TrFE)). Under the same strain level, piezoelectric ceramics can normally generate higher electric potentials than organic piezoelectric materials. However, much higher acoustic energy is required to activate ceramic materials. By contrast, piezoelectric polymers have the advantages of being lightweight, flexible, stretchable, and biocompatible, which make them much more sensitive to low-intensity acoustic vibrations for a much wider range of applications.

6.2.2.3 Triboelectric transducer

The most familiar example of triboelectric charge generation to us is rubbing a glass rod with a piece of silk fabric. In theory, the contact and separation of any two different insulating materials can generate triboelectric charges on material surfaces. For evaluating the charge generation capacity of a triboelectric material, the triboelectric series has been constructed. On this series, the materials that are placed further apart can generate more triboelectric charges than those that are placed closer together. The materials on the top end of the series are more likely to lose electrons and carry positive charges, while the materials at the other end are more easily negatively charged [21].

6.2.3 Evaluation of acoustoelectric conversion

Acoustoelectric energy conversion can be evaluated by various parameters, as summarized below.

6.2.3.1 Sound waves

- **Sound pressure** is the local pressure deviation from the ambient (average or equilibrium) atmospheric pressure, caused by a sound wave. In air, sound pressure can be measured using a microphone. The SI unit of sound pressure is the Pascal (Pa).
- **SPL** is a logarithmic measure of the effective pressure of a sound relative to a reference value. SPL (L_p) is measured in dB and defined by

$$L_p = 20\,\lg(P_1/P_0)$$

where P_1 is the sound pressure, and P_0 is the reference sound pressure of 0.00002 Pa.

6.2.3.2 Acoustic sensors

- The **signal-to-noise ratio** (SNR) measures the signal quality of an acoustic sensor. This ratio can be obtained using the following equation:

$$SNR = 10\,\log_{10}(P_s/P_n)$$

where P_s and P_n are the signal power and noise power, respectively. A higher signal-to-noise ratio indicates better sensing signal quality.

- **Sensitivity** is another key characteristic of sensor devices. The sensitivity (S) of the sensor device can be calculated using the following equation:

$$S = \frac{P(\text{Pa})}{V} = \frac{P_o \cdot 10^{\frac{L_p(\text{dB})}{20}}}{V}$$

where P(Pa) is the acoustic pressure in Pascals, V is the voltage output of the sensor, P_o is the reference sound pressure of 0.000 02 Pa, and L_p(dB) is the SPL in decibels.

- **Working frequency** is the range of sound frequency that the device can respond to, which decides the usage range of an acoustic sensor.
- **Working sound pressure** is the range of sound pressure that the device can respond to with electric signal outputs.

6.2.3.3 Acoustic energy harvester

- **Output voltage** and **output current** are the basic performance measurements of an acoustic conversion device. The maximum voltage and maximum output current are the peak values of the voltage signal and the current signal, respectively.
- The **instantaneous power** of an acoustic energy harvester can be calculated by the equation

$$P = UI$$

where U and I represent the voltage and current outputs on the optimum load, respectively.

- **Instantaneous area power density** can be calculated by the equation

$$P_A = P/A$$

where P is the instantaneous power, and A is the area of the energy conversion element.

- **Instantaneous volume power density** can be calculated by the equation

$$P_V = P/V$$

where V is the volume of the energy conversion element.

- **Integral power** can be calculated by

$$P_{e(Int)} = \int_{t_1}^{t_2} U(t)I(t)dt \times f_a$$

where U and I are the instantaneous voltage and current outputs, respectively. f_a is the frequency of the sound wave.

- The **effective value of power** P_{RMS} can be calculated by the equation

$$P_{RMS} = U_{RMS} \times I_{RMS}$$

where U_{RMS} and I_{RMS} are the effective values of voltage and current on the optimum load, respectively, when the instantaneous power is maximized.

- **Integral and effective energy** can be calculated by the following equations ($t = 3600$ s):

$$E_{e(Int)} = P_{e(Int)} \times t$$

$$E_{e(RMS)} = P_{e(RMS)} \times t.$$

- The **sound power** P_a of a sound source can be calculated according to the equation

$$P_a = \frac{Ap^2}{\rho c}\cos\theta$$

where A is the surface area of the device, p is sound pressure, ρ is the air density of 1.292 kg m^{-3}, c is sound speed in air (343.2 m s^{-1}), and θ is the angle between the direction of the sound propagation and the normal to the surface (zero degrees in this case).

- **Energy conversion efficiency** is calculated as the ratio of output electric power and input acoustic power ($\eta = \frac{P_e}{P_a}$):

$$\eta_{e(Int)} = P_{e(Int)}/P_a \times 100\%$$

$$\eta_{e(RMS)} = P_{e(RMS)}/P_a \times 100\%.$$

6.3 Acoustic sensors

6.3.1 Electrospun piezoelectric polymer nanofibers

Acoustic sensors made of piezoelectric polymer films such as PVDF, PVDF–HFP, and PVDF–TrFE can now be found on the commercial market. However, highly controlled multistep processes are often involved in the production of these piezoelectric films, leading to high prices [22]. The electrospinning process applies simultaneous mechanical stretching and electric poling effects during fiber formation, and electrospun nanofibers have been reported to have strong piezoelectricity without extra stretching and poling treatments [23–33], opening up promising potential for applications in sensing, energy conversion, and harvesting.

A PVDF nanofibrous acoustic sensor with incredible acoustic-to-electric conversion ability was reported by Lang *et al* [34]. The sensor device was prepared by sandwiching a piece of electrospun PVDF nanofiber web with two transparent PET films gold sputter-coated on one side. The gold-coated surfaces were contacted with the nanofiber layer (shown in figure 6.2(a)), and they functioned as electrodes to collect electrical signals. A hole was cut on each plastic film to allow the nanofibers to directly receive sound waves, as schematically shown in figure 6.2(b). A photo of the actual device is shown in figure 6.2(c). To eliminate the noise signals from the

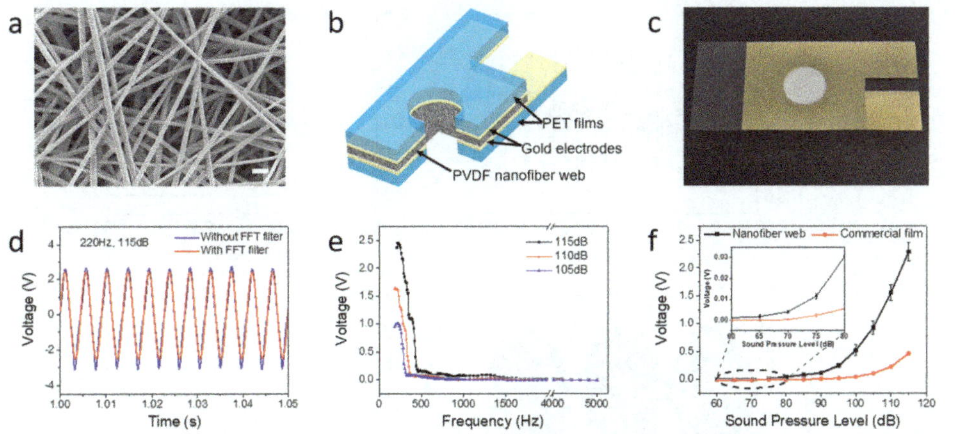

Figure 6.2. (a) SEM image of the electrospun PVDF nanofibers (scale bar, 1 μm), (b) schematic illustration of the acoustic sensor structure, (c) digital photo of the device, (d) voltage outputs of the device under sound with and without FFT treatment, and the effects of (e) sound frequency and (f) SPL (sound frequency = 220 Hz) on device voltage outputs [34].

background sound in the testing environment, electronic instrument, and connecting cables, a fast Fourier transform (FFT) technique was used to process the output signals [35]. After FFT filtering, the voltage output was very similar to the raw data, suggesting that the voltage response mainly comes from the sound wave (figure 6.2(d)). The devices can precisely distinguish sound waves in the low to middle frequency region, and show higher sensitivity to sound with a pressure level above 100 dB, which makes them very suitable for detecting noise (figure 6.2(e)). The nanofiber sensor devices are able to detect low-frequency sound with a sensitivity as high as 266 mV Pa^{-1}, more than five times higher than that of a commercial piezoelectric PVDF dense film (figure 6.2(f)). The authors also studied the effect of device structure (including the hole, hole size, nanofiber web thickness, and nanofiber web size) and the nanofiber's physical properties (including the orientation and fiber diameter) on the voltage output of the nanofiber device, and optimized these parameters.

To prove the ability to differentiate multiple frequencies of sound, two acoustic sources were used to generate bifrequency sound waves. From FFT treatment results, the nanofiber device is able to distinguish two sound waves, even with just a 0.05 Hz difference in frequency (figures 6.3(a) and (b)). Figure 6.3(c) schematically illustrates the sound sensing mechanism proposed by the authors. When acoustic waves hit the piezoelectric nanofibers, vibration takes place on the fibers and plastic films because of sound absorption. Both nanofibers and plastic films vibrate in the in-plane (V_1) and cross-plane (V_2) directions. The directly exposed nanofibers through the hole vibrate more intensely than those covered with the electrodes. Such asymmetric vibrations on the fibers and vibration propagation along the fibers enhance nanofiber deformation. The high β crystal phase content and oriented macromolecular chain dipoles facilitate piezoelectric conversion under compressive

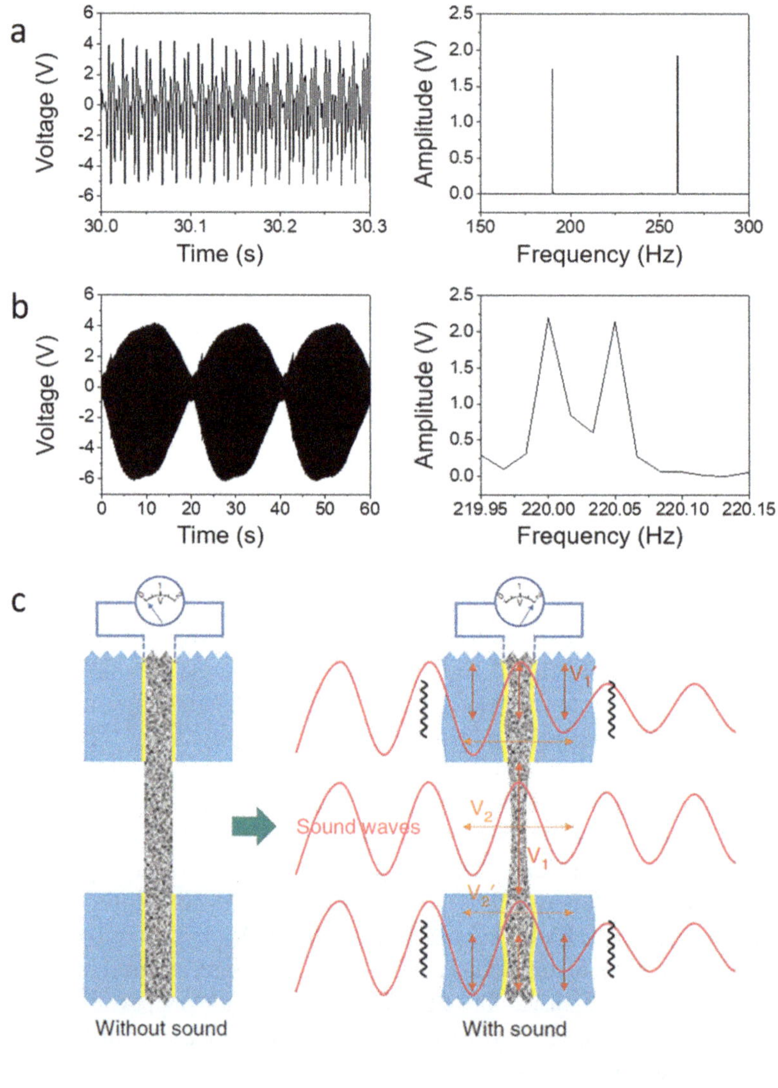

Figure 6.3. Voltage outputs of the nanofiber sensor device under bifrequency sound waves and the FFT-processed frequency spectrum of (a) 190 and 260 Hz and (b) 220.00 and 220.05 Hz (SPL = 115 dB for both loudspeakers); (c) proposed sound vibration modes and piezoelectric conversion mechanism of the nanofiber device [34].

impact. Power output could be generated by the cross-plane vibrations (V_2 and V_2') of the nanofibers and plastic films under sound waves. The in-plane vibrations (V_1 and V_1') assist vibration propagation throughout the entire nanofiber web.

With the same device structure, other piezoelectric polymers such as polyacrylonitrile, PVDF–HFP, and PVDF–TrFE were also investigated in this same study. The results show these piezoelectric nanofiber devices can have even higher

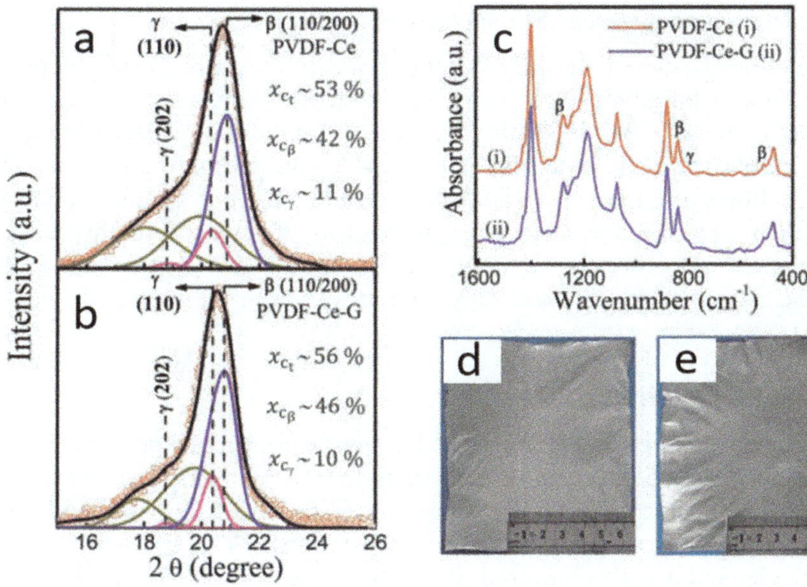

Figure 6.4. XRD patterns of (a) Ce^{3+}-doped PVDF and (b) Ce^{3+}-doped PVDF/graphene nanofibers. (c) FTIR spectra of Ce^{3+}-doped PVDF and Ce^{3+}-doped PVDF/graphene nanofibers. Photos of (d) Ce^{3+}-doped PVDF and (e) Ce^{3+}-doped PVDF/graphene nanofibers (10 × 13 cm^2) [36].

sensitivity than the PVDF nanofiber device: the highest sensitivity was achieved by PVDF–TrFE, with 750 mV Pa^{-1} at 100 dB. This suggests that the high acoustic-to-electric conversion property of electrospun nanofibers can be generally achieved by all piezoelectric polymers.

For enhancing the piezoelectric effect, different additives have been introduced into PVDF or its copolymer nanofibers. Garain *et al* [36] reported a Ce^{3+}-doped PVDF/graphene composite electrospun nanofibers for making piezoelectric acoustic transducers. The addition of Ce^{3+} and graphene increases the β phase content and crystallinity of the nanofibers, as shown in figure 6.4. They validated the practical application of the nanofiber device as an acoustic sensor; the device is sensitive to sound in the frequency range of 90–185 Hz, delivering a maximum output voltage of 3 V at an SPL of 88 dB.

Adhikary *et al* [37] fabricated composite electrospun nanofibers of Eu^{3+}-doped PVDF–HFP/graphene for an ultrasensitive wearable piezoelectric acoustic transducer. The addition of Eu^{3+} and graphene also increases the β phase content and crystallinity of the nanofibers. They demonstrated a sensitivity of 11 V Pa^{-1} to musical sound at 82 dB. Maity *et al* [38] reported a sensitive and efficient piezoelectric acoustic transducer made of a few layers of MoS_2 nanosheets incorporated into electrospun PVDF nanofiber webs. The FTIR, XRD, and XPS results show that the PVDF–MoS_2 composite nanofibers present a higher β phase content and crystallinity than neat PVDF nanofibers. As a result, the acoustic transducer shows an ultrahigh sensitivity of 19 V Pa^{-1} to musical sound at 90 dB, which is superior to most of the recently reported piezoelectric acoustic sensors.

6.3.2 Electrospun nonpiezoelectric polymer nanofibers

Inspired by the flow-based auditory system of mosquitoes, Zhou *et al* [39] prepared a directional sensor based on electromagnetic induction using a 2D nanofiber mesh as the sensing material. By depositing Au on a freestanding electrospun PMMA nanofiber mesh, they fabricated an Au–PMMA nanomesh. Within the 2D mesh, most of the fibers align between the two electrodes. Output signals are induced by the motion of the nanomesh in a magnetic field. Compared to the noisy signal recorded by the reference microphone, the nanomesh flow sensor can exclusively capture the signals at 400 Hz and reject the unwanted noise signals.

6.4 Acoustoelectric harvesters

Piezoelectric acoustic energy harvesters, also known as acoustoelectric generators, produce higher level of electric outputs than acoustic sensors, and are targeted by researchers so that the generated electricity can be directly or indirectly used for powering other microelectronic devices or systems. A diaphragm and energy conversion element are also required for an acoustoelectric harvester. Besides these, in some cases, a resonator is used to increase the vibrations caused by sound and an energy storage device is used to store the electric energy.

6.4.1 Piezoelectric harvesters

6.4.1.1 Electrospun piezoelectric polymer nanofibers

An acoustoelectric conversion device [40] based on a PVDF–TrFE nanofiber web was used for power generating purposes. As shown in figures 6.5(a)–(c), the device consists of eight holes on each electrode and a nanofiber web as a middle layer. During the experiment, a sound-collecting cone was set on the speaker to increase the sound pressure (figure 6.5(d)). Under a sound (SPL 115 dB, frequency 210 Hz) activation, it generated periodic voltage and current outputs with peak values as high as 14.5 V and 28.5 μA (figure 6.5(e)). Figure 6.5(f) shows the dependency of voltage outputs on sound frequency for the nanofiber device. The large voltage outputs were generated mainly from low-frequency sound. Figure 6.5(g) shows the effect of an external resistance on the electric outputs of the nanofiber device. The output power of the device calculated based on the voltage and current outputs is shown in figure 6.5(h). The device showed the maximum power output (141.3 μW) at the external resistance of 470 kΩ.

Table 6.2 lists the power generation features of the acoustoelectric devices made from electrospun PVDF–TrFE nanofibers, commercial dense piezoelectric PVDF–TrFE film, and electrospun PVDF nanofibers, which are estimated according to the results discussed in section 6.3.1. All of the calculated volume power density, power, energy, and energy conversion efficiency of the eight-hole PVDF–TrFE nanofiber device were much higher than those of the one-hole PVDF–TrFE nanofiber device, eight-hole commercial dense film device, and eight-hole PVDF nanofiber device. The calculated energy conversion efficiency of the eight-hole PVDF–TrFE nanofiber

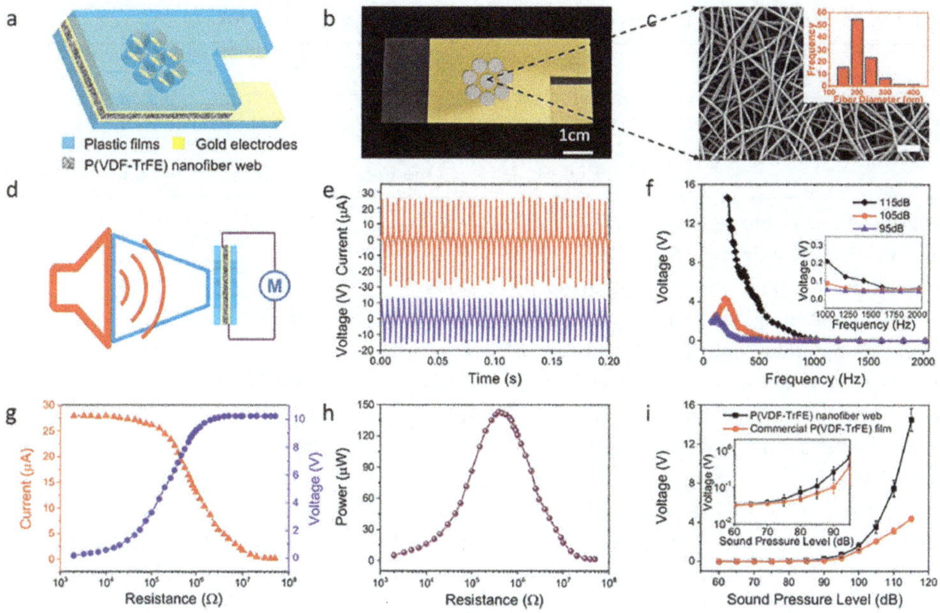

Figure 6.5. (a) Schematic illustration of the acoustic energy harvester device, (b) digital photo of the device, (c) SEM image of the electrospun PVDF–TrFE nanofibers (scale bar, 2 μm), (d) schematic illustration of the acoustoelectric conversion testing, (e) voltage and current outputs of the device under sound, (f) effect of sound frequency on the voltage output of the device, (g) dependence of the peak voltage and current output on the resistance of the external load, (h) dependence of the peak power output on the external load resistance, (i) effect of SPL on devices made of nanofiber web and commercial dense film (eight-hole device, hole diameter = 4.9 mm; web thickness = 20 μm; film area = 12 cm^2; (a)–(e) and (g)–(i): SPL = 115 dB and sound frequency = 210 Hz) [40].

device was as high as 60.3%, which is very promising for practical energy harvesting applications.

The holes in the electrodes were found to have significant effects on the output signals of the device. The effect of hole number on device outputs is shown in figure 6.6(a). By maintaining the overall hole area, the voltage output increases with the increasing number of holes until there are six. When the number of holes continues to increase, the voltage output remains at around 14 V. The device with different numbers of holes showed slightly different output responses to SPL (figure 6.6(b)). The results indicate devices with less holes are fit for making sensors due to their sensitivity to low SPL. Meanwhile, those with more holes are more suitable for making energy harvesters, as they are able to convert more energy when the SPL is high. A finite element method was used to analyze the vibration characteristics of the nanofiber device in order to understand the effect of holes on the acoustoelectric response of the devices. Figures 6.6(c) and (d) show the vibration energy of the nanofiber web and PET film in the eight-hole and one-hole devices, respectively. Figures 6.6(e) and (f) illustrate the vibration energy profiles of the nanofiber web in the one-hole and eight-hole devices. The central part of the hole shows a higher vibration energy density than the periphery of the hole. The eight-hole

Table 6.2. Power generation features of nanofiber acoustoelectric harvesters [40].

		PVDF–TrFE			
		Nano-nonwoven[a]		Commercial film[b]	PVDF Nano-nonwoven[c]
		Eight-hole	One-hole		
Peak open-circuit voltage (V)		14.5	7.8	4.6	5.3
Peak short-circuit current (μA)		28.5	11.2	5.8	6.6
Optimum load R (kΩ)		470	2500	2500	3000
Voltage on optimum load (V)		7.4	6.0	3.4	4.3
Current on optimum load (μA)		19.1	8.6	4.2	5.6
Instantaneous power (μW)		141.3	51.6	14.4	24.1
Area power density (μW cm^{-2})		11.8	4.3	1.2	2.0
Volume power density (μW cm^{-3})		5900	2150	600	1000
Power (μW)	Integral	25.8	12.5	5.2	7.6
	RMS[d]	25.6	12.6	5.0	7.6
Energy (Wh)	Integral	9.29×10^{-2}	4.50×10^{-2}	1.87×10^{-2}	2.72×10^{-2}
	RMS	9.22×10^{-2}	4.54×10^{-2}	1.80×10^{-2}	2.73×10^{-2}
Energy conversion efficiency (%)	Integral	60.3	29.2	12.1	17.8
	RMS	59.8	29.4	11.7	17.8

[a] Size 3 × 4 cm^2; thickness, 20 μm; nanofiber diameter, 240 ± 40 nm.
[b] Size 3 × 4 cm^2; thickness, 20 μm.
[c] Size 3 × 4 cm^2; thickness, 20 μm; nanofiber diameter, 310 ± 60 nm (calculated based on the active layer).
[d] RMS root mean square.

nanofiber web and PET film both have much higher vibration energy than the one-hole nanofiber web and PET film. This is because nanofibers in the hole vibrate more strongly than those covered with electrodes. Therefore, a multihole structure should include multiple vibration spots in the fibrous web, which would facilitate transport of vibration energy from the exposed areas to the entire device.

6.4.1.2 Nanowires

Other than continuous electrospun nanofibers, other forms of 1D nanomaterials have also been used in acoustic energy harvesting. Xudong Wang *et al* [41] developed a nanowire energy harvester that can be driven by an ultrasonic wave to generate continuous DC output. The energy harvester was based on vertically aligned ZnO nanowire arrays with a zigzag metal electrode on the top and a conductive electrode on the bottom. The zigzag electrode is used to create, collect, and output electricity generated from the nanowires. The output power of the energy harvester is about 1 pW per 2 mm^2 area substrate.

Cha *et al* [42] reported an energy harvesting device based on piezoelectric ZnO nanowires driven by 100 Hz frequency sound. The energy harvester was made on a polyethersulfone (PES) substrate with a PdAu coating as the electrode. Well-aligned ZnO nanowires were grown on the top of the PES electrode, and covered by a

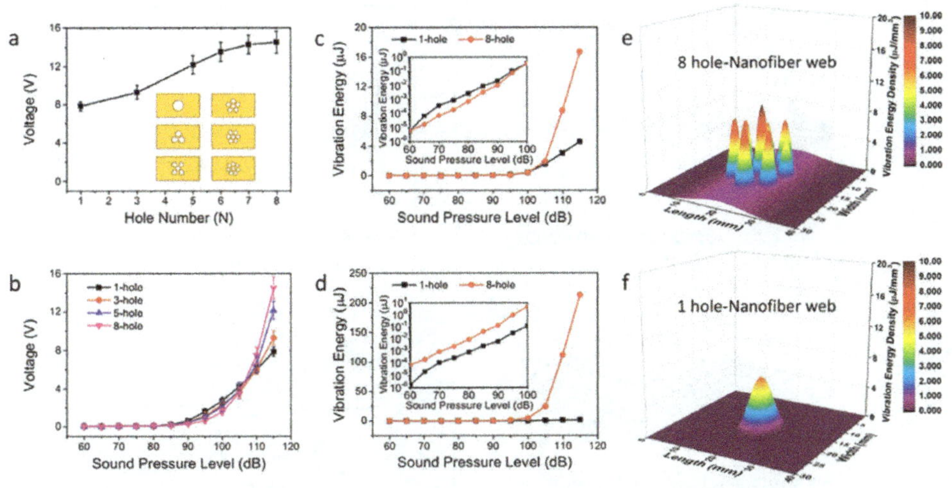

Figure 6.6. (a) Voltage output as a function of the number of holes (sound frequency = 210 Hz; SPL = 115 dB; web area = 12 cm^2; web thickness = 20 μm), (b) effect of SPL on voltage output of the nanofiber devices with different numbers of holes (sound frequency = 210 Hz). FEM modeled vibration energy of (c) nanofiber web and (d) PET film in one-hole and eight-hole devices (sound frequency = 210 Hz). Modeled vibration energy density of nanofiber webs in (e) eight-hole and (f) one-hole device (sound frequency = 210 Hz; SPL = 115 dB) [40].

vibration plate. When a sound wave irradiates the device, it causes the vibration of the top contact electrode and the compression of the ZnO nanowires, which results in the generation of an electric potential. The device can generate 50 mV of output voltage under sound with an intensity of 100 dB.

Lee *et al* [43] fabricated a hybrid cell that harvests both acoustic and solar energy. The sound wave-driven energy harvesting was demonstrated using both laterally bonded single wires and vertically aligned nanowire arrays, in the frequency range of 35–1000 Hz. Figure 6.7(a) schematically shows the experimental setup for single nanowires. An alumina rod was firmly attached to the sound wave generator at one end, and its other end was mounted to the energy harvesting device. The sound wave generated by the transducer was directly transmitted to the single nanowires. The vibration of the end surface of the alumina rod was measured with an accelerometer. As shown in figure 6.7(b), the voltage output is relatively uniform over the entire frequency range under investigation. The voltage output does not significantly depend on the sound frequency.

The aligned nanowires were combined with CdS (n-type)/CdTe (p-type) nanoparticles to make a hybrid device to harvest both acoustic and solar energy. Figure 6.7(c) presents a diagram of the hybrid device. Simulated sunlight illumination and acoustic vibration are supplied through the top PET/ITO layer and the bottom Au electrode, respectively. Figure 6.7(d) shows the energy band diagram of the hybrid device. The upper and lower diagrams correspond to the solar energy harvester and nanowire energy harvester, respectively. The solar energy harvester is supposed to generate DC output (figure 6.7(e)), while the sound harvester generates

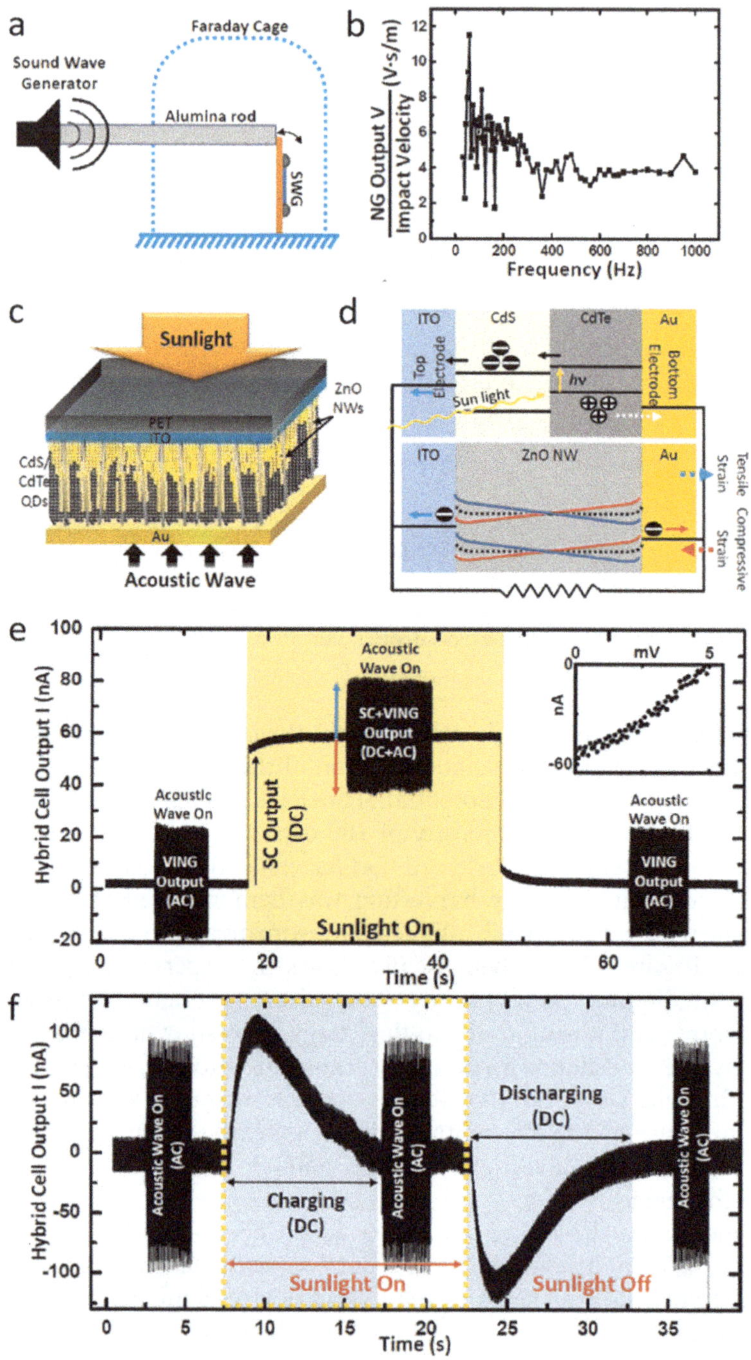

Figure 6.7. (a) Schematic of a single nanowire energy harvesting setup, and (b) effect of sound frequency on the ratio of single nanowire energy harvester voltage output to the applied mechanical impact velocity. (c) Schematic of a hybrid device with two different energy sources, (d) energy band diagram of the hybrid device showing electron flow under sunlight illumination and acoustic vibration, (e) current output of the hybrid device, and (f) current output of the hybrid device after connecting a capacitor [43].

AC voltage. A capacitor was added to allow the DC output to be stored (figure 6.7(f)).

6.4.2 Triboelectric acoustic generators

6.4.2.1 Electrospun PVDF nanofibers

In recent years, acoustic triboelectric energy harvesters have also been was investigated. Due to the wave characteristics of sound propagation, a resulting acoustic pressure separates and pushes back the two films at a certain frequency. On account of the triboelectric effect, an electrical signal with the same frequency is generated between the two contact surfaces of the two different materials, which have different electronegativities. The surface of the material with a higher electronegativity is negatively charged after it is separated, and the other material's surface has an equal amount of positive charges. Generally, polyamide and metal are used as positively charged materials in a triboelectric energy harvester, while negatively charged materials are mostly fluorinated polymers, such as PTFE, FEP, and PVDF [21].

Cui *et al* [17] reported a mesh-web-based triboelectric energy harvester with a sandwich structure of two substrates and one vibrating web. As shown in figure 6.8(a), a piece of aluminum (Al)-covered stainless steel mesh acts as the contact surface and bottom electrode, i.e. substrate B. Another piece of stainless steel mesh with a layer of electrospun PVDF nanofibers acts as the top electrode, i.e. substrate A (figure 6.8(b)). A piece of polyethylene film with Al coated on one surface and PVDF nanofibers on the other surface was used as the vibrating membrane of the device. The PVDF nanofiber and Al act as negatively and positively charged materials, respectively. The mesh structure allows air circulation and benefits the penetration of sound waves. The thin and lightweight vibrating film facilitates absorption of low-frequency sound wave energy, and displays a sensitive response to small sound wave pressure. Furthermore, the porous mesh substrate facilitates the spread of sound waves, and reduces the sound energy loss in the substrate. The triboelectric energy harvester device can be driven by sound waves in a broad bandwidth ranging from 50 Hz to 425 Hz, and it obtains a short-circuit current density as high as 45 mA m^{-2} and peak power density of 202 mW m^{-2}.

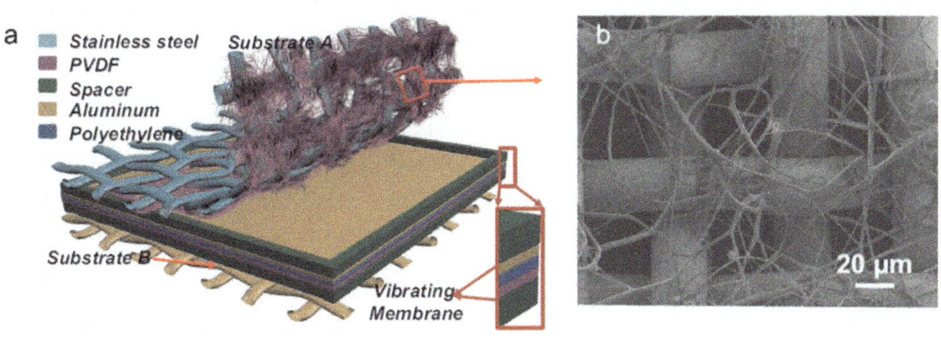

Figure 6.8. (a) Schematic of a sound-driven triboelectric energy harvester, and (b) SEM image of the PVDF nanofibers on the stainless steel mesh [17].

6.4.2.2 Nanowires

Yang *et al* [18] used a Helmholtz resonator with a size-tunable narrow neck to enhance the vibration and triboelectric effect of an energy harvester. They also optimized the parameters of the resonator to increase the output and broaden the frequency response. Figures 6.9(a) and (b) schematically show the structure of the device. The circular multilayer membrane on the surface of the cavity is the core part of the device. An Al thin film with a nanoporous surface plays the roles of an electrode and a contact surface. Figure 6.9(c) shows a SEM image of nanopores on the Al. A PTFE film with copper acts as another electrode. Inductively coupled plasma reactive ion etching was used to fabricate the aligned nanowires on the PTFE surface, which largely enhances the triboelectrification. Figure 6.9(d) shows a SEM image of the PTFE nanowires. A suitable initial prestress of 5.6 kPa and an optimum open ratio of 0.3, which can optimize the oscillation coupling between the air trapped in the cavity and PTFE thin film, are used to maximize the electrical output (figures 6.9(e)–(h)). Moreover, the natural frequencies of the devices can be designed by changing the parameters of the Helmholtz resonator. Four differently sized devices were used to broaden the working bandwidth from 10 to 1700 Hz.

Fan *et al* [19] reported a 125 μm thickness, rollable paper-based triboelectric energy harvester for sound wave energy harvesting. The device contains no Helmholtz resonator, which can achieve a maximum power density of 121 mW m^{-2} under an SPL of 117 dB. The paper substrate was coated with copper, which acts as a positively charged material that generates triboelectric charges upon contact with a thin PTFE nanowires membrane, as schematically shown in figure 6.10(a). Figure 6.10(b) shows a photograph of the multihole paper electrode. Figure 6.10(c) shows a SEM image of PTFE nanowires, which act as negatively charged materials. A digital photo of a paper-based acoustic energy harvester is presented in figure 6.10(d). ANSYS software was used to simulate the sound wave-induced PTFE membrane vibration under various frequencies, as shown in figure 6.10(e). From the simulation results, it can be seen that under sound with different frequencies, the PTFE membrane vibrated with different deformation regions and magnitudes. A cycle of electricity generation processes induced by vibration under sound is depicted in figure 6.10(f). Under sound, the PTFE will be separated from the copper. Then the altered inner dipole moment drives free electrons to flow from the electrode with PTFE nanowires to the multihole paper copper electrode until the maximum separation state is reached. The free electrons will flow in a reverse direction from the maximum separation state toward the maximum contact state. With that, a full cycle of electricity generation is completed. The authors studied the influence of the hole shape, central hole distribution, hole diameter, void-to-surface ratio, electrode thickness, and electrode substrate material on the electrical output and device frequency response; then, they optimized the output.

6.4.2.3 Nanopillars

Mehmet Kanik *et al* [44] constructed core–shell nanostructures, with PES as the core and As_2Se_3 as the shell, for building a 3D-printed triboelectric energy harvester

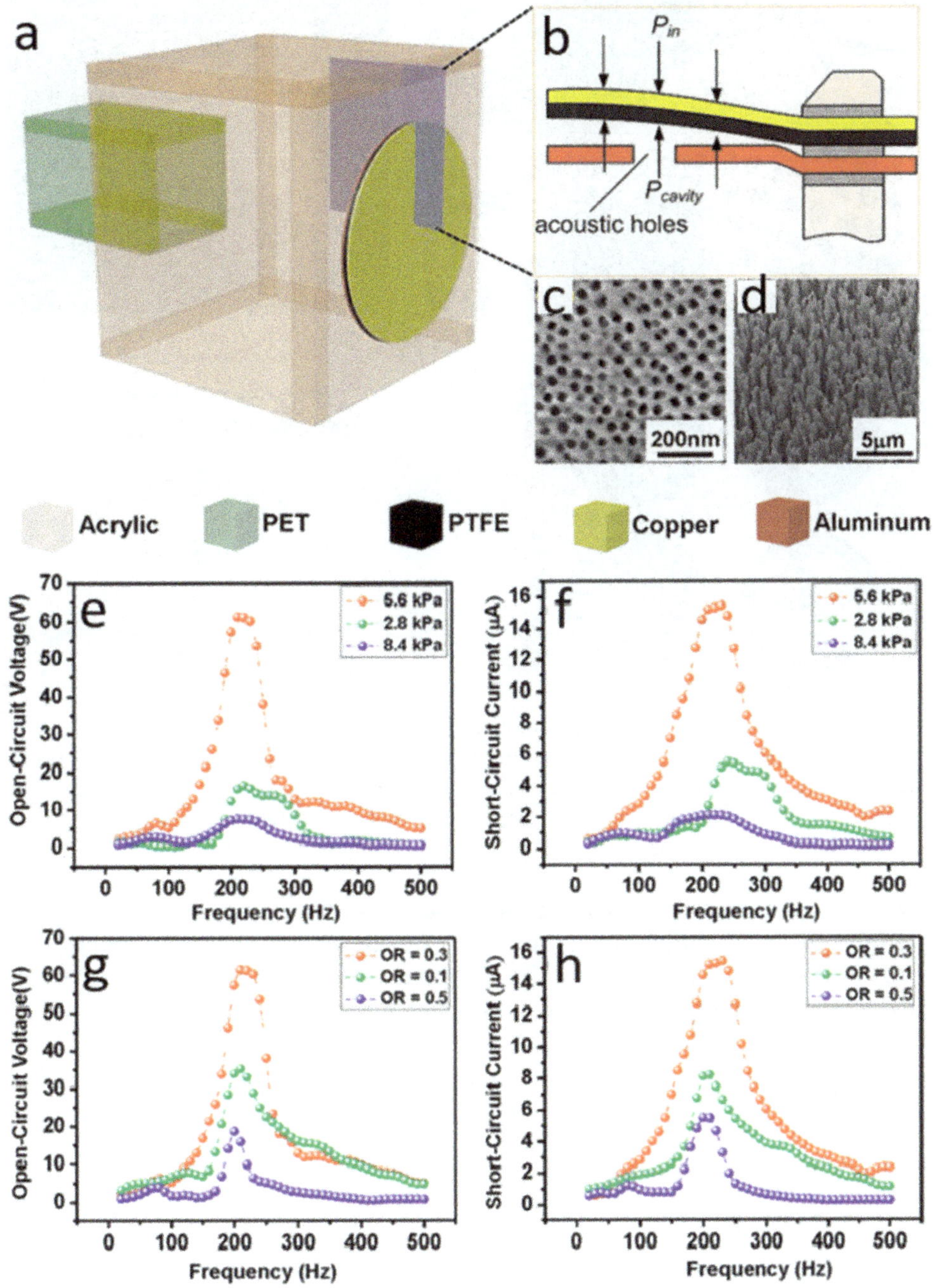

Figure 6.9. (a) Schematic and (b) cross-sectional view of the triboelectric energy harvester. (c) SEM image of nanopores on the Al electrode, and (d) SEM image of PTFE nanowires fabricated on the film surface by plasma etching. Effect of sound frequency on (e) voltage outputs and (f) current outputs with different prestresses. Effect of sound frequency on (g) voltage outputs and (h) current outputs with different open ratios [18].

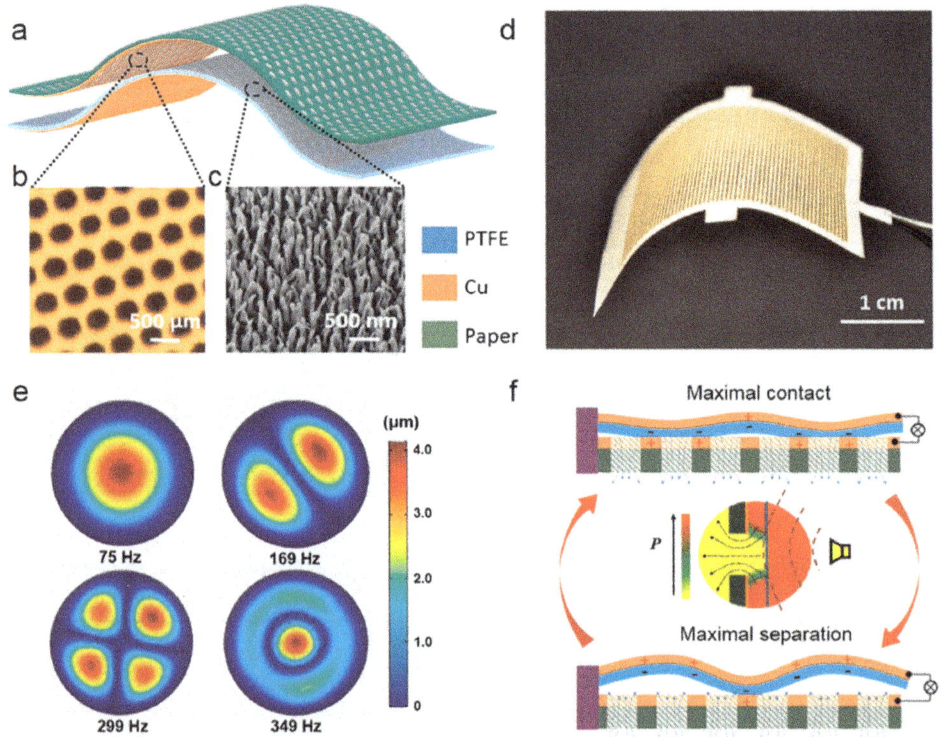

Figure 6.10. (a) Schematic illustration of the paper-based triboelectric energy harvester, (b) digital photo of the multihole paper electrode, (c) SEM image of the PTFE polymer nanowires, and (d) digital photo of the paper-based triboelectric energy harvester. (e) ANSYS analysis of PTFE membrane vibration under various sound frequencies, and (f) illustration to interpret the PTFE membrane vibration under sound and electricity generation [19].

device. In addition to an increase in the total surface area by using nanoscale materials, the application of a multilayer device structure is also important to improve the overall energy output. Measurements at sound frequencies between 10 and 100 Hz revealed that the device has a better response at 10 Hz with a peak-to-peak open-circuit voltage of 107.3 V and a peak-to-peak short-circuit current of 0.49 µA. The output power density (max. 5.47 mW m^{-2}) of the 3D triboelectric energy harvester decreases with increasing sound frequency.

6.5 Potential applications

6.5.1 Sound distinguishing

In addition to sensing the SPL and frequency of noise, nanofibrous sound sensors also have the ability to record human voices, for example, the pronunciation of ‘one, two, three, four, five’ [34] (figures 6.11(a) and (b)), ‘A, B, C, D, E’ [38], and ‘hello, bye’ [37]. Figures 6.11(a) and (b) show voice-activated voltage outputs of a sound sensor made of electrospun PVDF nanofiber web and a commercial microphone for comparison. The black, orange, blue, pink, and green lines represent voltage outputs and FFT profiles of the words ‘one’, ‘two’, ‘three’, ‘four’, and ‘five’, respectively.

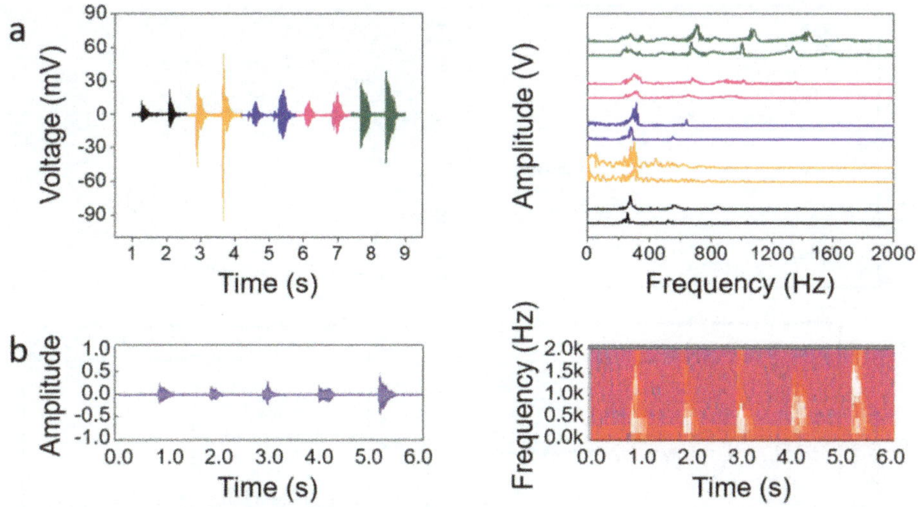

Figure 6.11. Voltage outputs of human voices saying 'one', 'two', 'three', 'four', and 'five', and the FFT-processed frequency spectrum recorded by (a) a sound sensor made of an electrospun PVDF nanofiber web and (b) a commercial microphone, respectively (SPL, 70–80 dB) [34].

From the voltage outputs and FFT-processed frequency spectra, the nanofibrous sound sensor exhibits an excellent ability to record human voices. This sensing ability effectively enables the sensors to distinguish between male and female voices [37]. The nanofiber sound sensor can also distinguish different musical instruments [38], such as a solo piano, violin, and flute.

The nanowire sound sensor with a Helmholtz resonator developed by Yang *et al* [18] can be used to localize acoustic sources with an error less than 7 cm in a plane (2×1.8 m^2). In their experiment, three nanowire sound sensors were arranged in an L shape and anchored in a 2D plane (figures 6.12(a) and 6.12(b)). When a balloon was burst on the 2D plane, the output voltages were detected by the nanowire sound sensor and the location of the sound source could be calculated by functions. Two correlation functions of the three output voltages were derived from the cross-correlation function (in figures 6.12(c) and (d)). The time lag at the highest peaks of their cross-correlation functions was used to estimate the time difference between the two output voltages. The time lags of the two cross-correlation functions were zero (AS_1 and AS_2, AS_2 and AS_3), which indicates the same distance from the acoustic source to the ASs. This calculation result is very consistent with experimental results.

With a broad working bandwidth, the multihole paper-based sound sensor produced by Fan *et al* [19] was capable of acting as an active self-powered microphone for human voice recording (figure 6.13(a)). Compared to other existing acoustic sensors, this paper-based sensor is rollable (figures 6.13(b) and (c)). The output of the rollable sensor with a symmetric circle pattern is independent of the sound wave's incident direction, as shown in figure 6.13(d). This feature makes the rollable sensor suitable to various sensing circumstances, such as theatric stage live recording, military surveillance, and omnibearing sensors.

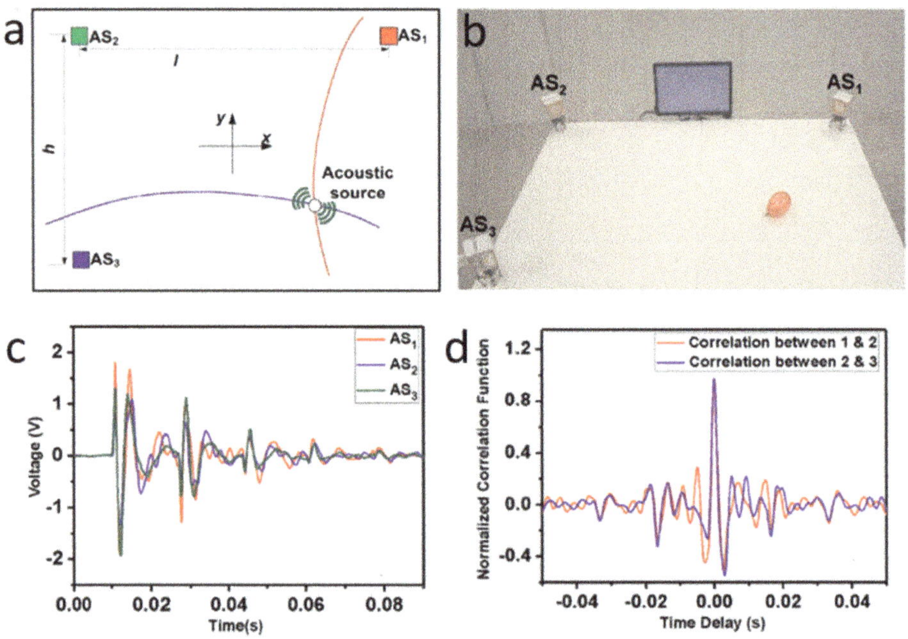

Figure 6.12. (a) Schematic illustration and (b) digital photo showing the working mechanism of sound sensors for sound localization, and (c) output signals from three sound sensors when a balloon burst. (d) Correlation functions of the output signals from different sound sensors [18].

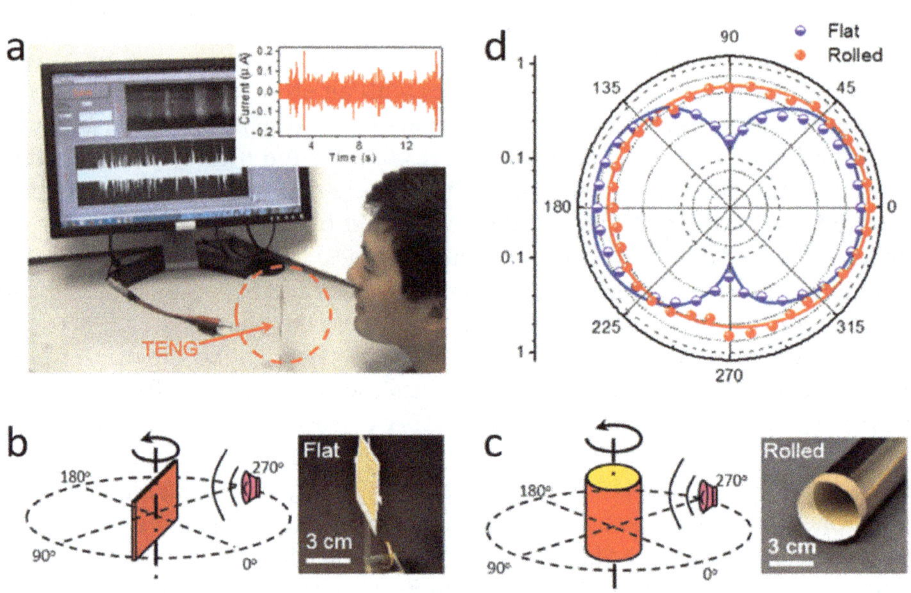

Figure 6.13. (a) Digital photo of a paper-thin triboelectric sound sensor as a self-powered microphone for sound recording. Inset is the output signals. Schematics showing the measurement of the directional patterns of the (b) flat and (c) rolled paper-thin triboelectric sound sensor. To the right are digital photos of the devices. (d) Directional patterns of the flat and rolled sound sensors [19].

6.5.2 Power supply

These acoustoelectric energy conversion devices that can generate a high level of electric signals can be directly used as power supplies for many microelectronic devices and electrochemical processes. Lang *et al* [40] reported a power generation device from a small piece of electrospun PVDF–TrFE nanofibers sandwiched with two pieces of PET electrode films, each of which had eight holes. Without prestorage in a capacitor, the voltage generated from this device can light up 27 commercial LEDs, run electrochemical polymerization reactions, and perform corrosion protection of metals (figure 6.14). Such an acoustoelectric generator may offer an effective solution to harvesting noise pollution into usable electricity for self-powered electronics.

Cui *et al* [17] reported a triboelectric sound energy harvester made of an electrospun PVDF nanofiber layer, two Al layers as triboelectric materials, and a copper mesh layer as the electrode. The energy harvester could instantaneously light up 138 LEDs under sound with an SPL 114 dB and a frequency of 160 Hz. After continuously working for 7 days and vibrating 100 million cycles, there was no decay in current outputs.

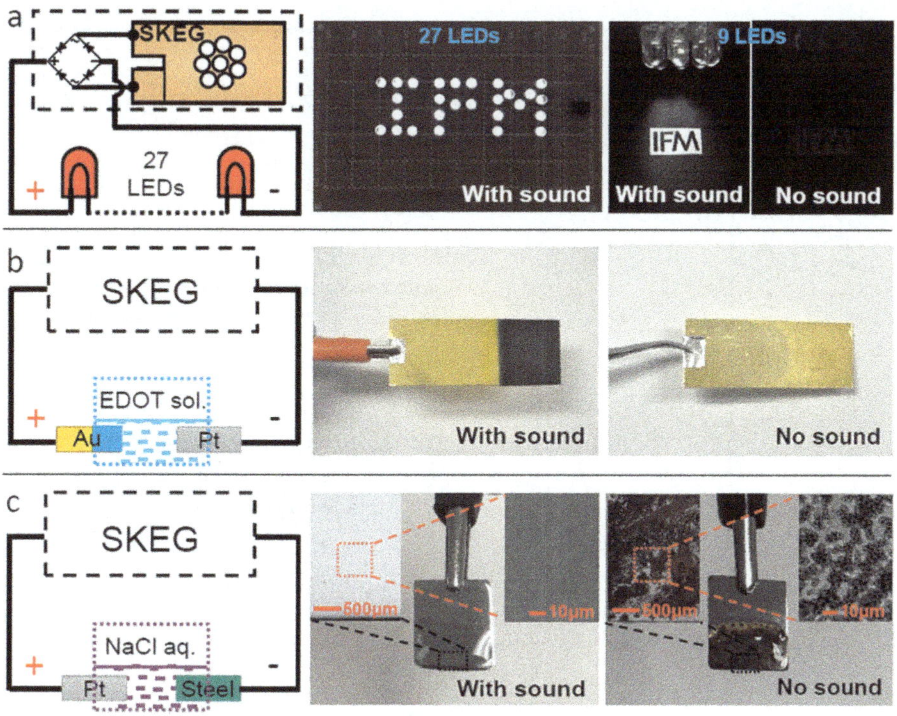

Figure 6.14. Electrospun nanofibrous acoustoelectric generators can (a) light up commercial LEDs, (b) run polymerization of 3,4-ethylenedioxythiophene (EDOT), and (c) perform metal corrosion protection (no energy storage device was used in any of the three demonstrations) [40].

Yang *et al* [18] used a Helmholtz resonator to enhance the vibration and triboelectric effect of the energy harvester. It produced a maximum electric power density of 60.2 mW m^{-2}, which directly and simultaneously lit up 17 commercial LEDs (figure 6.15).

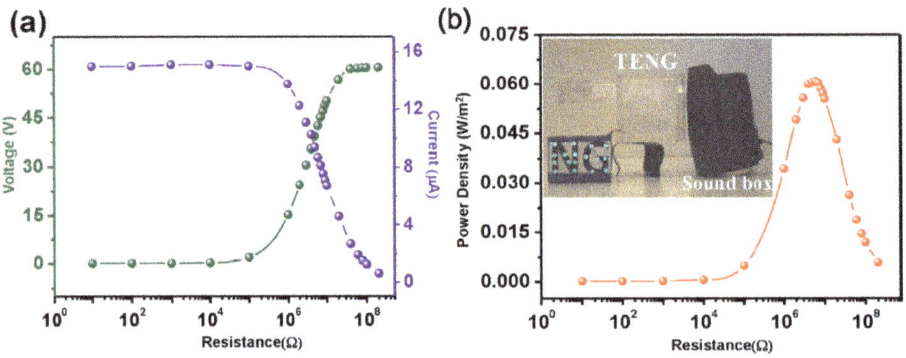

Figure 6.15. (a) Peak voltage and current output changes with the external load resistance, and (b) peak power output change with the external load resistance (inset: a sound energy harvester lighting up 17 LEDs) [18].

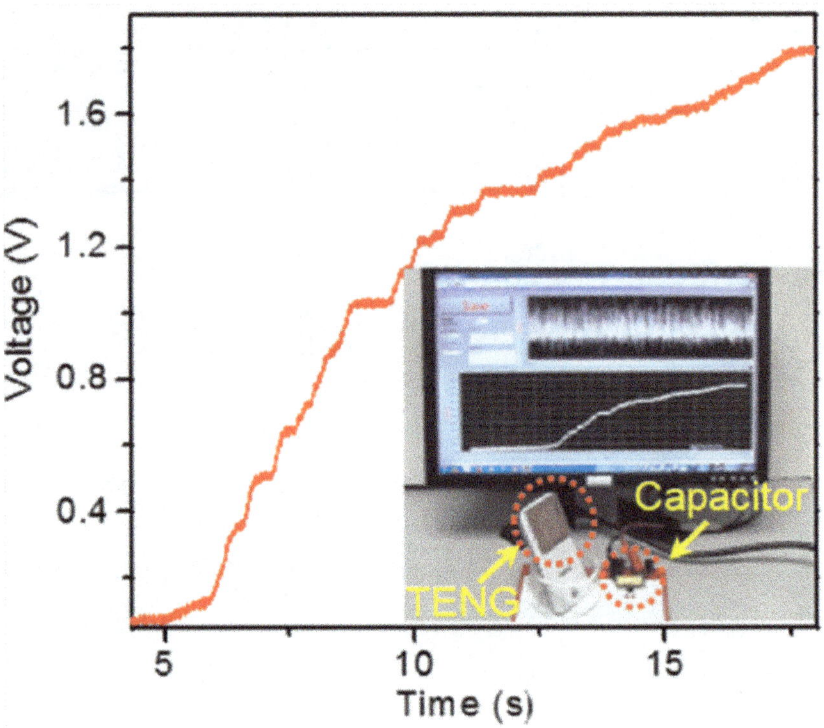

Figure 6.16. Capacitor charging response (insert: harvesting the sound energy from the cell phone via charging a 2 μF capacitor) [19].

Table 6.3. Comparison of the reported acoustoelectric transducers based on nanofibrous materials.

			Device parameters			Acoustic wave		Energy output						Ref.
Material	Design	Principle	Working unit size	D (cm^3)	R_{int} (kΩ)	SPL (dB)	f_r (Hz)	V_m (V)	I_m (μA)	Sensitivity (V Pa^{-1})	P (μW)	P_v (μW cm^{-3})	P_v/P_s (μW cm^{-3} · Pa^{-1})	
PVDF nanofibers	Electrospun PVDF nanofiber web clamped by two gold-coated PET films with one hole	Piezoelectric	$4 \times 3 \times 0.004$ cm^3	0.52		115	220	3.1		0.27				[34]
PVDF–TrFE nanofibers	Electrospun PVDF–TrFE nanofiber web clamped by two gold-coated PET films with eight holes	Piezoelectric	$4 \times 3 \times 0.002$ cm^3	0.46	470	115	210	14.5	28.5	1.3	141.3	306.5	27.3	[40]
PVDF composite nanofibers	Electrospun Ce^{3+}-doped PVDF/graphene composite nanofibers	Piezoelectric	3×4 cm^2			88	110	2.6						[36]
PVDF–HFP composite nanofibers	Electrospun Eu^{3+}-doped PVDF–HFP/graphene composite nanofibers	Piezoelectric				82				11				[37]
PVDF composite nanofibers	Electrospun PVDF–MoS_2 composite nanofibers	Piezoelectric	80 cm^2 × 0.015 cm	1.2		90		12	2×10^{-3}	19	2×10^{-4}	1.67×10^{-4}	2.64×10^{-4}	[38]
PVDF nanofibers	Electrospun PVDF nanofibers and Al layer with stainless steel mesh electrodes	Triboelectric	Vibrating membrane											
	$10 \times 10 \times 0.0005$ cm			114	160	45	450		20 200			[17]		
PMMA nanofibers	Au-coated aligned electrospun PMMA nanofiber mesh placed in a magnetic field	Electromagnetic	$0.8 \times 0.8 \times 0.0003$ cm^3											[39]

(*Continued*)

Table 6.3. *(Continued)*

			Device parameters			Acoustic wave		Energy output						Ref.
Material	Design	Principle	Working unit size	D (cm^3)	R_{int} (kΩ)	SPL (dB)	f_r (Hz)	V_m (V)	I_m (μA)	Sensitivity (V Pa^{-1})	P (μW)	P_v (μW cm^{-3})	P_v/P_s (μW cm^{-3} · Pa^{-1})	
ZnO nanowires	Aligned ZnO nanowires covered by a zigzag silicon electrode coated with platinum	Piezoelectric	Area: 0.02 cm^2; nanowires: density, 10^9 cm^{-2}; height, 0.0001 cm; diameter, 0.000 004 cm		3.56		Ultrasonic wave 41 000	7×10^{-4}	1.50×10^{-4}		1.00×10^{-6}			[41]
ZnO nanowires	Flexible PdAu-coated PES substrate acting as a top electrode and ZnO nanowire arrays on a GaN/ sapphire substrate	Piezoelectric	Nanowires: length, 0.001 cm; diameter, 0.000 015 cm			100	100	5×10^{-2}						[42]
ZnO nanowires	Laterally bonded single wires or vertically aligned nanowire arrays	Piezoelectric	Nanowires: length, 0.01 cm; diameter, 0.000 08 cm		100		100	5×10^{-4}	1.2×10^{-2}			5 pW cm^{-2}		[43]
PTFE nanowires	Nanowire-based PTFE thin film triboelectric energy harvester with a Helmholtz resonator and nanoporous Al electrode and copper electrode	Triboelectric	radius, 6.5 cm; nanowires: diameter, 0.000 0054 cm; length, 0.000 11 cm	574.72	6000	110	240	60.5	15.1					[18]
PTFE nanowires	Nanowire-based PTFE thin film triboelectric energy harvester with multihole copper electrode	Triboelectric	Thickness, 0.0025 cm	0.20	800	117	250	22	17		194	968.0	68.4	[19]
PES nanopillars	Core–shell nanostructures (PES in the core and As_2Se_3 in the shell) for building a 3D-printed multilayered device	Triboelectric	$17.5 \times 5 \times 4.9$ cm^3			87.98	10	53.65	0.245		52.5	5.47 mW m^{-2}		[44]

The multihole paper-based PTFE nanowire sound energy harvester reported by Fan *et al* [19] was demonstrated to harvest sound energy from a commercial cell phone when playing music or when a human is talking on the phone. As shown in figure 6.16, the recycled sound energy is capable of charging a commercial capacitor up to 1.8 V in about 12 s.

Table 6.3 lists most of the current reports on acoustoelectric transducers based on nanofibrous materials. Acoustoelectric transducers made of electrospun piezoelectric polymer nanofibers show high output voltage both based on piezoelectricity and triboelectricity principles. However, acoustoelectric transducers made of nanowires based on triboelectricity are better for energy conversion than those based on piezoelectricity.

6.6 Conclusions

Owing to their high aspect ratio, high specific surface area, and surface charge density, acoustoelectric sensors and energy harvesters based on piezoelectric nanofibrous materials are not only produced by simpler fabrication processes, but also show higher sensitivity and energy conversion efficiency than those based on conventional techniques and materials. Nanofibrous materials, especially for electrospun nanofibers, are lightweight and have great flexibility, which makes them particularly suitable for the emerging applications of wearable electronics.

Electrospinning is a one-step process to produce piezoelectric nanofibers with strong piezoelectricity. It is not a tedious multistep method involving mechanical stretching and high-voltage poling at an elevated temperature. At the same time, there is a wide range of materials on the triboelectric series that can be made into nanofibers for triboelectric energy harvesting. The research on acoustoelectric energy conversion is still at its early stage, with ongoing efforts in improving materials fabrication, transducer device structure, electrical outputs, and responsive frequency range, and it can be expected that many new applications will be explored with nanofibrous materials.

References

[1] Blumstein D T, Mennill D J and Clemins P *et al* 2011 Acoustic monitoring in terrestrial environments using microphone arrays: applications, technological considerations and prospectus *J. Appl. Ecol.* **48** 758–67

[2] Liang S Y, Hecker R L and Landers R G 2004 Machining process monitoring and control: the state-of-the-art *J. Manuf. Sci. Eng. Trans. ASME* **126** 297–310

[3] Stansfeld S, Haines M and Brown B 2000 Noise and health in the urban environment *Rev. Environ. Health* **15** 43–82

[4] Barnett G H, Kormos D W and Steiner C P *et al* 1993 Intraoperative localization using an armless, frameless stereotactic wand. Technical note *J. Neurosurg.* **78** 510–4

[5] Ghiurcau M V, Rusu C and Bilcu R C *et al* 2012 Audio based solutions for detecting intruders in wild areas *Signal Process.* **92** 829–40

[6] Lai T, Huang C and Tsou C 2008 Design and fabrication of acoustic wave actuated microgenerator for portable electronic devices *MEMS/MOEMS 2008. Symp. on Design, Test, Integration and Packaging of MEMS/MOEMS, 2008* pp 28–33
[7] Khan F U 2013 Electromagnetic-based acoustic energy harvester *2013 16th Int. Multi Topic Conf. (INMIC)* pp 125–30
[8] Bhat R 2014 Acoustic energy harvesting *Int. J. Sci. Res.* **3** 6
[9] Chen C-C, Yan W-Y and Wu Y-Y *et al* 2014 Investigating the optimum efficiency of acoustoelectric conversion plate devices *Int. J. Eng. Technol.* **4** 86–95
[10] Khan F U 2016 Electromagnetic energy harvester for harvesting acoustic energy *Sādhanā* **41** 1–9
[11] Scheeper P R, Van Der Donk A G H and Olthuis W *et al* 1994 A review of silicon microphones *Sens. Actuators,* A **44** 1–11
[12] Torkkeli A, Rusanen O and Saarilahti J *et al* 2000 Capacitive microphone with low-stress polysilicon membrane and high-stress polysilicon backplate *Sens. Actuators,* A **85** 116–23
[13] Consales M, Ricciardi A and Crescitelli A *et al* 2012 Lab-on-fiber technology: toward multifunctional optical nanoprobes *ACS Nano* **6** 3163–70
[14] Toda M and Thompson M L 2006 Contact-type vibration sensors using curved clamped PVDF film *IEEE Sens. J.* **6** 1170–7
[15] Xu J, Dapino M J and Gallego-Perez D *et al* 2009 Microphone based on polyvinylidene fluoride (PVDF) micro-pillars and patterned electrodes *Sens. Actuators,* A **153** 24–32
[16] Kim J 2009 Improvement of piezoelectricity in piezoelectric paper made with cellulose *Project Report* Inha University http://oai.dtic.mil/oai/oai?verb=getRecord&metadataPrefix=html&identifier=ADA510948
[17] Cui N, Gu L and Liu J *et al* 2015 High performance sound driven triboelectric nanogenerator for harvesting noise energy *Nano Energy* **15** 321–8
[18] Yang J, Chen J and Liu Y *et al* 2014 Triboelectrification-based organic film nanogenerator for acoustic energy harvesting and self-powered active acoustic sensing *ACS Nano* **8** 2649–57
[19] Fan X, Chen J and Yang J *et al* 2015 Ultrathin, rollable, paper-based triboelectric nanogenerator for acoustic energy harvesting and self-powered sound recording *ACS Nano* **9** 4236–43
[20] Khan F U and Izhar 2015 State of the art in acoustic energy harvesting *J. Micromech. Microeng.* **25** 023001
[21] Zhu G, Peng B and Chen J *et al* 2015 Triboelectric nanogenerators as a new energy technology: from fundamentals, devices, to applications *Nano Energy* **14** 126–38
[22] Lovinger A J 1983 Ferroelectric polymers *Science* **220** 1115
[23] Fang J, Wang X and Lin T 2011 Electrical power generator from randomly oriented electrospun poly(vinylidene fluoride) nanofibre membranes *J. Mater. Chem.* **21** 11088
[24] Mandal D, Yoon S and Kim K J 2011 Origin of piezoelectricity in an electrospun poly (vinylidene fluoride-trifluoroethylene) nanofiber web-based nanogenerator and nano-pressure sensor *Macromol. Rapid Commun.* **32** 831
[25] Persano L, Dagdeviren C and Su Y *et al* 2013 High performance piezoelectric devices based on aligned arrays of nanofibers of poly(vinylidenefluoride-co-trifluoroethylene) *Nat. Commun.* **4** 1633
[26] Wang Y R, Zheng J M and Ren G Y *et al* 2011 A flexible piezoelectric force sensor based on PVDF fabrics *Smart Mater. Struct.* **20** 045009

[27] Ren G, Cai F and Li B *et al* 2013 Flexible pressure sensor based on a poly(VDF-TrFE) nanofiber web *Macromol. Mater. Eng.* **298** 541–6
[28] Fang J, Niu H and Wang H *et al* 2013 Enhanced mechanical energy harvesting using needleless electrospun poly(vinylidene fluoride) nanofibre webs *Energy Environ. Sci.* **6** 2196–202
[29] Li B, Xu C and Zheng J *et al* 2014 Sensitivity of pressure sensors enhanced by doping silver nanowires *Sensors* **14** 9889–99
[30] Mandal D, Henkel K and Schmeisser D 2014 Improved performance of a polymer nanogenerator based on silver nanoparticles doped electrospun P(VDF-HFP) nanofibers *Phys. Chem. Chem. Phys.* **16** 10403–7
[31] Adhikary P, Garain S and Mandal D 2015 The co-operative performance of a hydrated salt assisted sponge like P(VDF-HFP) piezoelectric generator: an effective piezoelectric based energy harvester *Phys. Chem. Chem. Phys.* **17** 7275–81
[32] Chang C, Tran V H and Wang J *et al* 2010 Direct-write piezoelectric polymeric nanogenerator with high energy conversion efficiency *Nano Lett.* **10** 726–31
[33] Hansen B J, Liu Y and Yang R *et al* 2010 Hybrid nanogenerator for concurrently harvesting biomechanical and biochemical energy *ACS Nano* **4** 3647–52
[34] Lang C, Fang J and Shao H *et al* 2016 High-sensitivity acoustic sensors from nanofibre webs *Nat. Commun.* **7** 11108
[35] Duhamel P and Vetterli M 1990 Fast Fourier transforms: a tutorial review and a state of the art *Signal Process.* **19** 259–99
[36] Garain S, Jana S and Sinha T K *et al* 2016 Design of *in situ* poled Ce^{3+}-doped electrospun PVDF/graphene composite nanofibers for fabrication of nanopressure sensor and ultrasensitive acoustic nanogenerator *ACS Appl. Mater. Interfaces* **8** 4532–40
[37] Adhikary P, Biswas A and Mandal D 2016 Improved sensitivity of wearable nanogenerators made of electrospun Eu^{3+} doped P(VDF-HFP)/graphene composite nanofibers for self-powered voice recognition *Nanotechnology* **27** 495501
[38] Maity K, Mahanty B and Sinha T K *et al* 2017 Two-dimensional piezoelectric MoS_2-modulated nanogenerator and nanosensor made of poly(vinlydine fluoride) nanofiber webs for self-powered electronics and robotics *Energy Technol.* **5** 234–43
[39] Zhou J, Li B and Liu J *et al* 2018 Highly-damped nanofiber mesh for ultrasensitive broadband acoustic flow detection *J. Micromech. Microeng.* **28** 095003
[40] Lang C, Fang J and Shao H *et al* 2017 High-output acoustoelectric power generators from poly(vinylidenefluoride-co-trifluoroethylene) electrospun nano-nonwovens *Nano Energy* **35** 146–53
[41] Wang X, Song J and Liu J *et al* 2007 Direct-current nanogenerator driven by ultrasonic waves *Science* **316** 102–5
[42] Cha S N, Seo J S and Kim S M *et al* 2010 Sound-driven piezoelectric nanowire-based nanogenerators *Adv. Mater.* **22** 4726–30
[43] Lee M, Yang R and Li C *et al* 2010 Nanowire-quantum dot hybridized cell for harvesting sound and solar energies *J. Phys. Chem. Lett.* **1** 2929–35
[44] Kanik M, Say M G and Daglar B *et al* 2015 A motion- and sound-activated, 3D-printed, chalcogenide-based triboelectric nanogenerator *Adv. Mater.* **27** 2367–76

Chapter 7

Flexible and stretchable nanofibrous piezo- and triboelectric wearable electronics

Roohollah Bagherzadeh, Mohammad Sajad Sorayani Bafqi, Nikoo Saveh Shemshaki, Zohreh Moarref, S Ghasemi-Nezhad, F Maleki and Parisa Fakhri

7.1 Introduction

Energy can be scavenged from a variety of sources from the ambient environment, including natural sources, such as wind, ocean waves, and sunlight; vibration from mechanical energy; Joule heating of electronic devices in the form of the thermal energy; and electromagnetic energy from inductor coils and electronic devices. Moreover, a great amount of energy is available for harvesting, such as mechanical energy due to vibrations from body movement, respiration, and blood flow in vessels; thermal energy from body heat; and biochemical energy generated during physiological processes and metabolic reactions. These all have potential utilization in self-powered devices [1]. Renewable power can be obtained by generating electrical energy from different types of natural sources, including light, thermal, and dynamic energy. Thermal energy can be conveniently transduced into electrical energy by the Seebeck effect but solar power is probably the most well-known method in this context. Solar cells provide excellent power density in direct sunlight, whereas they perform poorly in indirect sunlight, indoor lights or in conditions where cells are infected with contamination [2, 3]. Therefore, thermal energies are highly dependent on time, weather conditions, and location, which may not all be accessible. Therefore, other types of natural sources of energy, such as human motion-based energy harvesting, have attracted tremendous attention due to growing popularity as independently sustainable power sources for portable smart electronics applications [4]. The kinetic energy is present in the form of vibrations, and random displacements or forces can be converted into electrical energy using electromagnetic, piezoelectric, or triboelectric mechanisms [2, 5, 6]. The amount of electrical energy generated by these approaches depends on both the quantity of energy in the environment and the efficiency of the power conversion devices [2, 7].

Based on the findings of recent studies on flexible mechanical and electrical sensing devices, it has been shown that there is a great and wide application potential, such as in displays, robotics, and energy harvesting [8–10]. To achieve these advances, various solutions have been considered through the use of electrical active materials on flexible polymeric-based materials to obtain different functions, such as high bendability [11], ultra-sensitivity [12], transparency [13], and human–device interfaces [14]. In general, a number of elements, such as nanotubes and nanofibers, need to be integrated in flexible textiles materials so that mechanical and electrical signals can be measured [8]. Moreover, various fabrication methods can be utilized with polymeric materials to achieve high flexibility in sensing devices, such as low-temperature deposition, solution processing, transfer printing, electrospinning, and nano/micro molding [6, 8]. Here, polymeric materials such as nanofibers play an important role as a biocompatible, less irritating support layer [8]. Furthermore, the proper use of a polymer becomes more important when the electronic device becomes more complex and multifunctional [15]. In general, mechanical flexibility or stretchability is the key to these technologies; for the device to be easily bent, stretched, and adapted into a favorable shape are some of the final goal in this area [16]. Less irritating conformal contact on the skin [17] and biocompatible sealing for implantable devices are only some of the most important issues for the application of these materials [18].

Electronic textiles are fabrics that feature electronics and interconnections woven into them in such a way that their physical flexibility cannot be replaced with other existing electronic manufacturing techniques [19]. It is expected that in the near future textile-based electronics will contain both functionality and connectivity in a flexible and stretchable substrate. These devices will leave the laboratory scale and enter the market place, and will be utilized to achieve a broad range of outcomes in different environments [20]. To achieve full textile-based electronic systems, various components, such as transistors, sensors, displays, and batteries, should be developed and integrated into textile substances. It is also well documented that using kinetic energy harvesting systems can be very helpful to provide the required electrical energy for these systems. Different techniques and technologies have been investigated to power these systems:

- solar energy harvesting systems;
- electrostatic energy harvesters;
- thermoelectric-based structures;
- pyroelectric-based harvesters;
- piezoelectric and triboelectric energy harvesters.

7.1.1 Solar energy harvesting systems

Solar energy is the most abundant exploitable renewable source of energy. A photovoltaic (PV) process should be utilized by using semiconductor materials with the PV effect to convert solar radiation into electricity. Despite the considerable capacity of solar energy, PV technology only accounts for 0.1% of electricity generation so far. PV technology has been growing and expanding rapidly [1]. The development of new-generation PV technologies that operate with enhanced efficiency at lower costs is a hot research topic. The development of nanotechnology has also enabled researchers to develop advanced materials and to create innovative solutions to the drawbacks related to traditional materials. Electrospun polymer and ceramic nanofibers are a specific classification of materials used in these applications [21]. Considerable studies have been conducted to develop not only flexible but also stretchable nanofibrous solar energy harvester. For instance, flexible electrospun TiO_2–SiO_2 mats can achieve a 1.3–3.4 mm radius of curvature even after heat treatment [22]. In addition, hybrid ceramic/polymer fibers can be easily fabricated by different techniques, such as the electrospinning method. The advantages of electrospinning methods aid, for instance, the fabrication of composite nanofibers in which the ceramic nanoparticles such as hydroxyapatite can be randomly decorated on PA6 fibers without agglomeration [23]. Nevertheless, due to the fragility and rigidity of the component, this method cannot be used to produce flexible energy harvesters.

7.1.2 Electrostatic energy harvesters

In general, a capacitor consists of two plates that are electrically isolated from each other by an insulator such as air or vacuum. The charging of the plates by a battery of voltage V creates equal but opposite charges on the plates, Q, leading to a storage of the charge when the voltage source is disconnected [2].

The definition of the capacitance of such a capacitor is given by

$$C = \frac{Q}{V}$$

where C is the capacitance in Farads, Q is the charge on the plate in Coulombs, and V is the voltage in volts. For a parallel plate capacitor, C can also be given by the following equation:

$$C = \varepsilon \frac{A}{d}$$

where ε is the permittivity (F m^{-1}) of the insulator material, A is the area of the plates (m^2), and d is the gap distance between two plates (m).

The above equation can be also expressed in terms of the dielectric constant, $\kappa = \varepsilon/\varepsilon_0$, of the insulator material:

$$C = k\varepsilon_0 \frac{A}{d}$$

where the ε_0 is the permittivity of free space.

The voltage across a parallel plate capacitor is given by

$$V = \frac{Qd}{\varepsilon_0 A}.$$

The energy stored in a capacitor, with plate charge Q and potential difference V, is given by

$$E = 0.5QV = 0.5CV2 = 0.5Q2/C.$$

The perpendicular force between the plates is given by following equation if the charge on the plates is held constant:

$$F = 0.5Q2d/\varepsilon A.$$

Furthermore, the perpendicular force between the plates is given as follows if the voltage between the plates is held constant:

$$F = 0.5\varepsilon AV2/d2.$$

The work done against the electrostatic force between the plates provides the harvested energy.

Electrostatic generators can be classified into three types [2]:
(1) in-plane overlap varying;
(2) in-plane gap closing;
(3) out-of-plane gap closing.

These three approaches can be operated in either charge-constrained or voltage-constrained cycles. It has been proved that the voltage-constrained cycling provides more energy than the charge-constrained approach. However, by incorporating a capacitor in parallel with the energy harvesting capacitor, the energy from the charge-constrained system can approach that of the voltage-constrained system as the parallel capacitance approaches infinity. This parallel capacitor efficiently constrains the voltage on the energy harvesting capacitor [2]. Roundy [24] states that in-plane gap closing offers the highest power output with an optimized design producing 100 μW cm^{-3}; out-of-plane gap closing produces the next highest, followed by in-plane overlap varying. Maximum power generation occurs for very small dielectric gaps. Recently, Lue *et al* [25] created a flexible electrostatic energy harvester based on an electret made of an electrospun PVDF–PTFD nanofibrous scaffold covered by chemical vapor deposition (CVD)-deposited parylene C. They also propose to cover electrospun nanofibrous material with a parylene C layer in order to offer stable charge storage. The maximum peak power obtained is 45.6 μW, corresponding to 48 nJ/cycle.

7.1.3 Thermoelectric-based structures

The thermoelectric phenomenon provides the opportunity for energy harvesting from the environment and particularly from the body to generate electricity that can power medical devices, implants, and other electrical devices. The thermoelectric phenomenon was first observed in 1821 when Seebeck found that a voltage was generated at room temperature as a result of the temperature gradient between two ends of bismuth wires and copper (Cu) wires [26]. Afterward, in 1834, Peltier observed that by passing an electric current through bismuth and Cu wires at room temperature, a temperature difference would be generated [26, 27]. The Seebeck effect and Peltier effect provide a basis for thermoelectric devices. Figure 7.1(a) shows the working principle of thermoelectric equipment based on the Seebeck effect. In this setup, a temperature gradient will occur between two junctions of dissimilar conductive or semiconductive materials when the top ones are heated and the bottom junctions are cooled. Subsequently, electron–hole pairs are generated at the heated end by absorbing heat, then the electron–hole pairs recombine and release heat at the cold end. Due to the mobility of electron–hole pairs, the Seebeck voltage generates between the two ends, resulting in current flow. Figure 7.1(b) presents the Peltier effect as the process of thermoelectric cooling. By applying a voltage between the junctions of two different conductive or semiconductive materials, electron–hole pairs are generated adjacent to the junctions and flow away, resulting in heating of the end of one junction and cooling on the other end.

To evaluate the efficiency of thermoelectric devices, a dimensionless formula is used:

$$ZT = S^2\sigma T/\kappa$$

where S is the Seebeck coefficient, σ is electrical conductivity, T is the absolute temperature, and κ is the thermal conductivity. Accordingly, a higher ZT value leads to higher performance of thermoelectric materials at a specific temperature. The reduced dimension of electrospun nanofibers will decrease the thermal conductivity without any change in electrical conductivity, which would increase the ZT value [28, 29].

In 2010, polyaniline nanotubes doped by β-naphthalene sulfonic acid were shown to have a significant high Seebeck coefficient compared to the reference sample without a nanostructure [30]. This result has been related to the concurrent decrease in the thermal conductivity and increase in the electrical conductivity, which can be

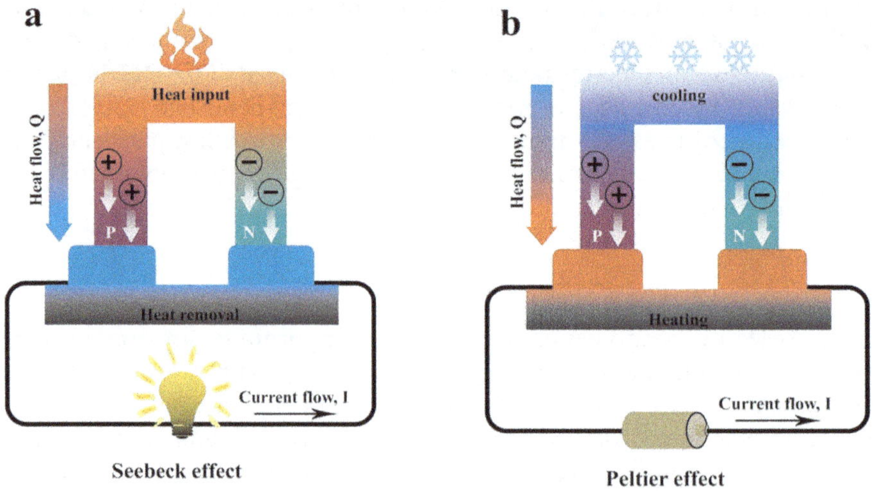

Figure 7.1. Schematic of the working principle of thermoelectric devices. (a) Seebeck effect; and (b) Peltier effect [26].

attributed to the elongated and nanostructured shape of the conductive structure [30]. Similar observations have been reported by Taggart *et al* [31]. In this study, long poly(3,4-ethylene dioxythiophene) (PEDOT) nanowires have been compared to thin films in terms of their thermoelectric properties. The fabricated nanowires were found to exhibit a higher Seebeck coefficient value than films [31].

Recently, the usage of organic conductive polymers with favorable thermoelectric properties has increased [32]. The unique properties of these polymers such as facile processability, low thermal conductivity, low density, and good transport properties, can address some limitations of inorganic materials including the high cost of low-abundance materials, toxicity, and difficult fabrication process [33]. The most commonly investigated conductive polymers include polypyrrole, polythiophene, PEDOT, and polyaniline. Although the electrical conductivity of conductive polymers is comparable with inorganic thermoelectric materials, their Seebeck coefficients are lower than those of inorganic thermoelectric materials [34].

The progress of thermoelectric materials has developed significantly. However, few studies have been conducted on thermoelectricity of polymeric nanofibers. In 2010, Yin *et al* [35] fabricated $Ca_3Co_4O_9$/poly(vinyl pyrrolidone) electrospun nanofibers followed by annealing to achieve nanocrystalline thermoelectric ceramics. The results showed a lower thermal conductivity, higher electrical conductivity, and enhanced Seebeck coefficient. In order to increase the electrical conductivity of conductive polymers and subsequently enhance their Seebeck coefficients, some studies investigated the incorporation of carbon-based composites with conductive polymers. In 2012, enhanced thermoelectric properties were found in composite nanofibers made of CNTs and polyaniline, fabricated by electrospinning and *in situ* polymerization [34]. Slobodian *et al* showed that the flexible thermoelectric film composed of carbon nanofiber (with a fiber diameter of 150 nm and length of 10 μm) and ethylene–octene copolymer (EOC) exhibits a higher power value ($S^2\sigma$) compared to MWCNTs/EOC films [33].

The effect of poly(3-hexylthiophene) (P3HT) nanofibers on decreasing the thermal conductivity of the structure in comparison with P3HT film was investigated by Hiura *et al* [29]. The study showed no significant difference in power value and electrical conductivity between the P3HT nanofibrous mat and the P3HT film. However, the results exhibited a lower thermal conductivity of the nanofibrous mat [29]. In a recently conducted study, Zhang *et al* fabricated lead telluride hollow nanofibers by electrospinning with different concentrations of lead [36]. Based on the results, by decreasing the content of lead, the electrical conductivity decreased. Moreover, thermopower increased with temperature in nanofibers with higher lead contents.

7.1.4 Pyroelectric-based harvesters

At micro/nanoscale, where a spatial temperature gradient (dT/dx) of thermoelectric devices is more difficult to realize, pyroelectric materials that require a temporal temperature gradient (dT/dt) are considered more desirable for thermal to electrical energy conversion [37]. The stability of many pyroelectric materials at high temperature (~1200 °C) indicates their merit over thermoelectrics materials for energy harvesting from high-temperature sources. Moreover, pyroelectric devices require little or no maintenance since they do not include any moving parts compared to piezoelectric energy harvesting devices [38]. The pyroelectric effect is described as the generation of an electric current in materials with a noncentrosymmetric crystal structure when a temporal temperature gradient (dT/dt) is applied [39]. Some common pyroelectric materials include triglycine sulfate, PVDF, lithium tantalate, gallium nitride, and ZnO [39]. Nanofibers are attractive structures for flexible pyroelectric energy harvesting devices; however, few studies have investigated the pyroelectric application of electrospun nanofibers [40]. PVDF and its copolymer PVDF–TrFE are the most common pyroelectric materials that are produced using electrospinning [38, 41]. When PVDF is poled or mechanically stretched, the relative content of the β phase, which contributes to the pyroelectric and piezoelectric effects, increases [42]. Therefore, a higher content of the β phase is desirable for higher energy harvesting efficiency [38]. Recently, You *et al* [43] fabricated a flexible and lightweight nanogenerator based on pyroelectric and piezoelectric effects for harvesting thermal and mechanical energies. In this study, PVDF electrospun nanofibers were fabricated as an active layer. To improve the flexibility of the nanogenerator, thermoplastic polyurethane electrospun nanofibers used as a substrate and a conductive PEDOT:PSS–PVP nanofibrous layer along with a CNT layer used as the electrodes for the fabricated nanogenerator. Based on the results, by applying a thermal gradient and strain simultaneously, a piezoelectric–pyroelectric electrical current of the nanogenerator was obtained. This study shows the potential of the flexible hybrid piezoelectric–pyroelectric nanogenerators for self-powered electronic textiles.

7.1.5 Piezoelectric and triboelectric energy harvesters

Vibration-based mechanical energy is available in the environment and is more accessible than solar and thermal energy. These vibrations, with broad frequencies ranging from a few hertz to several kilohertz, are abundant in the human environment. The energy density from these vibrations ranges from a few microwatts to milliwatts per cubic centimeter [5, 44, 45]. Plenty of research has been performed in the last few years regarding different aspects of PENGs and TENGs to convert biomechanical energy into electrical energy [5, 46]. A few stretchable TENGs were successfully reported by different researchers, although their use is difficult in small-scale wearable systems due to their large dimension factor [47]. PENGs on the other hand, possess a small dimension factor, making them appropriate for nano-to-microsystems [48, 49]. PENGs also have a simple mechanism and modest design that can be applied in energy harvesting devices, particularly for biomedical applications. Despite their simple working mechanism, which makes them favorable for human kinematic energy harvesting and implantable devices, PENGs suffer from lower output power compared to TENGs [50]. Moreover, piezoelectric materials are brittle by nature and intrinsically not suitable for stretching. These types of nanogenerators also use two electrodes, which are not stable under stretching conditions. These issues are the foremost drawbacks affecting their practical use as a stretchable PENG [46]. Meanwhile, the harvesting of mechanical energy by piezoelectric materials has received enormous attention due to their ability to directly convert mechanical energy into electricity. Traditionally, PZT, has been the dominant material used for mechanical energy harvesting [1, 6]. The piezoelectric ZnO nanowire arrays were also demonstrated to convert mechanical energy into electrical energy [51]. Furthermore, a scalable sweeping–printing method was proposed to fabricate stretchable and high-output nanogenerators; an open-circuit voltage up to 2.03 V, close-circuit current of 100 nA, and peak output power density of $\sim$11 mW cm^{-3} were achieved [52]. Then, with the aim of developing flexible PENGs, PVDF was considered as an alternative material for piezoelectricity; nanofibers with diameters ranging from several nanometers to a few hundred nanometers were produced by electrospinning to form the fiber-based PENGs. Thanks to the polymer-based piezoelectric materials, fiber-based energy harvesters have particularly attracted more attention because of their light weight, low cost, and mechanical flexibility [53]. Of all the piezoelectric polymers, PVDF and its copolymers are most widely used [5, 54]. Domain switching and associated ferro/piezoelectric properties of the electrospun PVDF fibers have shown well-defined ferroelectric and piezoelectric properties [55]. Near-field electrospinning has also achieved the direct-write and precise patterning to deposit substantial nanofibers in a continuous and controllable process [56]. For functional material such as piezoelectric PVDF, a comparative study utilizing both continuous near-field electrospinning and conventional electrospinning processes of electrospun PVDF-based membranes was conducted in terms of fiber dimension, processing parameters, and crystalline structure [57]. Stretchability with good performance are the top two goals to achieve in this research. Various sophisticated designs have been studied using piezoelectric nanoparticles as fillers; the structural design of materials to achieve better performances and flexibility has also been studied. Chun *et al* demonstrated the fabrication of a stretchable piezoelectric composite based on piezoelectric hemispheres embedded in PDMS, but the final device was not stretchable due to the use of an indium tin oxide-coated poly(ethylene terephthalate) substrate [58]. Likewise, Duan *et al* reported a stretchable piezoelectric utilizing ZnO nanoribbons and the buckling of PVDF nanofibers, but the outputs were very low [59]. A wavy pattern of PZT nanoribbons was also transferred on a stretchable substrate to fabricate a stretchable piezoelectric nanogenerator (SPENG), but the complexity of their production and their low energy conversion efficiency make these methods not difficult to employ in practical applications [60]. Jeong *et al* also demonstrated an SPENG using piezoelectric fillers embedded into an elastomer-based substrate with silver nanowires as an electrode, showing output under stretching and high mechanical durability; the reported SPENG was operated at a high strain of 200%, but the toxicity of lead-based active materials may limit its application for a body-attachable and implantable harvester [61]. Additionally, under lower strains similar to human body motions, where the stretching range is up to 30%, the output is very low (0.7 V and 50 nA at 50% stretching), making practical biomechanical application difficult [62]. In general, except for the work by Jeong *et al* [61], these methodologies showed conceptual stretchable piezoelectric devices with no practical operation under stretching modes. Furthermore, the insufficient stretchability of the reported SPENGs, less than 15%, limits their use for human kinematic energy harvesting, specifically from the motion of joints like elbows or knees, where large strains are generated and high stretchability, conformability, and efficiency are demanded [46, 60].

It is well known that the performance of SPENGs is strongly influenced by the stretchability of their two main components, the electrode and the active piezoelectric materials [63]. Both the electrode and active materials must be stretchable to fabricate a fully stretchable PENG with acceptable performance under stretching mode. Keen research has been performed to make stretchable, stable, and mechanically durable electrodes. These efforts can be mainly categorized into two different classes. In the first category, different types of stretchable electrodes have been reported, in which specifically conductive particles or fillers have been loaded in polymer-based substrates [63, 64]. The remaining challenge here is the limited mechanical and electrical stability of these types of composite electrodes under the stretching mode [65]. The second approach relies on conventional flexible electrode materials, such as those with waving [66], buckling [67], nano-mesh [68], and serpentine structures [69]; this approach produces both stretchability and reasonable performance [70]. The main drawbacks of the second approach is the issue of sophisticated fabrication processes such as vacuum processes, etching, lithography, and the use of expensive materials. One of the best examples of these modification processes to achieve the stretchable PENG is reported by Saddique *et al* [46]. They reported an efficient approach to make a stretchable and robust SPENG based on a stretchable graphite electrode and a stacked mat of piezoelectric nanofibers. They fabricated stretchable electrodes by placing the graphite on a 3D micropatterned stretchable substrate. They believed that this fabrication process would make this device suitable for self-powered wearable sensing systems.

7.2 Recent progress in electrospun nanofibers and piezo–triboelectric nanogenerators

Electrospun nanofibers with their specific characteristics and properties have recently gained lots of attention, especially in driving the advent and growth of wearable electronics, which in turn resulted in a wide range of functional devices: piezo–triboelectric nanogenerators, sensors, batteries, capacitors, actuators, transistors, etc. The process of energy harvesting and storage comprises charge generation, transport, and collection, as well as diffusion. In this sense, electrospun nanofibers can function as an electron expressway, shorten the carrier transportation path, and enhance the carrier collection ability. The combination of these characteristics provides these materials with great potential to be used in the next generation of energy harvesters [71–73].

Nanofiber-based generators are a promising power supply for flexible micro/nanoscale devices, capable of directly harvesting energy from ambient sources. Since the invention of the first nanogenerator by Wang *et al* in 2006 [51], several works have been performed with great efforts to design, develop, and improve wearable nanogenerators. They may be used for a wide variety of applications due to their exclusive properties such as flexibility, sensing susceptibility, and their ability of integration with other related devices. The possibility to use elastic and deformable materials leads to broadening device design options. PENGs and TENGs have a great potential in flexible and stretchable electronics due to their inherent mechanical flexibility, lightness, and compactness [64]. In this context, by combination of electrospun PVDF–TrFE and a PDMS/MWCNT composite layer, a small-scale flexible piezo–triboelectric hybrid nanogenerator can be fabricated and used as a finger-motion sensor [74]. Due to the faster electron flow of MWCNTs and the proper microstructure of the composite layer, a better output performance is expected [74]. It was also claimed that MWCNT-doped PDMS contributes to tuning the internal resistance and increasing the output voltage. Electronic devices possessing stretchability in response to high strains are in frequent demand for the detection of biomechanical motion and realizing self-powered sensing [75, 76]. Using electrospun nanofibers of polyurethane (PU) loaded with barium titanate (BT) nanoparticles stacked together with PVDF–TrFE nanofibers and a graphite electrode on the 3D micropatterned PDMS substrate, an omnidirectionally stretchable, robust, and environmentally friendly nanogenerator nanofibrous stretchable piezoelectric nanogenerator (nf SPENG) was produced to harvest mechanical energy under stretching, bending, tapping, crumpling, and biomechanical motion of the human body [75]. The 3D micropatterned substrate provides the device with adequate stretchability along with stability and mechanical robustness to the conventional electrode materials. Meanwhile, the combination of a PU–BT nanocomposite with PVDF–TrFE nanofibers results in a high piezoelectric output. According to Siddiqui *et al*'s [75] investigation, such a designed structure is capable of withstanding mechanical deformation during stretching, and maintains its performance up to 9000 vigorous stretching cycles at a strain of 30%.

The high surface area-to-volume ratio, flexibility in materials selection, and ease of incorporating nanoparticles make electrospun nanofibers suitable for sensor applications. Sensors made from electrospun nanofibers often have more susceptibility than those made of the same material with the film casting method [77]. A hybrid nanogenerator, made of PVDF–TrFE electrospun nanofibers as the piezoelectric active layer, electrospun PU nanofibers coated by silver nanowires and conductive CNTs as the fiber-based electrodes, and micropatterned PDMS as the triboactive layer and encapsulating agent, can be attached to a soft surface to harvest tapping or continuous pressure energies. Owing to the high sensitivity of piezoelectric materials, the device can also be used as a self-powered sensor for real-time monitoring of human physiological signals, such as respiratory signals and the radial artery pulse. The separate structure of the device enables it to generate high piezoelectric and triboelectric outputs simultaneously (figure 7.2) [76].

Another method used to produce nonwrinkled, highly stretchable piezoelectric devices, such as sensors and energy converters/providers, is mechano-electrospinning (MES). This method was presented by Duan *et al* [59]. The noticeable difference between MES and near-field electrospinning [78]/traditional electrospinning [79] is that MES has a high-speed motion platform exerting a large dragging force on the fiber, which helps attain positioning. There are some advantages of MES for

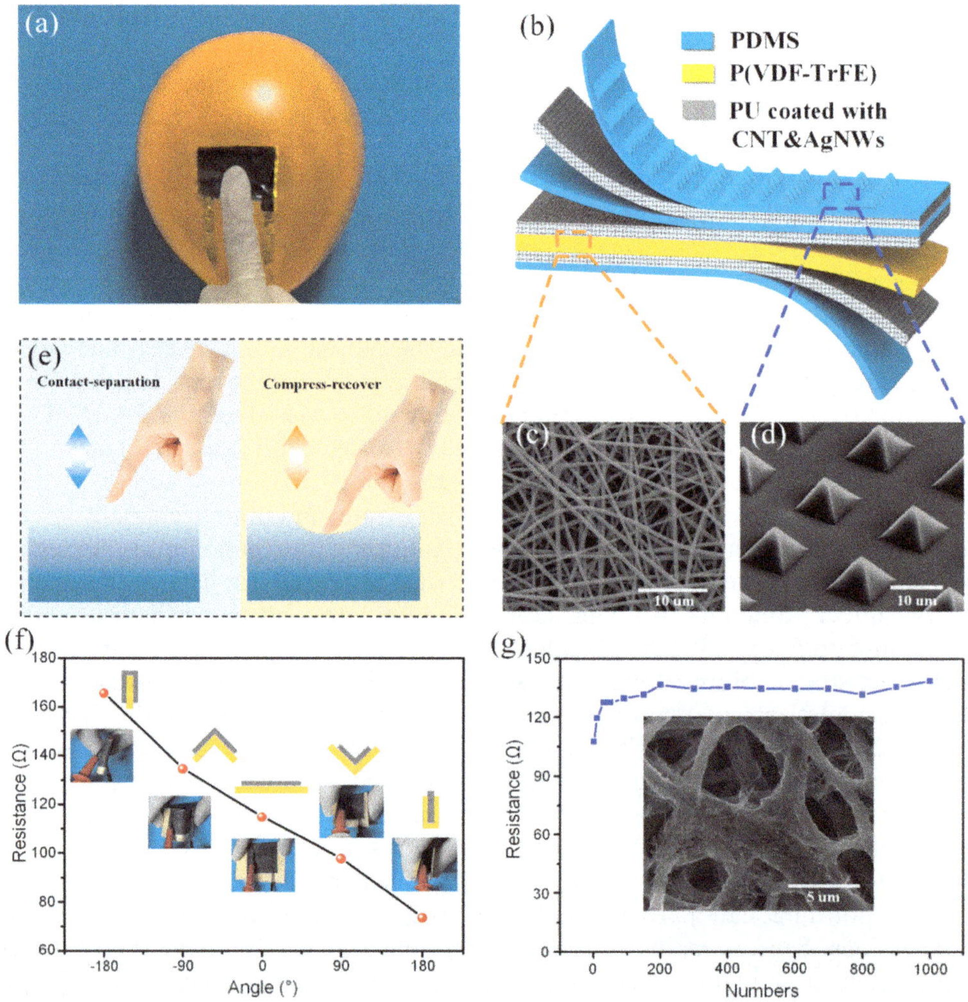

Figure 7.2. Overview of the hybrid nanogenerator. (a) Photo of the device attached to the surface of a balloon. (b) Schematic diagram of the device. (c) SEM image of the electrospun PVDF–TrFE nanofiber mat. (d) SEM image of the pyramid structure of PDMS. (e) Two mechanical processes when a soft body is pressed. (f) Conductivity measurement of the PU fiber-based electrode under different bending angles. The insets show the bent state of the electrode. (g) Stability test of the electrode under fold/unfold deformation. The inset is a SEM image of the fiber-based electrode [76].

fabricating stretchable piezoelectric devices, such as the ability to precisely direct-write piezoelectric nanofibers onto the surface of elastomers like PDMS, and the ability to improve the polymer polar phase in one step [80]. This can happen because the relatively large mechanical drawing force during production can overcome the electric repulsion resulting from the insulated elastomer substrate [81]. Being able to tune the morphology of electrospun nanofibers from a ribbon to a cylinder by adjusting the process parameters, MES can form controllable in/out-of-surface buckling. Accordingly, Duan *et al* [59] introduced a stretchable nanogenerator with

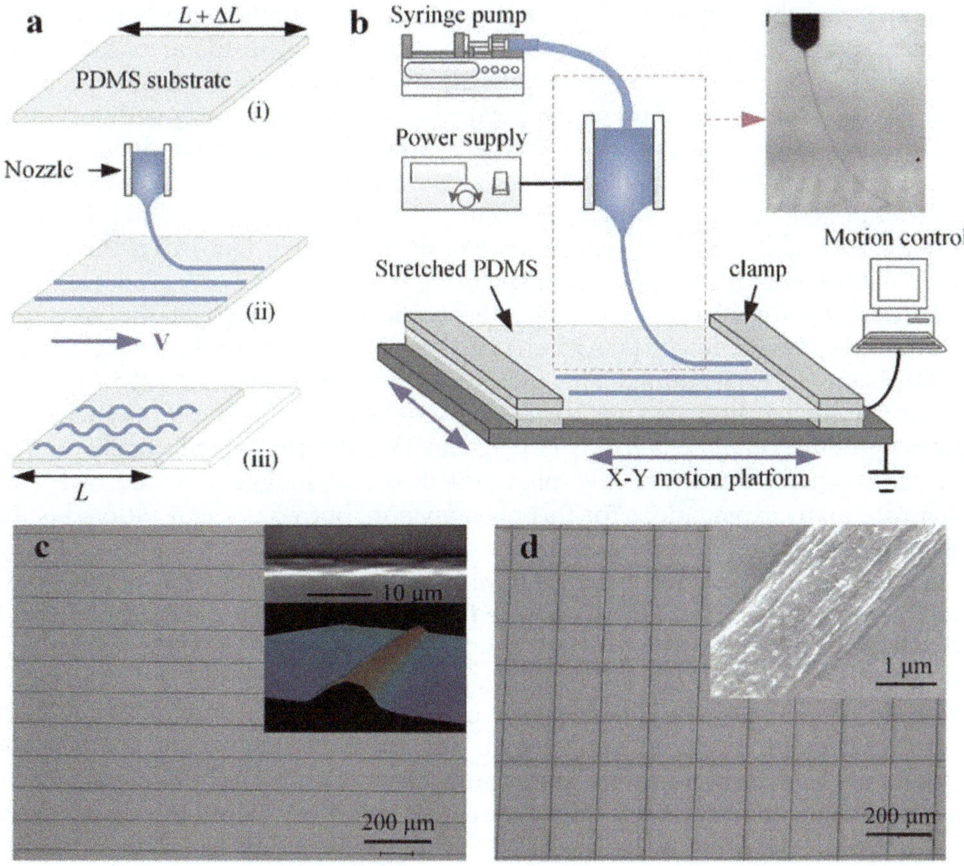

Figure 7.3. (a) Three steps for directly fabricating buckled fibers: (i) prestraining the elastomeric substrate, (ii) direct-writing straight fibers on the prestrained substrate, and (iii) releasing the prestrained substrate; the buckled fibers appear once the prestrained substrate is released. (b) Schematic diagram of MES. (c) Parallel and (d) latticed fibers on PDMS substrates. Inset SEM/LSCM images show the morphology of a single PVDF fiber [59].

highly uniform, controllable buckled fiber arrays produced using MES for direct-writing PVDF fibers onto a prestrained PDMS elastomeric substrate (figure 7.3). Owing to their excellent response at an extremely large applied strain, in-surface deformation of the stretchable pressure sensor with an in-surface buckled fiber array, and stable piezoelectric performance up to the failure strain of PDMS (~110% in their study), these stretchable piezoelectric generators and sensors have the potential to be integrated into stretchable electronics such as human monitoring devices and artificial skin [59].

Several different factors, such as electrospinning parameters, fiber alignments, nanofiber diameter distribution, electrode-related factors (like their specific surface area), as well as applied stress/strain, can affect the outputs of harvesting devices [82, 83]. The effect of applied stress and strain rate on the output of PVDF–TrFE electrospun nanofibrous flexible nanogenerators has been examined: the pulse peaks generated by the nanogenerator increases with the applied mechanical strain frequency, and the generated output is also proportional to the amplitude of the applied stress [82]. Studies also show that the electrical output of the nanogenerator is independent of the position of the electrodes but increases along with increasing electrode specific surface area and decreases when the results were divided by the surface area [83]. On the matter of fiber alignment and arrangement, according to Bafqi *et al* [84], applying a rotary collector and magnetic field is beneficial to the crystalline structure of the electrospun PVDF nanofibers, improving the PVDF-based nanogenerator's output and efficiency.

Particularly for PVDF nanofibers, the β phase percentage of the crystalline structure responsible for its piezoelectric properties can control the performance of PVDF-based nanogenerators. The addition of different fillers, LiCl [85], lead-free potassium sodium niobate (KNN) [86], Fe_3O_4–GO [87], ZnO nanoparticles [88], microcrystalline cellulose [89], CNTs [89], clay [89, 90], etc, can effectively improve the percentage of the piezoactive phase (β) in the crystalline structure of PVDF. There are many explanations for this phenomenon, but what seems rational is that fillers can act as the nucleating agent for creation and growth of the β phase [91].

As mentioned so far, most studies have been done on PVDF and its copolymer, and they are the most likely to be used as the piezoactive component [92–94]. However, recent advances provide new opportunities in the field of polymer piezoelectric materials [95]. For example, Street *et al* [96] studied the piezoelectric property of chitin by varying the type of electrospinning collector. Apparently, electrospun chitin produced using the air gap collector shows significantly higher crystallinity and piezoelectricity than chitin produced through flat-plate-collector electrospinning. They claimed that electrospun chitin can be promising for heavy metal filtration [96], but their piezoelectric and tensile results also suggest that this electrospun piezoelectric nanofiber can be applied in flexible energy harvesters and sensors.

Shear piezoelectricity is the coupling between a uniaxial polarization and a shear stress or strain. Due to the helical conformation of the polymer chain leading to an intrinsic shear piezoelectricity, PLLA can be a special selection among all the polymers possessing piezoelectric properties [97, 98]. Being a biocompatible and biodegradable polymer [99], PLLA has great potential as a functional material for *in vivo* energy harvesting or biomedical sensing, especially triboelectric application in biomedical devices [99–102].

There are studies involving PLLA nanofibers grown by electrospinning incorporated into self-powered strain sensors and energy harvesters [103]. A high electric field in the electrospinning fabrication process results in polarized and relatively long PLLA nanofibers along the fiber axis; this is found to be suitable for large-scale nanogenerator fabrication. In a work reported by Zhu *et al* [103], an energy harvester is fabricated using a polyimide substrate with comb-shaped gold electrodes and electrospun PLLA. The generated electric charge across the nanofibers was collected via the Au comb electrodes and showed reasonable linearity as a function of deformation with both open-circuit voltage and short-circuit current.

Using PLA as a biodegradable polymer in triboelectric applications, a biodegradable TENG was designed; PLA was paired with gelatin and magnesium as the counter electrode [104]. Compared with previously reported surface morphology-modified designs [105], electrospun PLA nanofibers paired with rough gelatin film shows the largest open-circuit voltage and short-circuit current.

7.3 Hybridization approaches for piezo–triboelectric devices

Due to flexibility in material selection, different designs in electronic devices (in particular in wearable and flexible devices), and various applications, electrospun piezo–triboelectric nanofibers constitute a broad and active subject.

The main goal of the combination of piezoelectric generators with triboelectric generators is the enhancement of electrical output (electrical power, voltage, and current) through utilizing the synergic effect of these generators. Hence, many efforts have been made to produce and optimize these generators so that in one cycle of use both can be used simultaneously. In other words, the difference between a hybrid generator and a piezoelectric generator or a triboelectric generator is higher electrical outputs when under the same mechanical stress.

A hybrid piezo–triboelectric generator consists of at least one piezoelectric section and a pair of triboelectric layers, such that the piezoelectric section can form one of the triboelectric layers. In this type of generator, the change in mechanical stress produces an electrical voltage in the piezoelectric part and the frictional contact of the triboelectric circuit creates an electrical output.

Different working mechanisms have been developed to create the above conditions in a hybrid piezo–triboelectric generator. Among them, one of the most common and widely used is the contact–separation of a pair of layers, where at least one of these layers has piezoelectric properties. The complementary layer is also selected with respect to the piezoelectric layer and the position of materials in the triboelectric series. In this working mechanism, also called *press–release*, the design of a suitable geometry or the use of an appropriate element for separating layers after loading is necessary.

In the latest research, an arc-shaped geometry is used to restore the initial state of the layers. In this type of design, an arc-shaped polymer section is usually employed as a supporter layer, and the piezoelectric layer is attached to it. The compressive stresses loading in the elastic range of this polymeric element causes bending in the support and consequently in the piezoelectric layer, thus creating a piezoelectric charge. The compressive stresses loading also cause physical contact between layers, resulting in the generation of a triboelectric charge.

After removal of the loading, the arc-shaped structure retrieves its initial geometry in order to separate the triboelectric layers, and the change in mechanical stress in the piezoelectric layer returns to its original state.

An example of this type of design was used in the research by Suk Jung *et al* [106]. The hybrid generator consists of two different layers: an arch-shaped piezoelectric generator with an Au/PVDF/Au structure on the top, and on the bottom, and a triboelectric generator with a PTFE/Al structure with a press–release mode of operation. This hybrid generator operates by the vertical force on the top layer, which has a prestrained piezoelectric PVDF layer with an arch shape supported by a polyimide substrate. Utilizing the arch shape of the piezoelectric generator enhances the applied strain on the piezoelectric layer as well as drives the shape of the generator to the original state after release of the applied force. Ordinary Al electrodes were used in the bottom layer to harvest the generated charges from the surfaces. Therefore, to benefit from both the piezoelectric and triboelectric generator, the authors used the bottom electrode of the piezoelectric generator as an electrode for the triboelectric generator. With this assembly, the piezoelectric layer generates a positive piezoelectric potential between the Au_1 electrode and the Au_2 electrode; therefore, electrons flow from Au_2 to Au_1. Physical contact between the Au_2 electrode and PTFE due to the applied forces on the top layer (contact mode) results in a charge transfer between two layers. According to the triboelectric series, the PTFE tends to gain electrons, and electrons will transfer to the Au_2 electrode and form electrostatic charges. After releasing the applied load (separating mode), the arc structure starts to return to its initial shape; and since the electrons flow back to Au_2 to neutralize it and the PVDF has a compressive force, the hybrid generator produces both triboelectric and piezoelectric voltages, and electrons flow from the Al and Au_1 electrode to the Au_2 electrode. Therefore, this working mechanism simultaneously obtains both piezoelectric and triboelectric output signals within one press–release cycle.

Another press–release mechanism involves the use of a suitable element for separating and restoring the initial state of the converter. Typically, a spring or sponge is used in the corners and sides of the double layer as a suitable element. In the studies of Hassan *et al* [107], such elements were used. Flat piezoelectric and triboelectric layers were vertically stacked on each other. A PDMS:$ZnSnO_3$ composite with Cu electrodes on the top and bottom formed the piezoelectric part, whereas a PDMS film and a Cu film comprised the complementary friction layers of the TENG. Here, the bottom Cu electrode of the piezoelectric part also served as one of the friction layers of the triboelectric layer. A Cu film attached on the bottom surface of the PDMS:ZnSnO3 layer acts as a triboelectric layer. To generate electrical charges by utilizing the contact–separation of the PDMS and Cu film, the spacer for the triboelectric layers was made of sponges, and the piezoelectric part was deposited on the top electrode of the TENG. This design effectively harvests any vertical stress upon the generator since the piezoelectric part remains under variable strain during the whole pushup cycle.

In the initial state, there is no charge on the rough PDMS surface and driving Cu electrode of the TENG. After a complimentary contact electrification during first few cycles, negative charges appear on the PDMS surface while positive charges appear on the driving Cu electrode. By applying an external force on the top surface, the PDMS:$ZnSnO_3$ composite is put under a lot of stress while the bottom surface experiences compressive strain upon touching the PDMS surface. In this way, piezoelectric potential energy is induced across the electrodes of the piezoelectric part. Meanwhile, the varying gap distance between the charged PDMS and Cu induces triboelectric output power; but upon contact, it reaches peak voltage and then suddenly falls to zero as the charges reach their equilibrium states. When the external force is released, the PENG returns to its original state by the application of stored stress in the elastic sponge (spacer), which is responsible for the reversal of stress on the piezoelectric part and the generation of reverse piezoelectric potential energy. Meanwhile, the bottom Cu of the piezoelectric part starts to separate from the PDMS of TENG, thus increasing the air gap. This changes the electric field and produces the reverse triboelectric charge in the external circuit. Finally, the hybrid generator reaches its original state.

7.4 Large-scale nanofibrous energy harvesting strategies

7.4.1 Introduction

For harvesting energy from human movement, fiber-based electrical power generators are highly desirable since they are comfortable, lightweight, and are very similar to conventional fabrics. Flexible textile structures can be designed to provide piezoelectric output under a large number of variable mechanical deformation and loading systems [108]. By using active materials as different structures on the surface of a fabric, or as yarns woven or knitted into the textile, energy harvesting textiles can be fabricated in a reasonable scale [109]. Since the energy scavenging from textile-based PENGs and TENGs can provide the sustainable power system required for wearable electronic devices, it has attracted tremendous attention by many researchers and manufacturers who see these materials as suitable to integrate with other wearable technologies [110]. Intelligent textiles can also be easily adapted to work with these power systems [111]. This section discusses various types of energy harvesting textiles on a large scale, and their applications as viable energy sources. The method of manufacturing various piezo fabrics as woven, knitted, and yarn structures is also scrutinized.

7.4.2 Classification of large-scale piezo–triboelectric energy harvesters

There are many different piezoelectric devices, such as actuators, sensors, frequency controllers, and transducers [112]. Piezo- and triboelectric textile-based energy harvesters could be developed using various types of fibers of different structures, such as woven or knitted fabrics, yarns, and composites, which have greater strength compared with traditional fibers [113]. There are different ways to scale up nanofibrous mats for energy harvesting; for example, using piezo and triboelectric mats in the fabrics or composites as a layer. Another way to scale up piezo–triboelectric energy harvesters is converting fibrous mats to yarns and then weaving these yarns into fabrics.

7.4.3 Piezoelectric and triboelectric structural layer

Guang Zhu *et al* [111] describe a simple, cost-effective, and scalable approach to manufacture a TENG with improved electric output. In their work, they developed a new TENG that has not only a greatly simplified structure but also substantially higher power output enabled by nanoparticle-based surface modification. The nanogenerator has a multilayered structure with two different substrates. This can be considered as one of the first attempts to experimentally show the scaling-up of the nanogenerator's power output, which enables the generator to power not just a single electronic device but many of them simultaneously. The fabricated nanogenerator was capable of producing an estimated output open-circuit voltage of ~1200 V, which was enough to instantaneously light up 600 commercial LED bulbs. This work has opened up the practicality of harvesting mechanical energy by the nanogenerator on a large scale. The working mechanism demonstrated here can be further applied to harvest other types of mechanical energy, such as that produced by rolling wheels, wind, and ocean waves, on a large scale.

In another study, Zeng *et al* [114] produced a novel all-in-one fiber power nanogenerator for wearable application, consisting of a PVDF–$NaNbO_3$ nanofiber nonwoven fabric as an active piezoelectric layer, and an elastic conducting knitted fabric as the top and bottom electrodes (figure 7.4). Since these electrodes should be conductive and flexible, segmented polyurethane and silver-coated polyamide multifilament yarns were utilized to fabricate them in a simple procedure. Since the two main components of the nanogenerator in this approach are made based on fibers, the scale of the device can be expanded to the hundreds of cm^2. Three advantages can be seen from this strategy to fabricate energy harvesting devices. First, the knitted fabric-based electrodes provide the function of inducing a localized multimode deformation field in the piezoelectric fabric. Secondly, the integration process of the nanogenerator devices can be simplified without the size and shape

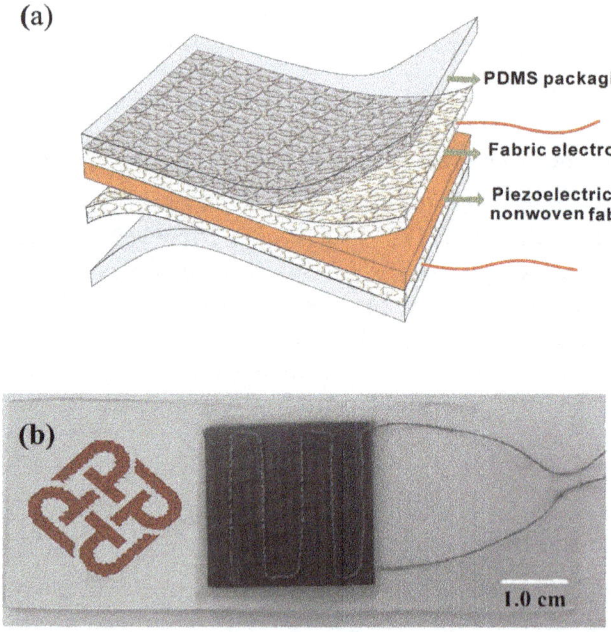

Figure 7.4. (a) Schematic of the all-fiber electric power nanogenerator. (b) Photo of a fully packaged generator device [114].

limit due to the use of a fibrous active layer and electrodes. Thirdly, the whole generator device, including the active piezoelectric and fabric electrodes, are very flexible and lightweight; this enables us to fabricate wearable devices and integrate them into intelligent textiles.

7.4.4 Composite laminate with embedded piezo and triboelectric films

Fabrication of composite laminate can also be considered for scaling up the power harvesters. For example, phenyl salicylate is casted upon the matrix to prevent the metallic connector from getting embedded in it. Then a glass epoxy laminate is fabricated and the prepared film sensors are placed between the 2nd and 3rd layers during the fabrication process. After curing, the phenyl salicylate on the metal connectors is removed by heating it at 60 °C. Glass epoxy tabs are later bonded on the ends of the specimens. The dimensions of the specimens with tabs can be easily set as required [115].

7.4.5 Piezoelectric and triboelectric woven and knitted structures

Compared to other structures, a woven or knitted structure offers greater strength and more stability. There are different types of woven or knitted structures, such as single, double, or multilayered fabrics. Therefore, it is easier for a fabric to tune its structure to achieve desired characteristics, such as tear strength, tensile strength, shear strength, drapeability, air permeability, water absorbency, and crease resistance [113]. As an example, Elias *et al* [108] demonstrated 3D spacer technology-based all-fiber piezoelectric fabrics as power generators and energy harvesters. In this strategy for scaling up, they produced a high β-phase (~80%) piezoelectric PVDF using melt spinning under a high electric field (0.6 MV m^{-1}), and then produced yarns knitted into a fabric structure. The Ag-coated PA66 yarns were also used as the top and bottom electrodes. This structure provides an output power density in the range of 1.10–5.10 μW cm^{-2} at applied impact pressures of 0.02–0.10 MPa. The enhanced performance of this type of all-fiber piezoelectric fabric is due to the advantage of efficient charge collection of electrodes as well as the uniform distribution of pressure on the surface of the fabric.

7.4.6 Piezoelectric and triboelectric yarns

Yarns can be considered as engineering tools to produce fabric structures. Therefore, piezo–triboelectric yarns can be considered as another scaling-up approach for power harvesting devices.

In the work presented by Elias *et al* [108], a substantial increase in piezoelectric output of the PVDF yarns was detected using $ZnSnO_3$-based perovskite. This led to doubling the piezoelectric constant from 60 pm V^{-1} to nearly 130 pm V^{-1}. They believe their method of producing large quantities of high-quality piezoelectric yarn and piezoelectric fabric can be considered as an ideal option for the development of high-performance energy harvesting textile-based structures for wearable electronic devices.

Metal wires or common spun yarns with conductive metals or carbonaceous material coatings have been utilized as the electrodes in TENGs. Dielectric PDMS films can also be coated on conductive yarns or metal wires as triboelectrification layers to fabricate flexible large-scale energy harvesting devices. These modified yarns or wires can be woven or knitted into a textile-based TENG or PENG. It is believed that weaving yarns into fabric is one of the oldest handicrafts. In this technique, long yarns can weave together and create supermechanical properties through well-designed weaving methods. In this context, active yarns can be fabricated separately with predefined properties and then woven or knitted into the textile alongside traditional yarns. This approach places fewer constraints on functional materials and can be considered as one of the promising approaches for scaling up the energy harvesting devices [109]. Moreover, these structures are very flexible by nature and are easy to incorporate and integrate with conventional clothing. It will be very interesting to produce smart textiles and clothing integrated with self-chargeable electronic devices.

Yu *et al* [110] propose a strategy for TENG textiles as power harvesting cloths in which core–shell yarns were used to form this structure. They suggest core conductive fibers wrapped or twisted with polymeric or natural fibers. The resulting TENG textiles are comfortable, flexible, and fashionable, and their production processes are compatible with industrial, large-scale manufacturing. In one work, TENG textiles woven from core–shell yarns were successfully fabricated to harvest various types of energies, such as mechanical energy during patting, walking, and running, and then transform these energies into electrical energy [116]. The core–shell yarns, with core conductive fibers as electrodes and shell dielectric fibers as electrification layers, enables the versatility of the core–shell-yarn-based TENG in material choices and structure designs. The use of this textile structure as an energy harvesting device which is able to wash and tailor were further demonstrated, which can be considered as a strong feature for a TENG [110]. These types of TENG textiles have demonstrated capability to effectively scavenge mechanical energy for powering small personal electronics. However, the main challenge of selecting the most practical way to incorporate TENGs into garments still needs to be addressed.

7.5 Future trend of energy harvesting methods for portable medical devices

In the past decade, wearable electronics have received much interest due to their attractive properties such as flexibility, stretchability, conformability, and huge potential for applications in medical devices harvesting energy from body movements. With growing attention in wearable devices, various energy harvesters have been studied for scavenging energies from different sources within the ambient environment; for example, PENGs and TENGs to harvest mechanical energy, pyroelectric and thermoelectric nanogenerators to harvest thermal energy, and PV cells to harvest solar energy.

Recently, the hybridization of individual energy harvesters has become a new trend with the aim to enhance power output. First, the hybrid cell for harvesting mechanical and solar energy was presented by Chen *et al* in 2009 [117]. They showed that two energy harvesters can operate simultaneously or individually, and can also be integrated in parallel and serial to enhance output performance.

After that, various hybrid nanogenerators, such as the piezo–tribo nanogenerator [118, 119], piezo–pyro nanogenerator [120], piezo–tribo–pyro nanogenerator [121], and tribo-solar nanogenerator [122], were developed. In such hybridized devices, various energies can be harvested from a single hybrid device. Mechanical, thermal, and solar energy are available in our surroundings and can be utilized wherever and whenever one or all of them are available. The integration of various energy harvesters into one device illustrates a newly materializing field to fabricate multi-energies scavenging devices. Figure 7.5 shows combinations of various energy harvesters.

The available hybrid nanogenerators are prepared based on individual energy harvesters stacked together in series or in parallel. These distinct energy harvesters have independent electrodes, which is undesirable because of the device dimensions.

An attractive future trend of smart wearable electronics is to design multifunctional hybrid devices to achieve a one-structure-based nanogenerator to integrate multiple types of energy and thus attain better electrical performance. To date, very few reports have described one-structure-based piezo–tribo–pyro–photoelectric multiple-function nanogenerators for harvesting mechanical, thermal, and solar energy [123–125].

The hybrid nanogenerator described by Zhang *et al* [123] is mainly based on PZT as a piezoelectric, pyroelectric, and photoelectric material. In the approach presented by Ji and coworkers [124], a multi-effect coupled nanogenerator based on ferroelectric barium titanate was reported. Zhang and his colleagues [125] review

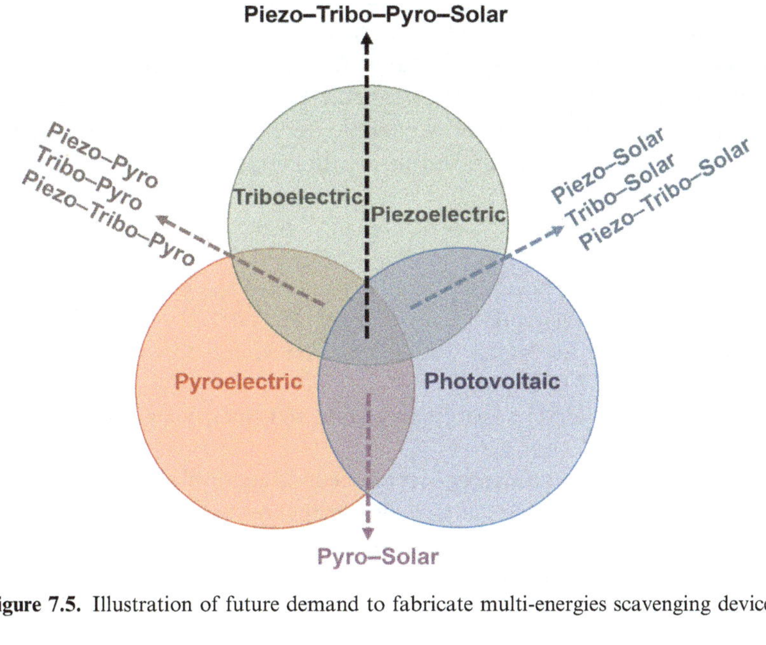

Figure 7.5. Illustration of future demand to fabricate multi-energies scavenging devices.

the feasibility of 'all-in-one' self-powered flexible microsystems by introducing the technology of TENGs. In these devices, a multi-effects coupled nanogenerator that integrates a PENG, TENG, and PV cell into a single device with two electrodes has been demonstrated. Multi-effect coupled nanogenerators are capable of individually/simultaneously harvesting mechanical, thermal, and solar energy. Development of versatile nanogenerators and sensors that use these hybrid and flexible structures can profoundly revolutionize the power harvesting and pressure sensing applications in smart clothing.

7.6 Conclusion

Flexibility and stretchability of nanofibrous piezo- and triboelectric wearable electronics are considered as the two main functions of wearable power-supplying and self-powered sensing systems. These two abilities of wearable electronics are considered as the sustainable development needs of future wearable electronics. Fiber-based nanogenerator technology is another means to harvest human motion energy in a continuous, convenient, and environmentally friendly way and transform these environmental energy into the electrical energy required to power wearable electronics. Moreover, flexibility and stretchability directly contradict the conductivity and overall performance of these electronic devices; this can be considered as the main challenge for the development of these devices.

With the rapid improvement of PENGs and TENGs, and their expanding application fields, electrical output tuning will gradually become an unavoidable problem in their utilization and commercialization. So far, engineering tools developed to control the energy conversion capability of these energy harvesting devices address fiber diameter, distribution, alignment, etc. However, there is also a need to find greater control over this issue. It is firmly believed that many major issues of the current textile-based nanogenerators and all other wearable electronics will be solved with the progress of material science and technology, particularly nanoscience and technology. Fibrous wearable power harvesting devices are bound to shine brightly in the future of wearable technologies due to their excellent properties and ability to be integrated into other materials and technologies.

References

[1] Wang Z L and Wu W J 2012 Nanotechnology-enabled energy harvesting for self-powered micro-/nanosystems *Angew Chem. Int. Ed. Engl.* **51** 11700–21

[2] Beeby S P *et al* 2006 Energy harvesting vibration sources for microsystems applications *Meas. Sci. Technol.* **17** R175

[3] Lee J-H *et al* 2016 All-in-one energy harvesting and storage devices *J. Mater. Chem.* A **4** 7983–99

[4] Fuh Y-K *et al* 2015 Hybrid energy harvester consisting of piezoelectric fibers with largely enhanced 20 V for wearable and muscle-driven applications *ACS Appl. Mater.* **7** 16923–31

[5] Bafqi M S S, Bagherzadeh R and Latifi M J 2015 Fabrication of composite PVDF-ZnO nanofiber mats by electrospinning for energy scavenging application with enhanced efficiency *J. Polym. Res.* **22** 130

[6] Sorayani Bafqi M S, Bagherzadeh R and Latifi M J 2017 Nanofiber alignment tuning: an engineering design tool in fabricating wearable power harvesting devices *J. Polym. Res.* **47** 535–50
[7] Khaligh A, Zeng P and Zheng C J 2010 Kinetic energy harvesting using piezoelectric and electromagnetic technologies—state of the art *IEEE Trans. Ind. Electron.* **57** 850–60
[8] Pang C, Lee C and Suh K Y 2013 Recent advances in flexible sensors for wearable and implantable devices *J. Appl. Polym. Sci.* **130** 1429–41
[9] Fan F R, Tang W and Wang Z L J 2016 Flexible nanogenerators for energy harvesting and self-powered electronics *Adv. Mater.* **28** 4283–305
[10] Ostfeld A E *et al* 2016 High-performance flexible energy storage and harvesting system for wearable electronics *Sci. Rep.* **6** 26122
[11] Kim D *et al* 2012 Mechanical bending of flexible complementary inverters based on organic and oxide thin film transistors *Org. Electron.* **13** 2401–5
[12] Pang C *et al* 2012 A flexible and highly sensitive strain-gauge sensor using reversible interlocking of nanofibres *Nat. Mater.* **11** 795
[13] Ramuz M *et al* 2012 Transparent, optical, pressure-sensitive artificial skin for large-area stretchable electronics *Adv. Mater.* **24** 3223–7
[14] Boland J J J 2010 Flexible electronics: within touch of artificial skin *Nat. Mater.* **9** 790
[15] Golmohammadi Rostami S *et al* 2015 Multi-layer electrospun nanofiber mats with chemical agent sensor function *J. Ind. Text.* **45** 467–80
[16] Fan F-R *et al* 2012 Transparent triboelectric nanogenerators and self-powered pressure sensors based on micropatterned plastic films *Nano Lett.* **12** 3109–14
[17] Bae W G *et al* 2013 Enhanced skin adhesive patch with modulus-tunable composite micropillars *Adv. Healthc. Mater.* **2** 109–13
[18] Kim R-H *et al* 2010 Waterproof AlInGaP optoelectronics on stretchable substrates with applications in biomedicine and robotics *Nat. Mater.* **9** 929
[19] Stoppa M and Chiolerio A J 2014 Wearable electronics and smart textiles: a critical review *Sensors* **14** 11957–92
[20] Heo J S *et al* 2018 Recent progress of textile-based wearable electronics: a comprehensive review of materials, devices, and applications *Small* **14** 1703034
[21] Esfahani H, Jose R and Ramakrishna S J M 2017 Electrospun ceramic nanofiber mats today: synthesis, properties, and applications *Materials* **10** 1238
[22] Biswas A, Park H and Sigmund W M 2012 Flexible ceramic nanofibermat electrospun from TiO_2–SiO_2 aqueous sol *Ceram. Int.* **38** 883–6
[23] Esfahani H *et al* 2016 Electrospun nylon 6/zinc doped hydroxyapatite membrane for protein separation: mechanism of fouling and blocking model *Mater. Sci. Eng. C Mater. Biol. Appl.* **59** 420–8
[24] Roundy S J 2003 *Energy Scavenging For Wireless Sensor Nodes With a Focus on Vibration to Electricity Conversion* (Berkeley, CA: University of California)
[25] Lu Y *et al* 2017 A flexible electrostatic kinetic energy harvester based on electret films of electrospun nanofibers *Smart Mater. Struct.* **27** 014001
[26] Gao C and Chen G 2016 Conducting polymer/carbon particle thermoelectric composites: emerging green energy materials *Compos. Sci. Technol.* **124** 52–70
[27] Peltier J C A 1834 Investigation of the heat developed by electric currents in homogeneous materials and at the junction of two different conductors *Ann. Chim. Phys.* **56** 371

[28] Kamarudin M A *et al* 2013 A review on the fabrication of polymer-based thermoelectric materials and fabrication methods *Sci. Word. J.* **2013** 713640
[29] Hiura S *et al* 2017 Thermoelectric properties of poly (3-hexylthiophene) nanofiber mat with a large void fraction *Materials* **10** 468
[30] Sun Y *et al* 2010 A three-in-one improvement in thermoelectric properties of polyaniline brought by nanostructures *Synth. Met.* **160** 2371–6
[31] Taggart D K *et al* 2010 Enhanced thermoelectric metrics in ultra-long electrodeposited PEDOT nanowires *Nano Lett.* **11** 125–31
[32] Persano L, Camposeo A and Pisignano D 2015 Active polymer nanofibers for photonics, electronics, energy generation and micromechanics *Prog. Polym. Sci.* **43** 48–95
[33] Slobodian P *et al* 2013 Thermoelectric properties of carbon nanotube and nanofiber based ethylene-octene copolymer composites for thermoelectric devices *J. Nanomater.* **2013** 5
[34] Wang Q *et al* 2012 Enhanced thermoelectric properties of CNT/PANI composite nanofibers by highly orienting the arrangement of polymer chains *J. Mater. Chem.* **22** 17612–8
[35] Yin T *et al* 2010 Nanocrystalline thermoelectric $Ca_3Co_4O_9$ ceramics by sol–gel based electrospinning and spark plasma sintering *J. Phys. Chem.* C **114** 10061–5
[36] Zhang M *et al* 2018 Synthesis and thermoelectric characterization of lead telluride hollow nanofibers *Front. Chem.* **6**
[37] Sebald G *et al* 2009 On thermoelectric and pyroelectric energy harvesting *Smart Mater. Struct.* **18** 125006
[38] Lingam D *et al* 2013 Nano/microscale pyroelectric energy harvesting: challenges and opportunities *Int. J. Smart Nano Mater.* **4** 229–45
[39] Lang S B 2005 Pyroelectricity: from ancient curiosity to modern imaging tool *Phys. Today* **58** 31
[40] Persano L *et al* 2013 High performance piezoelectric devices based on aligned arrays of nanofibers of poly (vinylidenefluoride-co-trifluoroethylene) *Nat. Commun.* **4** 1633
[41] Lee M *et al* 2012 A hybrid piezoelectric structure for wearable nanogenerators *Adv. Mater.* **24** 1759–64
[42] Sencadas V, Gregorio R Jr and Lanceros-Méndez S J 2009 α to β phase transformation and microestructural changes of PVDF films induced by uniaxial stretch *J. Macromol. Sci.* B **48** 514–25
[43] You M-H *et al* 2018 A self-powered flexible hybrid piezoelectric–pyroelectric nanogenerator based on non-woven nanofiber membranes *J. Mater. Chem.* A **6** 3500–9
[44] Roundy S *et al* 2005 Improving power output for vibration-based energy scavengers *IEEE Perv. Comput.* **4** 28–36
[45] Wang X J 2012 Piezoelectric nanogenerators—harvesting ambient mechanical energy at the nanometer scale *Nano Energy* **1** 13–24
[46] Siddiqui S *et al* 2018 An omnidirectionally stretchable piezoelectric nanogenerator based on hybrid nanofibers and carbon electrodes for multimodal straining and human kinematics energy harvesting *Adv. Energy Mater.* **8** 1701520
[47] Wu Y *et al* 2015 Hybrid energy cell for harvesting mechanical energy from one motion using two approaches *Nano Energy* **11** 162–70
[48] Fuh Y-K *et al* 2015 Self-powered sensing elements based on direct-write, highly flexible piezoelectric polymeric nano/microfibers *Nano Energy* **11** 671–7
[49] Hu Y and Wang Z L J 2015 Recent progress in piezoelectric nanogenerators as a sustainable power source in self-powered systems and active sensors *Nano Energy* **14** 3–14

[50] Dagdeviren C *et al* 2014 Conformal piezoelectric energy harvesting and storage from motions of the heart, lung, and diaphragm *Proc. Natl. Acad. Sci. U.S.A.* **111** 1927–32
[51] Wang Z L and Song J J S 2006 Piezoelectric nanogenerators based on zinc oxide nanowire arrays *Science* **312** 242–6
[52] Zhu G *et al* 2010 Flexible high-output nanogenerator based on lateral ZnO nanowire array *Nano Lett.* **10** 3151–5
[53] Sengupta D *et al* 2018 Optimized polyvinylidene fluoride nanofiber webs for flexible energy harvesters *Multidisciplinary Digital Publishing Institute Proc.*
[54] Zandesh G *et al* 2017 Piezoelectric electrospun nanofibrous energy harvesting devices: influence of the electrodes position and finite variation of dimensions *J. Ind. Text.* **47** 348–62
[55] Baji A *et al* 2011 Electrospinning induced ferroelectricity in poly (vinylidene fluoride) fibers *Nanoscale* **3** 3068–71
[56] Sun D *et al* 2006 Near-field electrospinning *Nano Lett.* **6** 839–42
[57] Fuh Y-K *et al* 2012 Comparative study of polyvinylidene fluoride nanofibrous membranes prepared by continuous near-field and conventional electrospinning processes *Micro Nano Lett.* **7** 376–9
[58] Chun J *et al* 2015 Highly anisotropic power generation in piezoelectric hemispheres composed stretchable composite film for self-powered motion sensor *Nano Energy* **11** 1–10
[59] Duan Y *et al* 2014 Non-wrinkled, highly stretchable piezoelectric devices by electrohydrodynamic direct-writing *Nanoscale* **6** 3289–95
[60] Qi Y *et al* 2011 Enhanced piezoelectricity and stretchability in energy harvesting devices fabricated from buckled PZT ribbons *Nano Lett.* **11** 1331–6
[61] Jeong C K *et al* 2015 A hyper-stretchable elastic-composite energy harvester *Adv. Mater.* **27** 2866–75
[62] Chortos A and Bao Z J 2014 Skin-inspired electronic devices *Mater. Today* **17** 321–31
[63] Trung T Q and Lee N E 2017 Recent progress on stretchable electronic devices with intrinsically stretchable components *Adv. Mater.* **29** 1603167
[64] Wu H *et al* 2016 Energy harvesters for wearable and stretchable electronics: from flexibility to stretchability *Adv. Mater.* **28** 9881–919
[65] Ahn J-H and Je J H 2012 Stretchable electronics: materials, architectures and integrations *J. Phys. D: Appl. Phys.* **45** 103001
[66] Xu F *et al* 2012 Wavy ribbons of carbon nanotubes for stretchable conductors *Adv. Funct. Mater.* **22** 1279–83
[67] Jin Y *et al* 2016 Buckled Au@ PVP nanofiber networks for highly transparent and stretchable conductors *Adv. Electron. Mater.* **2** 1500302
[68] Guo C F *et al* 2015 Enhancing the scratch resistance by introducing chemical bonding in highly stretchable and transparent electrodes *Nano Lett.* **16** 594–600
[69] Xu S *et al* 2013 Stretchable batteries with self-similar serpentine interconnects and integrated wireless recharging systems *Nat. Commun.* **4** 1543
[70] Trung T and Lee N-E 2017 Materials and devices for transparent stretchable electronics *J. Mater. Chem.* C **5** 2202–22
[71] Bhardwaj N and Kundu S C 2010 Electrospinning: a fascinating fiber fabrication technique *Biotechnol. Adv.* **28** 325–47
[72] Thavasi V, Singh G and Ramakrishna S 2008 Electrospun nanofibers in energy and environmental applications *Energy Environ. Sci.* **1** 205–21

[73] Peng S *et al* 2016 Multi-functional electrospun nanofibres for advances in tissue regeneration, energy conversion & storage, and water treatment *Chem. Soc. Rev.* **45** 1225–41
[74] Wang X *et al* 2016 A flexible triboelectric-piezoelectric hybrid nanogenerator based on P (VDF-TrFE) nanofibers and PDMS/MWCNT for wearable devices *Sci. Rep.* **6** 36409
[75] Siddiqui S *et al* 2018 An omnidirectionally stretchable piezoelectric nanogenerator based on hybrid nanofibers and carbon electrodes for multimodal straining and human kinematics energy harvesting *Adv. Energy Mater.* **8** 1701520
[76] Chen X *et al* 2017 Flexible fiber-based hybrid nanogenerator for biomechanical energy harvesting and physiological monitoring *Nano Energy* **38** 43–50
[77] Chinnappan A *et al* 2017 An overview of electrospun nanofibers and their application in energy storage, sensors and wearable/flexible electronics *J. Mater. Chem.* C **5** 12657–73
[78] Sun D *et al* 2006 Near-field electrospinning *Nano Lett.* **6** 839–42
[79] Reneker D H and Yarin A L 2008 Electrospinning jets and polymer nanofibers *Polymer* **49** 2387–425
[80] Chang C *et al* 2010 Direct-write piezoelectric polymeric nanogenerator with high energy conversion efficiency *Nano Lett.* **10** 726–31
[81] Han T, Reneker D H and Yarin A L 2008 Pendulum-like motion of straight electrified jets *Polymer* **49** 2160–9
[82] Serairi L *et al* 2017 Flexible piezoelectric nanogenerators based on PVDF-TrFE nanofibers *Eur. Phys. J.: Appl. Phys.* **80** 30901
[83] Zandesh G *et al* 2017 Piezoelectric electrospun nanofibrous energy harvesting devices: influence of the electrodes position and finite variation of dimensions *J. Ind. Text.* **47** 348–62
[84] Sorayani Bafqi M S, Bagherzadeh R and Latifi M 2017 Nanofiber alignment tuning: an engineering design tool in fabricating wearable power harvesting devices *J. Ind. Text.* **47** 535–50
[85] Mokhtari F, Shamshirsaz M and Latifi M 2016 Investigation of β phase formation in piezoelectric response of electrospun polyvinylidene fluoride nanofibers: LiCl additive and increasing fibers tension *Polym. Eng. Sci.* **56** 61–70
[86] Abdullah I Y *et al* 2016 Enhancement piezoelectricity in poly (vinylidene fluoride) by filler piezoceramics lead-free potassium sodium niobate (KNN) *Opt. Quantum Electron.* **48** 149
[87] Samadi A, Hosseini S M and Mohseni M 2018 Investigation of the electromagnetic microwaves absorption and piezoelectric properties of electrospun Fe_3O_4-GO/PVDF hybrid nanocomposites *Org. Electron.* **59** 149–55
[88] Bafqi M S S, Bagherzadeh R and Latifi M 2015 Fabrication of composite PVDF-ZnO nanofiber mats by electrospinning for energy scavenging application with enhanced efficiency *J. Polym. Res.* **22** 1–9
[89] Bodkhe S *et al* 2014 Beta-phase enhancement in polyvinylidene fluoride through filler addition: comparing cellulose with carbon nanotubes and clay *J. Polym. Res.* **21** 434
[90] Neppalli R *et al* 2013 The effect of clay and of electrospinning on the polymorphism, structure and morphology of poly (vinylidene fluoride) *Eur. Polym. J.* **49** 90–9
[91] Georgousis G *et al* 2015 Strain sensing in polymer/carbon nanotube composites by electrical resistance measurement *Composites, Part B* **68** 162–9
[92] Xin Y *et al* 2018 A brief review on piezoelectric PVDF nanofibers prepared by electrospinning *Ferroelectrics* **526** 140–51

[93] Wang X *et al* 2018 Tactile-sensing based on flexible PVDF nanofibers via electrospinning: a review *Sensors* **18** 330

[94] Xin Y *et al* 2016 The use of polyvinylidene fluoride (PVDF) films as sensors for vibration measurement: a brief review *Ferroelectrics* **502** 28–42

[95] Jing Q and Kar-Narayan S 2018 Nanostructured polymer-based piezoelectric and triboelectric materials and devices for energy harvesting applications *J. Phys. D: Appl. Phys.* **51** 303001

[96] Street R M *et al* 2018 Variable piezoelectricity of electrospun chitin *Carbohydr. Polym.* **195** 218–24

[97] Smith M *et al* 2017 Direct observation of shear piezoelectricity in poly-l-lactic acid nanowires *APL Mater.* **5** 074105

[98] Ochiai T and Fukada E 1998 Electromechanical properties of poly-L-lactic acid *Jpn. J. Appl. Phys.* **37** 3374

[99] Jamshidian M *et al* 2010 Poly-lactic acid: production, applications, nanocomposites, and release studies *Compr. Rev. Food Sci. Food Saf.* **9** 552–71

[100] Zhao G *et al* 2017 Electrospun poly (l-lactic acid) nanofibers for nanogenerator and diagnostic sensor applications *Macromol. Mater. Eng.* **302** 1600476

[101] Zhao C *et al* 2017 A poly (l-lactic acid) polymer-based thermally stable cantilever for vibration energy harvesting applications *Adv. Sustain. Syst.* **1** 1700068

[102] Sun J *et al* 2015 Triboelectric nanogenerator based on biocompatible polymer materials *J. Phys. Chem.* C **119** 9061–8

[103] Zhu J, Jia L and Huang R 2017 Electrospinning poly (l-lactic acid) piezoelectric ordered porous nanofibers for strain sensing and energy harvesting *J. Mater. Sci., Mater. Electron.* **28** 12080–5

[104] Pan R *et al* 2018 Fully biodegradable triboelectric nanogenerators based on electrospun polylactic acid and nanostructured gelatin films *Nano Energy* **45** 193–202

[105] Zhao L *et al* 2016 A size-unlimited surface microstructure modification method for achieving high performance triboelectric nanogenerator *Nano Energy* **28** 172–8

[106] Jung W-S *et al* 2015 High output piezo/triboelectric hybrid generator *Sci. Rep.* **5** 9309

[107] Hassan G *et al* 2017 A flat-panel-shaped hybrid piezo/triboelectric nanogenerator for ambient energy harvesting *Nanotechnology* **28** 175402

[108] Siores E 2015 Soft fibre based piezoelectric energy harvesting textiles *5th World Congress on Materials Science & Engineering*

[109] Torah R, Lawrie-Ashton J, Li Y, Arumugam S, Sodano H A and Beeby S 2018 Energy-harvesting materials for smart fabrics and textiles *MRS Bull.* **43** 214–9

[110] Yu A, Pu X, Wen R, Liu M, Zhou T, Zhang K, Zhang Y, Zhai J, Hu W and Wang Z L 2017 Core–shell-yarn-based triboelectric nanogenerator textiles as power cloths *ACS Nano* **11** 12764–71

[111] Zhu G L, Zong-Hong J, Qingshen B, Peng P, Caofeng Y, Zhou Y and Wang Z L 2013 Toward large-scale energy harvesting by a nanoparticle-enhanced triboelectric nanogenerator *Nano Lett.* **13** 847–53

[112] Uchino K 2010 *Advanced Piezoelectric Materials: Science and Technology* (Amsterdam: Elsevier)

[113] McCann J and Bryson D 2009 *Smart Clothes and Wearable Technology* (Amsterdam: Elsevier)

[114] Zeng W, Xiao-Ming T, Chen S, Shang S, Chan H L W and Choy S H 2013 Highly durable all-fiber nanogenerator for mechanical energy harvesting *Energy Environ. Sci.* **6** 2631–8

[115] Rathod V T, Hiremath S R and Mahapatra D R 2014 Strength and fatigue life evaluation of composite laminate with embedded sensors *Smart Sensor Phenomena, Technology, Networks, and Systems Integration* (Bellingham, WA: International Society for Optics and Photonics)

[116] Lund A, Rundqvist K, Nilsson E, Yu L, Hagström B and Müller C 2018 Energy harvesting textiles for a rainy day: woven piezoelectrics based on melt-spun PVDF microfibres with a conducting core *NPJ Flexible Electron.* **2** 9

[117] Xu C, Wang X and Wang Z L 2009 Nanowire structured hybrid cell for concurrently scavenging solar and mechanical energies *J. Am. Chem. Soc.* **131** 5866–72

[118] Karumuthil S C, Rajeev S P and Varghese S 2017 Piezo-tribo nanoenergy harvester using hybrid polydimethyl siloxane based nanocomposite *Nano Energy* **40** 487–94

[119] Singh H H and Khare N 2018 Flexible ZnO-PVDF/PTFE based piezo-tribo hybrid nanogenerator *Nano Energy* **51** 216–22

[120] Talemi P *et al* 2015 Flexible polymer-on-polymer architecture for piezo/pyroelectric energy harvesting *ACS Appl. Mater. Interfaces* **7** 8465–71

[121] Wang S, Wang Z L and Yang Y 2016 A one-structure-based hybridized nanogenerator for scavenging mechanical and thermal energies by triboelectric–piezoelectric–pyroelectric effects *Adv. Mater.* **28** 2881–7

[122] Zheng L *et al* 2014 Silicon-based hybrid cell for harvesting solar energy and raindrop electrostatic energy *Nano Energy* **9** 291–300

[123] Zhang K, Wang S and Yang Y 2017 A one-structure-based piezo-tribo-pyro-photoelectric effects coupled nanogenerator for simultaneously scavenging mechanical, thermal, and solar energies *Adv. Energy Mater.* **7**

[124] Ji Y, Zhang K and Yang Y 2018 A one-structure-based multieffects coupled nanogenerator for simultaneously scavenging thermal, solar, and mechanical energies *Adv. Sci.* **5** 1700622

[125] Zhang X-S *et al* 2018 All-in-one self-powered flexible microsystems based on triboelectric nanogenerators *Nano Energy* **47** 410–26

Chapter 8

Wearable triboelectric nanogenerators constructed from electrospun nanofibers

Zhaoling Li, Ahmed Elhassan, Ibrahim Abdalla, Miaomiao Zhu, Jianyong Yu and Bin Ding

Nowadays, wearable and flexible electronics are attracting increasingly widespread interests from both the industrial sector and the academic community. There is a huge number of emerging wearable products currently powered by traditional batteries, posing many unfavorable limitations and concerns. TENGs can be used to capture ambient mechanical energy from surrounding motions and then convert that energy to electricity for sustainably driving portable or wearable devices. This book chapter aims to give a comprehensive overview of the state-of-the-art developments in electrospun nanofibers for constructing wearable TENGs. The fundamental operation modes, basic device structures, electrospun triboelectric materials, standard matrix for high-performance nanofiber-based TENG, and main applications of nanofiber-based TENGs are systematically introduced. We also summarize the scientific achievements of world-class scientists who have been working in this area. Meanwhile, the development of nanofiber-based TENG and its global impact are elucidated by analyzing the statistical publication numbers and the geographic distribution of these publications.

8.1 Background

Recently, with the rapid development of wireless communication technology and other emerging fields like smart medicine, space, and construction, a massive amount of small electrical and electronic devices has been developed and widely used in our daily lives. These portable and wearable electronic devices are currently powered using traditional batteries, which have many unfavorable limitations and concerns [1, 2], such as relatively heavy weight, high cost, difficulties in maintenance and replacement, regular charging, and possible risk of health hazards as a result of electrolyte leakage [3]. In order to overcome these drawbacks, a sustainable power

doi:10.1088/978-0-7503-2005-4ch8

source that secures the independent and continuous operation of such portable devices is urgently needed. Among all environmental energy sources (chemical energy, solar energy, thermal energy, mechanical energy, etc) that have already been exploited, mechanical energy and its harvesting and conversion to electricity have attracted growing attention in the past decade. Due to their amazing ability to harvest various forms of ambient energy sources, these new energy harvesting devices can take advantage of nanomaterials and nanotechnology and effectively serve as sustainable and self-sufficient power sources. Nowadays, both PENGs and TENGs are used to capture the otherwise wasted mechanical energy from surrounding motions (such as the moving human body, vibrating objects, traveling automobiles, blowing wind, and ocean waves) and then convert them to electricity for sustainably driving portable devices. The PENG was first introduced in 2006 by Zhong Lin Wang [4, 5]; the basic theoretical framework was further developed in a later publication [6], and more details of fabrication and application are found in recent review articles [7–12]. In this chapter, the main objective is to give a systematic and comprehensive introduction to TENGs, specifically the wearable TENGs fabricated from electrospun nanofibers.

8.2 TENGs

As a phenomenon that has been known for thousands of years, triboelectrification is very common and can be easily observed during daily life. Most insulators or less conductive materials have a strong triboelectric effect, and any two different materials can be electrically charged through frequent friction. High-voltage static discharge accompanies these phenomena, and has mostly been considered as a negative effect as it may cause serious problems in manufacturing and transportation. Recently, the triboelectric effect has been used in desirable applications, especially after the great development of nanogenerators. In 2012, Zhong Lin Wang and his research team fabricated the first TENG by taking advantage of the triboelectric effect and electrostatic induction. When two different materials are brought to touch or rub each other, triboelectric charges can be created, which efficiently converts the abundant motion into usable electricity. Compared with other energy harvesting devices, the TENG exhibits many compelling features, such as wide material selection [13], simple device fabrication [14], high electric outputs [15, 16], and versatile device structures [17–19]. Currently, a great deal of research attention has been given to fabricate unique TENGs with high conversion efficiency [20–23], innovative structure [24–26], good resistance matching [27], and multiple energy harvesting applications [28–30]. In addition, many efforts have been dedicated to developing TENGs for harvesting large-scale electrical power from the ocean and wind, which is emerging as a promising field beyond green energy, called blue energy.

8.3 Fundamental operation modes and basic device structures

According to the triboelectric effect, any two different materials can be electrically charged through frequent friction due to the formation of chemical bonds between some parts of two surfaces (called adhesion) through exchanging electrons or ions/molecules to equalize their electrochemical potential during contact. When separated, triboelectric charges are produced by the accumulation of electrons on one surface, which can be forced into an external circuit using electrodes in order to balance the created electrical potential drop. Based on this principle, four different modes of TENGs have been developed, namely, vertical contact–separation, lateral sliding, single-electrode, and freestanding triboelectric layer modes. As illustrated in figure 8.1, each of these modes elaborates the interaction of triboelectric layers, and therefore describes the basic structure and required motion to achieve the power generation.

8.3.1 Vertical contact–separation mode

The working mechanism of this mode can be described as the potential difference between dissimilar dielectric surfaces resulting from the driven contact and separation of opposite surfaces, as can be seen in figure 8.1(a) [31]. When applying an external force to the device, the distance between the two surfaces will decrease and then cause charge exchange between the top and bottom surfaces of the stacked structure, generating positive charges on one surface and negative charges on another surface. This triboelectric effect is mainly determined by the intrinsic properties of selected materials [32–34]. As the force is released, the two surfaces are separated by a small gap, which induces a potential difference across the top and bottom surfaces, thus driving the free electrons flow between the two connected electrodes in order to balance the electrostatic field. Subsequently, when the force is loaded again, the transferred charges will flow back through the external load.

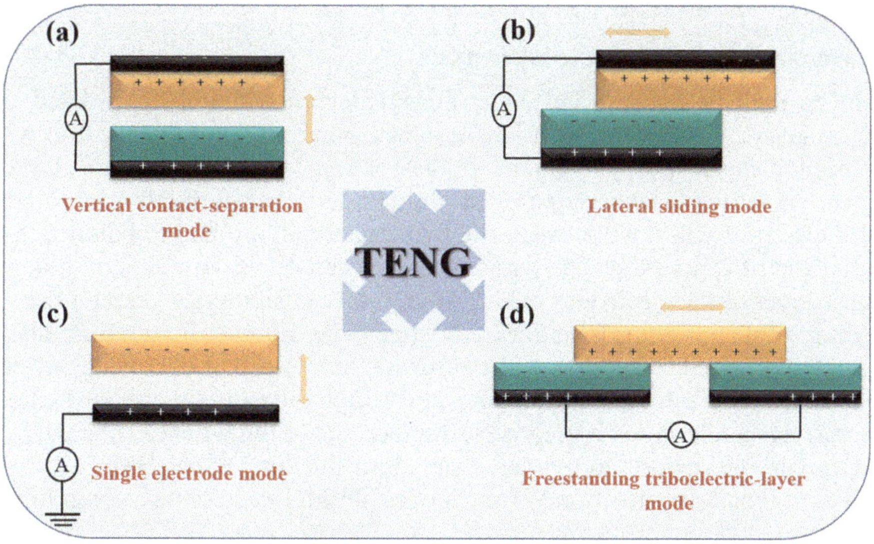

Figure 8.1. The four fundamental modes of TENGs: (a) vertical contact–separation mode, (b) lateral sliding mode, (c) single-electrode mode, and (d) freestanding triboelectric layer mode [31].

8.3.2 Lateral sliding mode

This mode has the same structure as the vertical contact–separation mode but involves horizontal motion. In the initial state, once the two dielectric contacted surfaces slide outward, as shown in figure 8.1(b), the charges on both surfaces are no longer balanced, leading to an increase in created triboelectric charges [19] as the contact area of the two surfaces decreases. Moreover, in order to balance the potential difference, these induced charges can be driven through an external circuit. Subsequently, when the two surfaces slide back to the original position, the charges will flow back from the previous electrode in the same manner. The electrical output can be obtained after periodic sliding between two dielectric materials [35]. Devices with a lateral sliding mode show high energy harvesting efficiency compared with the vertical contact–separation mode. Because of various sliding motions in human movement [36], this mode also shows quite promising prospects in wearable applications.

8.3.3 Single-electrode mode

Unlike other modes, this mode has one electrode and one dielectric material, as shown in figure 8.1(c). According to the triboelectric series, the local electrical field distribution of the TENG device (with limited size) will effectively change with the contact–separation movement of the freely moving object (such as human skin [37, 38]), resulting from the electron exchanges between the electrode and the ground in order to balance the potential at the electrode. Comparing with others modes, the output of this mode is lower, mainly due to the existence of only one electrode in the system [39]. However, the single-electrode TENG is more suitable for some applications such as fingertip-driven energy harvesting.

8.3.4 Freestanding triboelectric layer mode

As can be seen in figure 8.1(d), the TENG device is mainly composed of two dielectric layers as bottom electrodes (aligned horizontally or vertically) with the other friction layer. The movement of the friction layer (forward and backward) along the underneath layers creates an unbalanced charge distribution, and these induced electrons will flow between the two electrodes in order to balance the local potential distribution [40]. The moving object can be in contact, or not, with a dielectric layer of the electrodes, which, due to the possibility of keeping the surface charges on some materials for hours, can effectively decrease material abrasion and heat generation [35]. The cycled forward and backward movement between the paired electrodes produces electrical power through the external circuit. It should be noted that this mode shows a great performance due to the potential change between two stationary electrodes; moreover, there is no shielding effect [39].

Based on the above-mentioned four modes, TENG devices have been fabricated with diverse and composite structures in order to meet different applications. Table 8.1 summarizes the basic device structures of all four operation modes, which have been considered as the fundamental units of TENG technologies.

Table 8.1. Basic device structures of the four operation modes.

Basic modes	Basic device structures	Ref.
Vertical contact–separation mode	Spacer structure	[17, 41]
	Arch-shaped structure	[42]
	Spring-assisted separation structure	[15]
	Multiple-layer integration	[43]
	Microcavity–nanoparticle assembled structure	[44]
Lateral sliding mode	Plain-sliding structure	[19, 45]
	Linear-grating structure	[16, 36]
	Rotation-disk structure	[46]
	Rotation-cylinder structure	[47]
	Case-encapsulated structure	[48]
	Liquid–metal structure	[21]
Single-electrode mode	Contact–separation structure	[49, 50]
	Lateral sliding structure	[18]
Freestanding triboelectric layer mode	Linear-grating structure	[51]
	Rotation-disk structure	[52, 53]

8.4 Nanofiber-based wearable TENGs

With the rapid growth of portable electronics in modern society, wearable technologies have been recognized as the next generation of innovative technology and are attracting more attention from both academic and industrial areas. Owing to their wide applications in artificial intelligence, human–machine interfaces, personal sensing, and distributed sensors [54, 55], many efforts have been devoted to fabricating wearable devices with flexible, breathable, stretchable, and tailorable properties in order to meet the requirements of practical usage. Nowadays, due to the obvious limitations of traditional batteries as power supplies to drive electronic devices [1, 2], wearable TENGs have been considered as ideal replacements.

The enhancement of output performance of wearable TENGs is mainly dependent on two important parameters: sustainable material selection and structural design optimization. The constructing materials should be foldable, stretchable, and even washable in order to be part of human clothing [56]. Accordingly, many efforts have been made to fabricate textile-based TENGs because of their favorable wearable conditions [57–63]. Furthermore, with the development of nanotechnology, different methods have been successfully performed to maximize surface charge densities; this can be achieved by designing novel structures and modifying or functionalizing the contact surfaces of triboelectric materials. Due to their high surface area, rough morphologies and unique electronic properties [64, 65], nanoscale materials such as nanowires or nanoparticles have been used to provide more charge carrier sites [42, 66–69]. In addition, the charge density of the contact surfaces can be further enhanced through ion treatment [70–72], silicon templates [73], anodic aluminum, and oxide templates [74]. However, for these commonly used techniques there still exist some technical limitations, such as complicated and time-consuming procedures, expensive and sophisticated instruments, high operation cost, and unsatisfying practical performance [75], which thus pose a lot of challenges for developing wearable TENGs.

Among all the techniques used to fabricate wearable TENGs, more recent attention has been given to those used to fabricate nanofiber-based TENGs. Due to the significant advantages of simple processing, low cost, easy scalability, adjustable properties [76, 77], and its capability to process different materials (natural polymers, synthetic polymers, ceramics, inorganic nanoparticles, active agents, etc) the electrospinning technique can be effectively adopted as a highly efficient approach to produce multifunctional or multicomponent nanofibrous membranes. According to previous research, a high specific surface area and inherently rough structure of electrospun membranes can be significantly beneficial to increasing the electric outputs of wearable TENG devices, and as a result a higher charge density can be obtained. Moreover, through proper optimization of electrospinning parameters (temperature, humidity, voltage, collector, auxiliary electrode, etc), a wide range of fiber-based TENGs with unique features have been successfully constructed for wearable applications.

As shown in figure 8.2(a) [78], electrospun PES and cellulose acetate (CA) nanofibers were used as friction layers in a multilayered fiber-based TENG. In

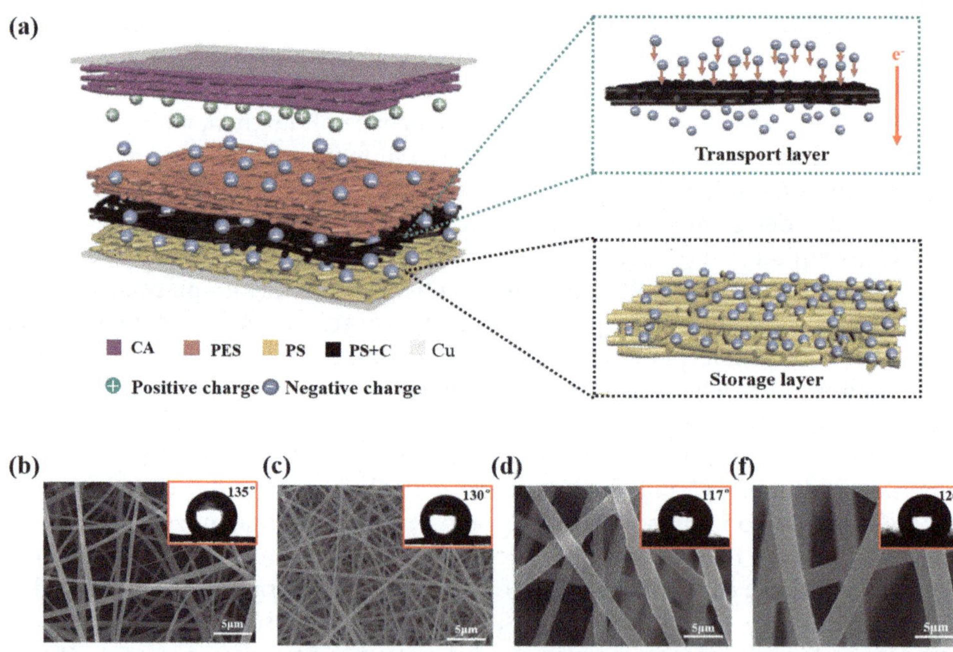

Figure 8.2. Working principle of the multilayer fiber-based TENG. (a) Illustration of triboelectric charge transport and storage process in the three-layer TENG. SEM images of (b) the electrospun CA nanofibers, (c) the electrospun PES nanofibers, (d) the electrospun PS nanofibers, (f) the electrospun PS nanofibers containing carbon black particles. Reproduced from reference [78]. Copyright 2018 Elsevier.

addition, a polystyrene (PS) nanofiber layer was chosen as the charge storage layer and a composite PS–carbon black nanofiber layer was chosen as the charge transport layer. The SEM images in figures 8.2(b)–(f) show the morphologies of the four membranes, namely the two friction layers, the charge storage layer, and the charge transport layer along with their water contact angles. As seen in figure 8.3, the presence of the charge transport layer and charge storage layer can greatly improve the output performance of the nanofiber-based TENG device. Accordingly, the short-circuit current and open-circuit voltage of the multilayered structure were distinctively 3.07 times and 2.56 times larger than those of the single-layer TENG. During the contact–separation process, the multilayered TENG showed a power density of 0.13 W m^{-2} with an effective area of 9 cm^2 at a load of 30 MΩ. Furthermore, the fabricated TENG possessed excellent durability, good stability (even after 600 cycles of continuous operation), and outstanding flexibility. It could even be integrated into a textile, and thus is very suitable for harvesting biomechanical energy from human motions.

Yin [79] obtained a high-porosity nanofiber-based TENG through the electrospinning technique. As shown in figure 8.4(a), this stretchable and tailorable arch-structured TENG has two friction layers, an electrospun PVDF nanofiber layer supported by a piece of conducting fabric as the negative triboelectric layer, and an electrospun thermoplastic polyurethanes (TPU) nanofiber layer supported by Ag elastic fabric as the positive triboelectric layer. Figures 8.4(b) and (c) show SEM images of the two friction layers. The nanofiber layers display a rough surface with nano/microholes between the fibers, providing breathability. Besides, due to the outstanding stretchability and flexibility, the nanofiber-based TENG can harvest

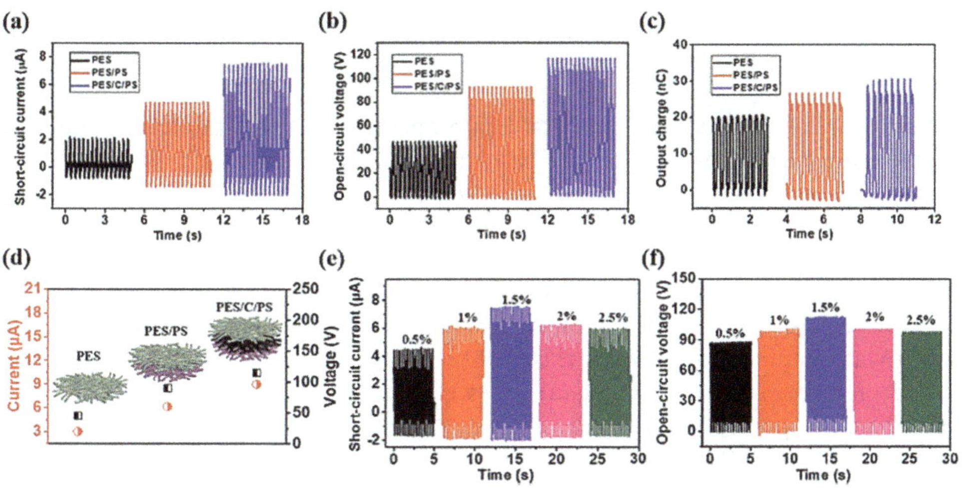

Figure 8.3. Electrical output of TENG with different negative friction layer structures. (a) Short-circuit current, (b) open-circuit voltage, and (c) charge generated from TENGs with different negative friction layer structures. (d) The trend of short-circuit current and open-circuit voltage of the TENGs. A comparison of (e) short-circuit current and (f) open-circuit voltage of the fabricated TENG with different concentrations of carbon black particles. Reproduced from reference [78]. Copyright 2018 Elsevier.

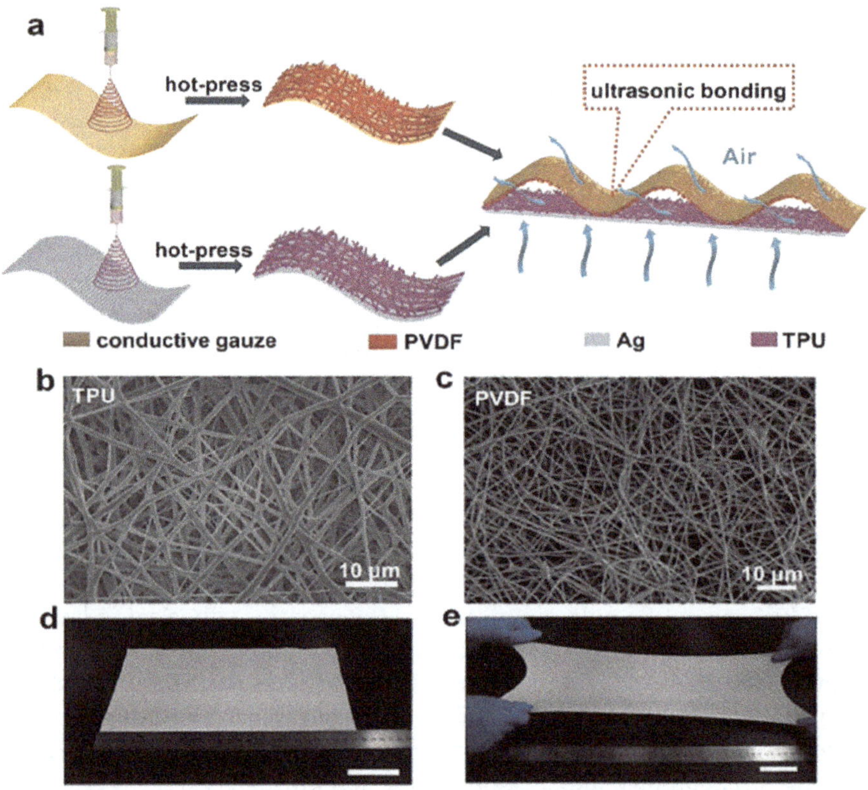

Figure 8.4. Structure diagram of a stretchable and tailorable TENG. (a) The schematic diagram of the fabrication processes. (b) SEM image of the TPU nanofibers. (c) SEM image of the PVDF nanofibers. Photographs to show (d) the TPU/Ag layer and (e) its stretching (scale bar, 5 cm). Reproduced from reference [79]. Copyright 2018 Wiley.

various types of irregular movements. Upon operation, the open-circuit voltage and short-circuit current of a single TENG unit can reach up to 170 V and 4.5 μA, respectively, and thus could be used as a self-powered motion sensor.

Figure 8.5 shows a biomimetic nanofiber-based triboelectric nanogenerator (BN-TENG) made of PHBV and PVDF nanofibers [80]. The PVDF nanofiber mat was subjected to press treatment to produce a honeycomb pattern (see the configuration in figures 8.5(b) and (d)). The nanofibers show a rough surface (figures 8.5(e)–(g)), which was obtained by mixing the PVDF solution with mineral oil during the electrospinning process. The hexagon structures of the pressure-treated PVDF nanofiber mat prevent charge dissipation. The BN-TENGs delivered a maximum output voltage of 1000 V with a short-circuit transfer charge density of 364 $\mu C \cdot m^{-2}$ and a load power density of 6 $W \cdot m^{-2}$, which are sufficient to power some wearable electronics and charge a commercial Li-ion battery.

In summary, electrospun nanofibrous membranes are quite suitable for fabricating wearable TENG devices for wide-range applications. The fabrication process of the nanofiber-based TENG is much simpler than that of other textile-based TENG

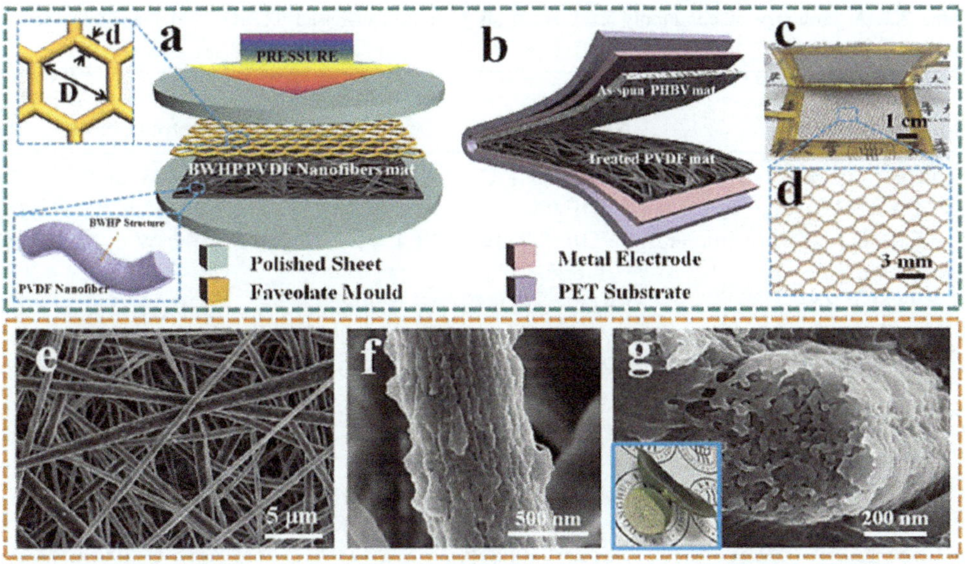

Figure 8.5. (a) Preparation of honeycomb mold pressure-treated PVDF nanofiber mats in the BN-TENG. D and d are defined as the double of side length and the width of the orthohexagonal mold, respectively (upper left). (b) Schematic diagram of the fabricated BN-TENG. (c), (d) Digital photographs of the fabricated BN-TENG and pressure-treated PVDF nanofiber mat. (e)–(g) SEM images of the PVDF nanofibers. Reproduced from reference [80]. Copyright 2018 Elsevier.

devices, since every single fiber is supposed to be considered in order to fabricate the electrode and friction layer. Owing to their fine fibrous structures with high surface area, electrospun membranes are ideal candidates to construct highly sensitive electronic devices that cannot be replaced by traditional textile-based TENG devices.

8.5 Electrospun triboelectric materials

According to triboelectrification, almost all materials can be electrically charged when put in contact with a different material, from natural materials to synthetic materials (organic and inorganic). Therefore, there is a wide range of materials to choose from when fabricating TENGs. As mentioned in the triboelectric series first published by John Carl Wilce, more negative charges will be achieved when a material from the bottom of the series is touching a material near the top of the series. Consequently, a huge range of polymers can be used to fabricate nanofiber-based TENG devices. According to their positions in triboelectric series, a high-efficiency positive triboelectric layer can be fabricated using metals such as Al, Ni, Cu, and Ag, as well as organic materials including nylon, cotton, silk, and polyurethane (PU). Typical negative triboelectric layers are PVDF, PDMS, PTFE, FEP, parylene, silicone, and rubber. Furthermore, different materials such as nanoscale metals and carbon-based materials can be included in triboelectric materials via electrospinning in order to improve the output performance of

Table 8.2. A summary of commonly used materials in nanofiber-based TENGs and their corresponding output performance.

No.	Positive layer	Negative layer	Supporting substrate	Output charge	Ref.
1	CA	PES + PS + CB	/	0.13 W m^{-2}	[78]
2	CA + PU	PVDF	/	1.3 W m^{-2}	[3]
3	External fabric or human skin	PVDF + PTFE + PET fabric	/	80 mW m^{-2}	[81]
4	PAN + PA6	PVDF + PDMS	PMMA	110 μA, 540 V	[82]
5	Conductive fabric	PVDF	PMMA	7 W m^{-2}	[83]
6	PA6	PVDF–HFP	RGO	80 V, 1.67 μA, 0.45 mW	[84]
7	Nickel-coated conductive fabric	PVDF + PU	/	150 V, 10.38 μA	[85]
8	Latex glove + bare hand	PVDF–TrFE	PDMS + Ag NW	217 W m^{-2}	[86]
9	TPU/Ag	PVDF/conducting fabric		170 V, 4.5 μA	[79]
10	PET/Al thin film	Fe_3O_4 NP/PVDF	Al	5.68 μA, 138 V	[87]
11	Nylon nanofiber	PVDF–Ag NW	Al	13 μA, 240 V	[88]
12	Nylon 6, soy protein, and lignin NFs	polyimide (PI) tape		1.1 μW	[89]

TENGs. As summarized in table 8.2, among all the polymers used to fabricate nanofiber-based TENGs, PVDF, PVDF copolymers, and PTFE are considered as the most effective negative triboelectric materials, which is mainly due to the high electron affinity of their fluorine side chains [13].

8.6 Standard matrix for high-performance nanofiber-based TENG

A key factor determining the energy conversion efficiency of TENG is surface charge density. Researchers have found that the final output performance of TENGs is mainly controlled by material compositions and device configurations. However, the electrospinning technique can create a large content of the crystalline β phase for ferroelectric polymers through simultaneous uniaxial stretching and electric field poling [90–93], which can significantly enhance the surface charge densities by delivering the greatest degree of spontaneous polarization per unit cell. Accordingly, along with their flexibility and stretchability, electrospun ferroelectric polymers are the best choice to fabricate wearable TENG devices [94, 95]. In addition, many efforts have been made to enhance the surface charge density through modifying or functionalizing the contact surfaces of triboelectric materials.

8.6.1 Polymer nanocomposites

Generally, most of the reported nanofiber-based TENGs can be classified into polymer composites, fabricated by embedding one or more components in a polymer matrix through electrospinning, which can effectively change the surface electrification and permittivity of the materials. Moreover, in order to further boost the output performance of energy harvesting devices, a wide range of materials (nano/micro metals and carbon-based materials [96–98]) have been explored for combination with nano/microfibers. For example, a nanofibrous membrane with a high power generation single-electrode triboelectric nanogenerator (SETENG) was fabricated by Kim [86] through incorporating Ag nanowire (NW) into PVDF–TrFE matrix during the electrospinning process. The output performance was enhanced as the Ag NW content increased from 0 to 2 wt%. As seen in figure 8.6, the output power of SETENG increased from 13 mW to 22 mW. In addition, the short-circuit current and open-circuit voltage increased up to 90 μA and 150 V by adding 2 wt% Ag NWs. This unique nanofiber-based SETENG can be used for wearable applications, trajectory monitoring, external physical sensing, and chemical stimuli sensing.

Ji-Su [87] prepared Fe_3O_4 nanoparticle/PVDF composite nanofibers (NFs) through electrospinning. As shown in figure 8.7, the incorporation of iron nanoparticles into PVDF nanofibers leads to improved output performance of TENG. It was confirmed that the Fe_3O_4 NP could be distributed uniformly within the nanofibers. When NP content was 11.3 wt%, the output voltage increased from 124 V to 138 V.

Cheon [88] prepared PVDF–Ag nanowire composite and nylon nanofibers through the electrospinning technique and then used them as the friction layers. The nanofiber-based TENG exhibited promising output performance. Benefiting from electrospinning, the electrospun membranes showed highly polar crystalline β

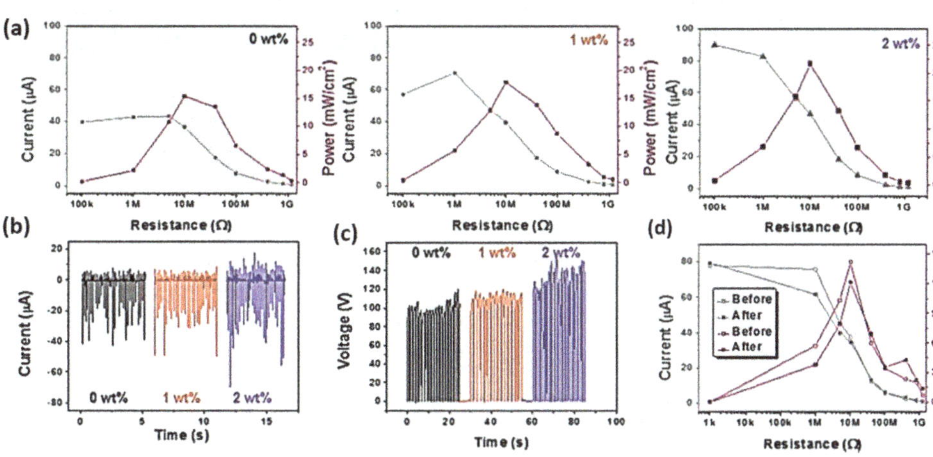

Figure 8.6. Performance of SETENGs: (a) calculated power increasing with the load resistance, for Ag NW contents of 0, 1, and 2 wt% in the PDMS-embedded composite NFs. (b) The short-circuit current and (c) open-circuit voltage. (d) Performance comparison of the PDMS-embedded composite NFs with 2 wt% Ag NWs before and after 10 000 cycles of a 10% stretching test. Reproduced from reference [86]. Copyright 2019 American Chemical Society.

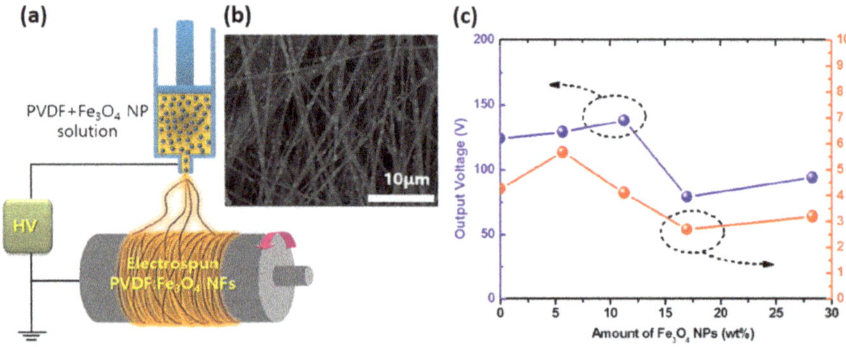

Figure 8.7. (a) Schematic diagram illustrating the fabrication of the Fe_3O_4 NP/PVDF composite nanofibers. (b) SEM images of Fe_3O_4 NP/PVDF composite nanofibers. (c) Output voltage and current of the TENG. Reproduced from reference [87]. Copyright 2018 American Chemical Society.

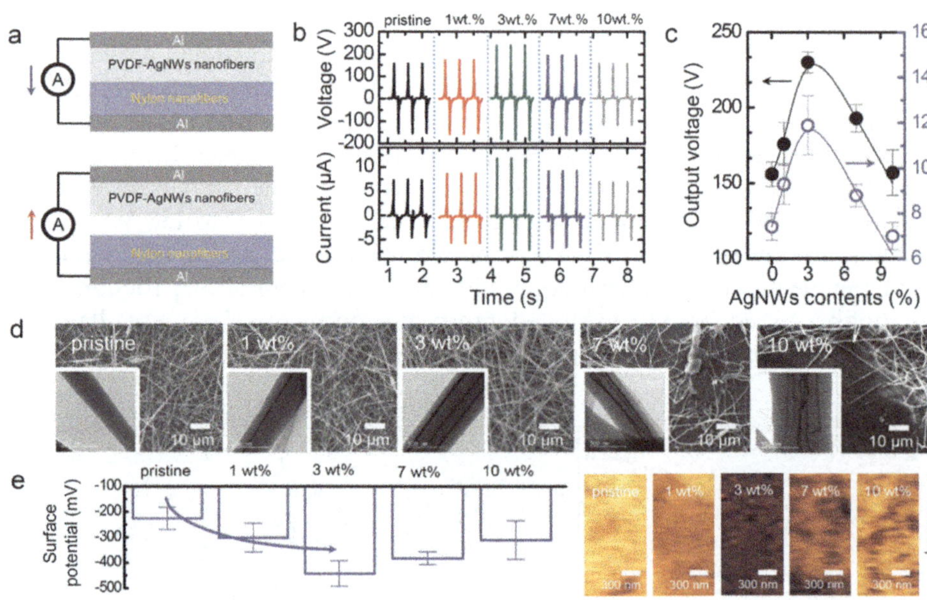

Figure 8.8. (a) Schematic illustration of the contact-mode TENGs based on PVDF–Ag NW composite and nylon nanofibers. (b) Output voltage and current of the TENGs. (c) Maximum output voltage and current extracted from panel (b). (d) SEM and TEM images of the PVDF–Ag NW composite nanofibers prepared with various Ag NW concentrations. (e) Surface potential of the PVDF–Ag NW composite nanofibers with various Ag NW concentrations. Reproduced from reference [88]. Copyright 2017 Wiley.

phase contents. As can be seen from figure 8.8, 3 wt% of Ag NWs led to a sharp increase in the output voltage (from 156 V to 240 V). Figure 8.8(d) shows the SEM and TEM images of the PVDF–Ag NW composite nanofibers prepared with various Ag NW concentrations. At a concentration up to 3 wt%, the PVDF–Ag NW exhibited a uniform and bead-free fibrous structure over a large area.

8.6.2 Surface modifications

Most friction materials for TENGs are subjected to surface modifications (either physical or chemical) to improve the interface contact properties [99, 100]. Physical surface modifications to increase the roughness of the friction layers can enhance the contact area of triboelectric materials. To enlarge surface roughness, micro/nano-structure materials have been incorporated into the surface of friction layers [101]. However, these methods have issues of complex procedures, long processing time, limited fabrication area, and high treatment cost [102].

In contrast to physical surface modification, chemical surface modification is simpler, and can better control the contact area, alter chemical properties, and even increase charge carrying sites without changing the substrate. In order to maximize the TENG's efficiency, many materials with the tendency of gaining or losing electrons (listed in several triboelectric series [27, 103, 104]) have been used to functionalize the triboelectric friction layers through various methods, such as fluorinating surfaces [23], ion injection [70], and molecular targeting [105]. Along with their ability to enhance the contact area of triboelectric materials, chemical modification can address many issues such as poor mechanical properties, low intrinsic hydrophobicity, and so on.

For example, Shin *et al* [27] used PET films as both friction layers. By functionalizing one PET surface with poly-L-lysine for positively charged amino groups ($-NH_3$) and another PET surface with trichloro (1H,1H,2H,2H-perfluorooctyl)silane for negatively charged $-CF_3$ groups, triboelectric polarities are obtained. As figure 8.9(a) shows, the contact–separation mode TENGs with four different functionalization pairs (named (i) PET:PET, (ii) PET:P-PET, (iii) PET:F-PET, and (iv) P-PET:F-PET) were readily achieved by dip-coating and self-assembling deposition methods. Figures 8.9(b) and (c) show under the forward connection mode the maximum $V_{\text{open-circuit}}$ (V_{oc}) of 330 V and $J_{\text{short-circuit}}$ (J_{sc}) of 270 mA m^{-2} for the P-PET:F-PET pair. The device had outstanding stability and could work for over 7200 cycles without losing performance. The approach demonstrated here is a simple, cost-competitive, and effective way to fabricate high-performance TENGs.

Shen *et al* [3] reported a high-performance humidity-resisting triboelectric nanogenerator (HR-TENG) through electrospinning and surface chemical amino modification. As figure 8.10(a) shows, the electrospun CA/PU membrane was firstly hydrolyzed through immersion in NaOH solution, and then the hydrolyzed membrane was immersed in a polyacrylamide (PAM) solution at different concentrations. Finally, the treated CA/PU membrane with a layer of amino groups was achieved after the electrostatic self-assembly process. The output of the TENG device increased with the increase in PAM concentration. In addition, the output remained at a relatively high level even when tested under a relative humidity up to 55%, as can be seen in figures 8.10(b)–(d); a current and voltage output up to 28 μA and 345 V, with a power density of 1.3 W m^{-2}, were obtained. The good output stability in high environmental humidity indicates that this device could be a sustainable power source to drive wearable electronics during human sports even with heavy perspiration.

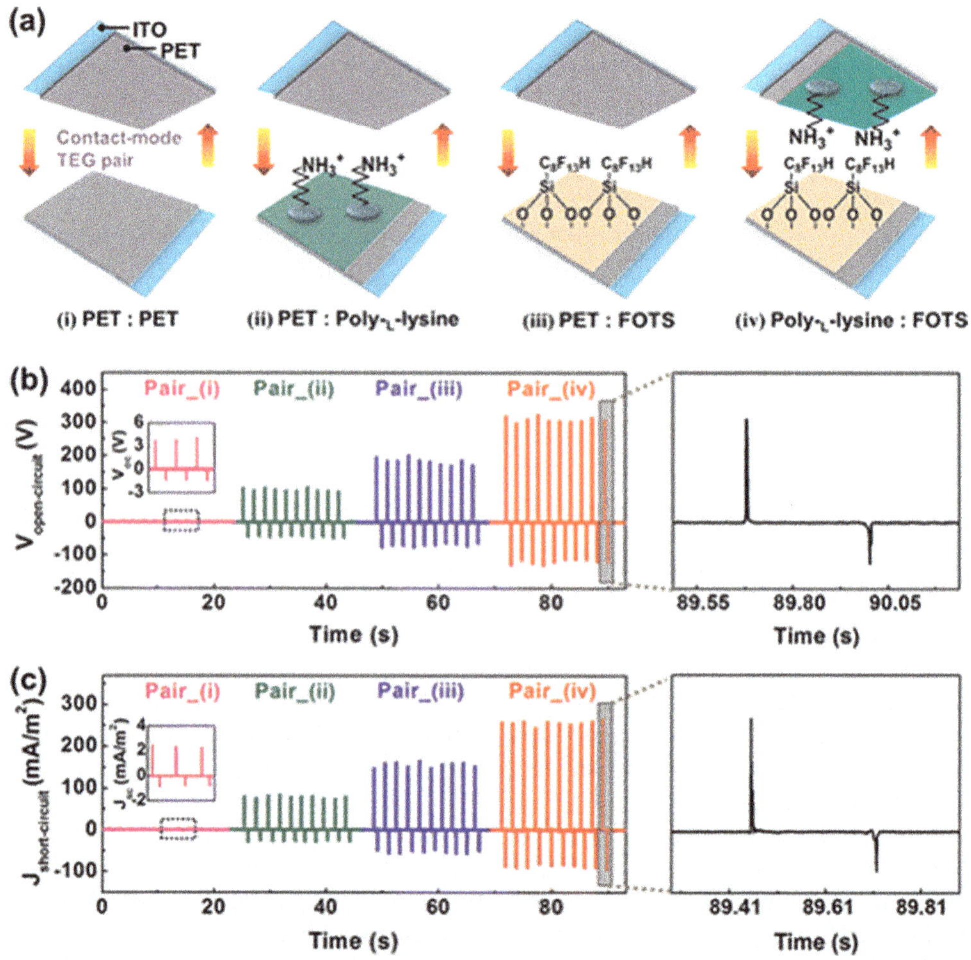

Figure 8.9. (a) The contact pairs of the TENGs. (i) PET:PET, (ii) PET:P-PET, (iii) PET:F-PET, and (iv) P-PET:F-PET. (b) Open-circuit voltages and (c) short-circuit current densities of the TENGs. Reproduced from reference [27]. Copyright 2015 American Chemical Society.

8.7 Applications of nanofiber-based TENGs

Electrospinning can prepare different kinds of membranes suitable to construct wearable TENGs [82]. The intrinsically rough structure of nanofibers can greatly enhance the output performance of TENGs, which are simple and cost-effective materials for self-powered wearable electronics [106]. TENGs can be used in many applications, with a particular advantage in harvesting mechanical energy, and can serve as self-powered active sensors. These devices can convert low-frequency mechanical energy of human motions, such as walking, shaking, jogging, stepping, typing, or deformation of clothes, into electricity for practical applications [107]. Moreover, the nanofiber-based TENGs can be widely applied in many emerging

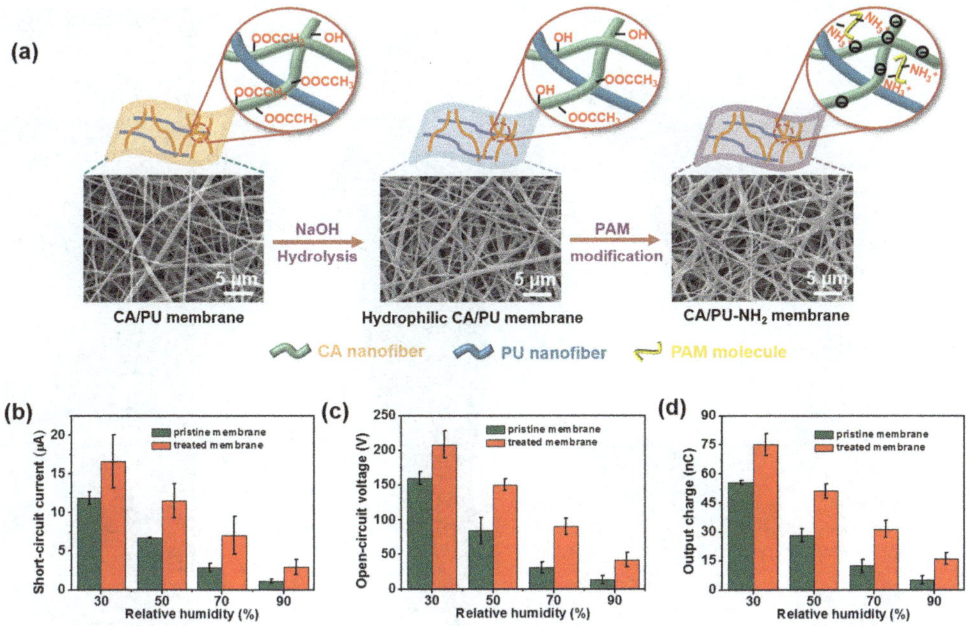

Figure 8.10. Surface amino modification onto the nanofibrous membranes as a strategy for hydrophobicity enhancement. (a) Schematic illustration of surface amino modification and SEM images of CA/PU nanofibrous membranes before and after chemical treatment. Electrical output of the HR-TENG under various humidity conditions. A comparison of (b) short-circuit current, (c) open-circuit voltage, and (d) output charge of the HR-TENG in different atmospheric humidities when the CA/PU membrane was treated by the amino-functionalized technique. Reproduced from reference [3]. Copyright 2017 Elsevier.

areas, such as implantable biomedical devices, wireless sensors, remote sensing of firefighters in hazardous situations, smart artificial skins, health monitoring, artificial intelligence, pressure detection, and wearable electronics [108–112]. The nanofiber-based TENGs can be directly used for powering commercial LEDs without using a storage or power regulation unit.

Figure 8.11(a) demonstrates that under normal hand tapping, a total of 500 LEDs, arranged in a pattern of the 'DONGHUA' logo, can be lit to full brightness simultaneously. The nanofiber-based TENGs can also be used for energy harvesting from human walking. A great number of LEDs were fully lit up when a human walked naturally with the insole, as presented in figure 8.11(b). Shaking the two cloths within a small area can generate enough electric power to light up many LEDs (figure 8.11(c)). The nanofiber-based TENGs could power a commercial thermometer (figure 8.11(d)), which showed an immediate finger temperature of 32.8 °C in a self-powered manner without an external power source. The power-supplying system had the capability to power an electronic watch, as illustrated in figure 8.11(e). In figure 8.11(f), the nanofiber-based TENG can realize the function of a finger motion detector. Figure 8.11(g) shows a photograph of the three-layer composite TENG combined with a commercial kneepad; this can be used for various human biomechanical energy harvesting purposes. Furthermore, 10 LEDs

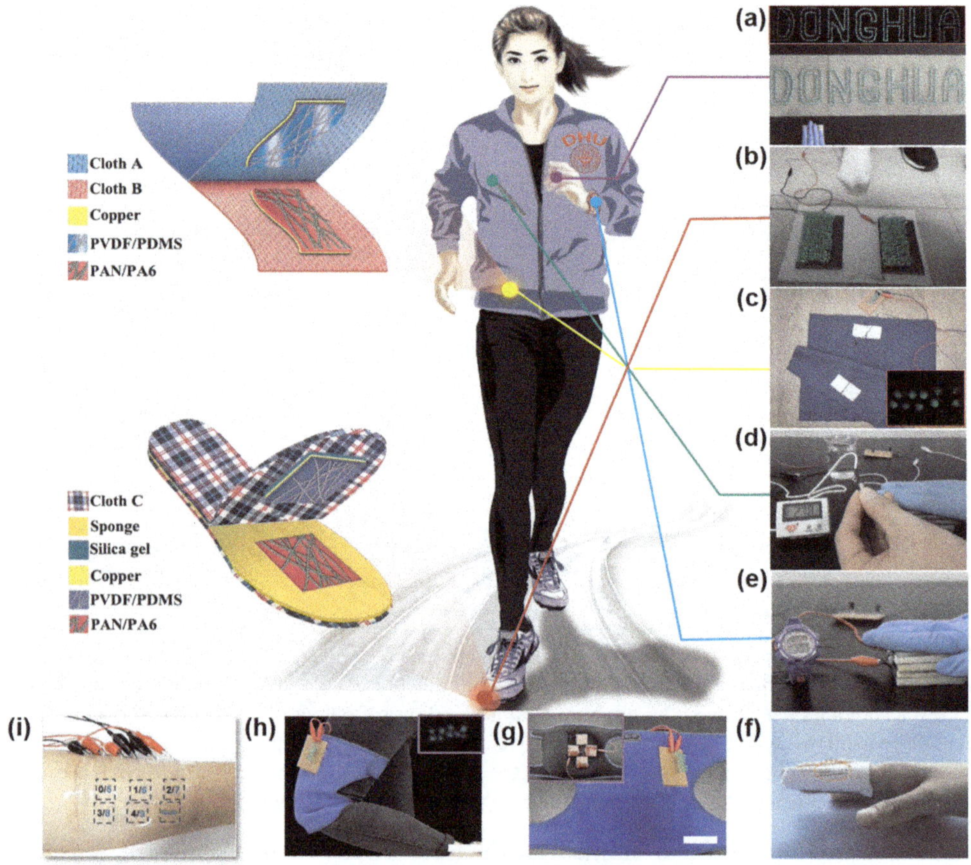

Figure 8.11. Photographs showing a nanofiber-based TENG designed to harvest various forms of biomechanical energy from the human body. (a) Photograph to show directly lighting up LEDs by gentle hand tapping of the TENG. The power-supplying system could sustainably power a (b) thermal meter and (c) an electronic watch. (d) Photograph showing that the TENG could be integrated into cloth for harvesting energy from normal human walking. (e) The TENG-based power generating insole could efficiently harvest energy from human walking and light up many LEDs. Reproduced from reference [82]. Copyright 2017 Elsevier. (f) An optical photograph of a finger motion detector. Reproduced from reference [112]. Copyright 2017 Springer Nature. (g) Photograph of three-layer composite TENG combined with a commercial kneepad. Inset shows the inner structure of the power kneepad. (h) Lighting up LEDs by human knee bending. Reproduced from reference [78]. Copyright 2017 Elsevier. (i) An image showing the key composition of the proposed panel. Reproduced from reference [86]. Copyright 2019 American Chemical Society.

can be lit up by a human knee bending (figure 8.11(h)). Figure 8.11(i) displays the image of the key composition consisted of six unit-cells, which could be used to write all the numbers.

8.8 Data collection and analysis

To express the growing research tendency of the nanofiber-based TENG field, herein we run search queries using common keywords that are related to the nanofiber-based TENG research in the Web of Science database with AND operator(s) between keywords. The keywords include electrospinning AND triboelectric, electrospinning AND triboelectricity, electrospun AND triboelectric, electrospun AND triboelectricity, electrospinning AND TENG, electrospinning AND TENG, electrospun AND TENG, electrospun AND TENG, nanofiber AND triboelectric, nanofiber AND triboelectricity, nanofibers AND triboelectric, nanofibers AND triboelectricity, nanofibre AND triboelectric, nanofibre AND triboelectricity, nanofibres AND triboelectric, nanofibres AND triboelectricity, nanofibrous AND triboelectric and nanofibrous AND triboelectricity. The results were retrieved on June 7, 2019 with the keywords appearing in the titles, abstracts, or keywords of the publications. As can be seen from figure 8.12, the number of publications increases each year, from 1 in 2013 to 46 in 2018. In the year 2019, there were already 25 publications by June 7, 2019. The above-mentioned search queries are relevant to the nanofiber-based TENG research situation worldwide, which clearly implies that this emerging field is becoming more and more popular and an increasing number of researchers are becoming involved in this line of research.

These research articles have been published in 35 different scientific journals indexed in the Web of Science database, and the related results are demonstrated in figure 8.13. These publications include *Advanced Materials, Advanced Energy Materials, ACS Nano, Advanced Functional Materials, Nano Energy, Journal of Materials Chemistry A, Small, Nano Research, ACS Applied Materials & Interfaces, Nanoscale, Analytical Chemistry, Journal of Materials Chemistry C, Sensors and Actuators B: Chemical, Advanced Healthcare Materials, Advanced Materials Interfaces, Bioconjugate Chemistry, Advanced Materials Technologies, Journal of Physical Chemistry C, APL Materials, Scientific Reports, Applied Physics Letters,*

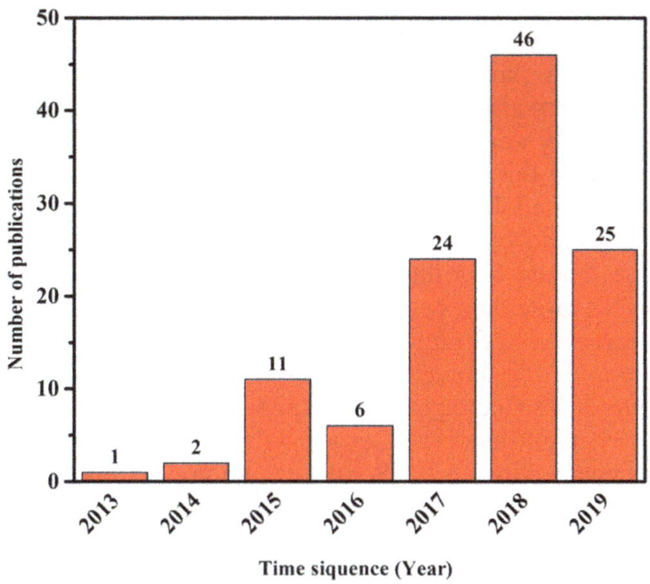

Figure 8.12. The number of publications on nanofiber-based TENG in each year (as of June 7, 2019).

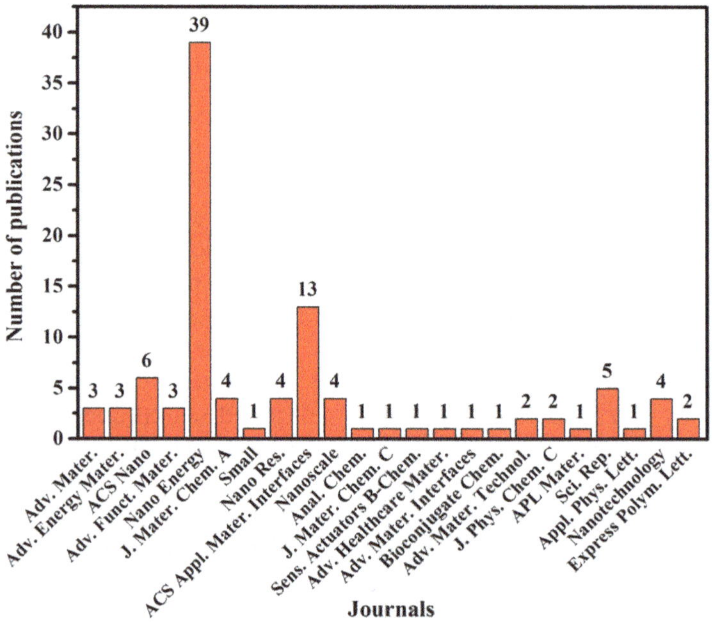

Figure 8.13. The number of publications on nanofiber-based TENGs in the top 23 scientific journals.

Nanotechnology, and *Express Polymer Letters*. Among these publications, almost 70% (81 out of 115) were published in journals with impact factors higher than 7.0 (as of 2019) according to Journal Citation Reports.

From the results obtained through the Web of Science database (figure 8.14), we can see that the nanofiber-based TENG research articles have been published by research institutions and universities across 12 countries: China, Korea, United States, Canada, Turkey, United Kingdom, Singapore, Switzerland, Bangladesh, Japan, Iran, and India. Among these countries, almost 62% (71 out of 115) of the articles were from China. Based on these results, it can be concluded that China is a fast-developing country in nanofiber-based TENG research and contributes greatly to the advancement of this field.

These nanofiber-based TENG research publications have been published by 57 research institutions and universities according to retrieval results from the Web of Science database. Figure 8.15 illustrates the research institutions and universities that have published more than two articles: Donghua University, Chinese Academy of Sciences, University of Strathclyde, University of Science and Technology of China, The Hong Kong Polytechnic University, National Central University, Georgia Institute of Technology, Peking University, Zhejiang University, Yonsei University, University of Illinois at Chicago, Ulsan National Institute of Science and Technology, Tsinghua University, Southwest Jiaotong University, Shanghai Jiao Tong University, Seoul National University of Science and Technology, Northwestern Polytechnical University, Nanyang Technological University, Hee University, and Advanced Institute of Science and Technology. Among those research institutions and universities, Donghua University and Chinese Academy

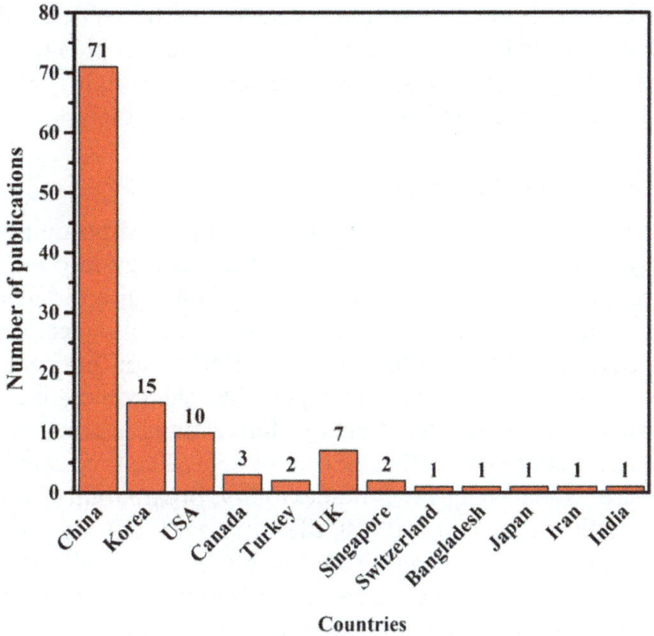

Figure 8.14. The number of publications on nanofiber-based TENGs in different countries.

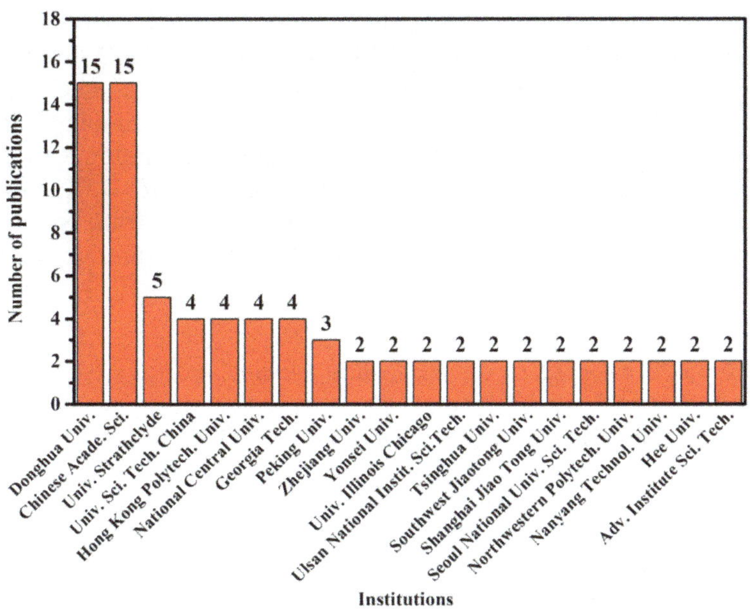

Figure 8.15. The number of publications on nanofiber-based TENGs from different research institutions.

of Sciences are two leading research agencies with well-established research foundations, and have constituted around 26% (30 out of 115) of all publications. Specifically, for Donghua University, four articles with high impact factors have been published from our research group.

8.9 Conclusions and perspectives

The invention of the nanofiber-based TENG has been an important breakthrough in the nano-energy field, as it can convert mechanical energy into electricity for self-powered electronics and systems. It provides a promising alternative approach to collect ambient ubiquitous energy using widely available organic and inorganic electrospun materials. The nanofiber-based TENG can not only be used for microscale power sources to drive some portable electronics, but it can also be scaled up to macroscale mechanical energy harvesting to produce a high-volume power density with a conversion efficiency up to 85%. The fabricated devices exhibit compelling features, such as light weight, flexibility, breathability, shape conformity, and even washability, and can harvest the abundant but wasted environmental energy. Such energy includes that produces by human walking, running, wind vibration, water flow, tire rotating, and so on. We expect that more nanofiber-based TENG devices will be increasingly and extensively applied in smart textiles and next-generation wearable electronics.

Acknowledgments

This work was financially supported by the National Natural Science Foundation of China (Grant Nos. 51703022 and 51873030), National Key R&D Program of China (2018YFC2000900), Natural Science Foundation of Shanghai (Grant No. 18ZR1402100), Shanghai Committee of Science and Technology (Grant No. 19QA1400100), Shanghai Municipal Science and Technology Committee of Shanghai Outstanding Academic Leaders Plan (Grant No. 18XD1400200), the Fundamental Research Funds for the Central Universities (Grant No. 16D310105), and the 'DHU Distinguished Young Professor Program' (Grant No. LZB2017002).

References

[1] Lee J *et al* 2014 Unlocking the potential of cation-disordered oxides for rechargeable lithium batteries *Science* **343** 519

[2] Steele B C H and Heinzel A 2001 Materials for fuel-cell technologies *Nature* **414** 345–52

[3] Shen J *et al* 2017 Humidity-resisting triboelectric nanogenerator for high performance biomechanical energy harvesting *Nano Energy* **40** 282–8

[4] Wang X *et al* 2006 Piezoelectric field effect transistor and nanoforce sensor based on a single ZnO nanowire *Nano Lett.* **6** 2768–72

[5] Wang Z L 2007 Nanopiezotronics *Adv. Mater.* **19** 889–92

[6] Zhang Y, Liu Y and Wang Z L 2015 Fundamental theory of piezotronics *Adv. Mater.* **23** 3004–13

[7] Wang Z L 2009 ZnO nanowire and nanobelt platform for nanotechnology *Mater. Sci. Eng. R-Rep.* **64** 33–71

[8] Wang Z L *et al* 2010 Lateral nanowire/nanobelt based nanogenerators, piezotronics and piezo-phototronics *Mater. Sci. Eng. R-Rep.* **70** 320–9
[9] Park J Y, Salauddin M and Rasel M S 2019 Nanogenerator for scavenging low frequency vibrations *J. Micromech. Microeng.* **29**
[10] Ali F *et al* 2019 Piezoelectric energy harvesters for biomedical applications *Nano Energy* **57** 879–902
[11] Chandrasekaran S *et al* 2019 Micro-scale to nano-scale generators for energy harvesting: Self powered piezoelectric, triboelectric and hybrid devices *Phys. Rep.-Rev. Sect. Phys. Lett.* **792** 1–33
[12] Ryu H, Yoon H-J and Kim S-W 2019 Hybrid energy harvesters: toward sustainable energy harvesting *Adv. Mater.* **31** e1802898
[13] Diaz A F and Felix-Navarro R M 2004 A semi-quantitative tribo-electric series for polymeric materials: the influence of chemical structure and properties *J. Electrostat.* **62** 277–90
[14] Zhu G *et al* 2015 Triboelectric nanogenerators as a new energy technology: from fundamentals, devices, to applications *Nano Energy* **14** 126–38
[15] Zhu G *et al* 2013 Toward large-scale energy harvesting by a nanoparticle-enhanced triboelectric nanogenerator *Nano Lett.* **13** 847–53
[16] Zhu G *et al* 2014 A shape-adaptive thin-film-based approach for 50% high-efficiency energy generation through micro-grating sliding electrification *Adv. Mater.* **26** 3788–96
[17] Zhu G *et al* 2012 Triboelectric-generator-driven pulse electrodeposition for micropatterning *Nano Lett.* **12** 4960–5
[18] Yang Y *et al* 2013 Single-electrode-based sliding triboelectric nanogenerator for self-powered displacement vector sensor system *ACS Nano* **7** 7342–51
[19] Wang S *et al* 2013 Sliding-triboelectric nanogenerators based on in-plane charge-separation mechanism *Nano Lett.* **13** 2226–33
[20] Hinchet R *et al* 2014 Performance optimization of vertical nanowire-based piezoelectric nanogenerators *Adv. Funct. Mater.* **24** 971–7
[21] Tang W *et al* 2015 Liquid-metal electrode for high-performance triboelectric nanogenerator at an instantaneous energy conversion efficiency of 70.6% *Adv. Funct. Mater.* **25** 3718–25
[22] Lin L *et al* 2015 Robust triboelectric nanogenerator based on rolling electrification and electrostatic induction at an instantaneous energy conversion efficiency of similar to 55% *ACS Nano* **9** 922–30
[23] Li H Y *et al* 2015 Significant enhancement of triboelectric charge density by fluorinated surface modification in nanoscale for converting mechanical energy *Adv. Funct. Mater.* **25** 5691–7
[24] Jiang T *et al* 2015 Structural optimization of triboelectric nanogenerator for harvesting water wave energy *ACS Nano* **9** 12562–72
[25] Seol M L *et al* 2015 Vertically stacked thin triboelectric nanogenerator for wind energy harvesting *Nano Energy* **14** 201–8
[26] Zhang C *et al* 2014 Rotating-disk-based direct-current triboelectric nanogenerator *Adv. Energy Mater.* **4**
[27] Shin S H *et al* 2015 Triboelectric charging sequence induced by surface functionalization as a method to fabricate high performance triboelectric generators *ACS Nano* **9** 4621–7
[28] Zheng L *et al* 2015 A hybridized power panel to simultaneously generate electricity from sunlight, raindrops, and wind around the clock *Adv. Energy Mater.* **5**

[29] Yang Y and Wang Z L 2015 Hybrid energy cells for simultaneously harvesting multi-types of energies *Nano Energy* **14** 245–56
[30] Jung W S *et al* 2015 High output piezo/triboelectric hybrid generator *Sci. Rep.* **5**
[31] Lee J P, Lee J W and Baik J M 2018 The progress of PVDF as a functional material for triboelectric nanogenerators and self-powered sensors *Micromachines* **9** 532
[32] Wu C *et al* 2016 Wearable electricity generators fabricated utilizing transparent electronic textiles based on polyester/Ag nanowires/graphene core–shell nanocomposites *ACS Nano* **10** 6449–57
[33] Bai P *et al* 2014 Membrane-based self-powered triboelectric sensors for pressure change detection and its uses in security surveillance and healthcare monitoring *Adv. Funct. Mater.* **24** 5807–13
[34] Fan F-R, Tian Z-Q and Lin Wang Z 2012 Flexible triboelectric generator *Nano Energy* **1** 328–34
[35] Wang Z L, Chen J and Lin L 2015 Progress in triboelectric nanogenerators as a new energy technology and self-powered sensors *Energy Environ. Sci.* **8** 2250–82
[36] Zhu G *et al* 2013 Linear-grating triboelectric generator based on sliding electrification *Nano Lett.* **13** 2282–9
[37] Zhang Q *et al* 2017 Service behavior of multifunctional triboelectric nanogenerators *Adv. Mater.* **29** 1606703
[38] Lai Y-C *et al* 2017 Single-thread-based wearable and highly stretchable triboelectric nanogenerators and their applications in cloth-based self-powered human-interactive and biomedical sensing *Adv. Funct. Mater.* **27** 1604462
[39] Niu S *et al* 2014 Theoretical investigation and structural optimization of single-electrode triboelectric nanogenerators *Adv. Funct. Mater.* **24** 3332–40
[40] Wang S *et al* 2014 Freestanding triboelectric-layer-based nanogenerators for harvesting energy from a moving object or human motion in contact and non-contact modes *Adv. Mater.* **26** 2818–24
[41] Fang H *et al* 2009 Controlled growth of aligned polymer nanowires *J. Phys. Chem.* C **113** 16571–4
[42] Wang S, Lin L and Wang Z L 2012 Nanoscale triboelectric-effect-enabled energy conversion for sustainably powering portable electronics *Nano Lett.* **12** 6339–46
[43] Bai P *et al* 2013 Integrated multilayered triboelectric nanogenerator for harvesting biomechanical energy from human motions *ACS Nano* **7** 3713–9
[44] Chun J *et al* 2015 Mesoporous pores impregnated with Au nanoparticles as effective dielectrics for enhancing triboelectric nanogenerator performance in harsh environments *Energy Environ. Sci.* **8** 3006–12
[45] Du W *et al* 2014 A three dimensional multi-layered sliding triboelectric nanogenerator *Adv. Energy Mater.* **4** 1301592
[46] Lin L *et al* 2013 Segmentally structured disk triboelectric nanogenerator for harvesting rotational mechanical energy *Nano Lett.* **13** 2916–23
[47] Bai P *et al* 2013 Cylindrical rotating triboelectric nanogenerator *ACS Nano* **7** 6361–6
[48] Jing Q *et al* 2014 Case-encapsulated triboelectric nanogenerator for harvesting energy from reciprocating sliding motion *ACS Nano* **8** 3836–42
[49] Yang Y *et al* 2013 A single-electrode based triboelectric nanogenerator as self-powered tracking system *Adv. Mater.* **25** 6594–601

[50] Meng B *et al* 2013 A transparent single-friction-surface triboelectric generator and self-powered touch sensor *Energy Environ. Sci.* **6** 3235–40
[51] Xie Y *et al* 2014 Grating-structured freestanding triboelectric-layer nanogenerator for harvesting mechanical energy at 85% total conversion efficiency *Adv. Mater.* **26** 6599–607
[52] Lin L *et al* 2014 Noncontact free-rotating disk triboelectric nanogenerator as a sustainable energy harvester and self-powered mechanical sensor *ACS Appl. Mater. Interfaces* **6** 3031–8
[53] Zhu G *et al* 2014 Radial-arrayed rotary electrification for high performance triboelectric generator *Nat. Commun.* **5** 3426
[54] Jin L *et al* 2016 Self-powered safety helmet based on hybridized nanogenerator for emergency *ACS Nano* **10** 7874–81
[55] Wang J *et al* 2015 A flexible fiber-based supercapacitor–triboelectric-nanogenerator power system for wearable electronics *Adv. Mater.* **27** 4830–6
[56] Wang J *et al* 2016 Extremely stretchable electroluminescent devices with ionic conductors *Adv. Mater.* **28** 4490–6
[57] Seung W *et al* 2015 Nanopatterned textile-based wearable triboelectric nanogenerator *ACS Nano* **9** 3501–9
[58] Zhou T *et al* 2014 Woven structured triboelectric nanogenerator for wearable devices *ACS Appl. Mater. Interfaces* **6** 14695–701
[59] Pu X *et al* 2016 Wearable self-charging power textile based on flexible yarn supercapacitors and fabric nanogenerators *Adv. Mater.* **28** 98
[60] Pu X *et al* 2015 A self-charging power unit by integration of a textile triboelectric nanogenerator and a flexible lithium-ion battery for wearable electronics *Adv. Mater.* **27** 2472–8
[61] Li S L *et al* 2015 Cloth-based power shirt for wearable energy harvesting and clothes ornamentation *ACS Appl. Mater. Interfaces* **7** 14912–6
[62] Kim K N *et al* 2015 Highly stretchable 2D fabrics for wearable triboelectric nanogenerator under harsh environments *ACS Nano* **9** 6394–400
[63] Wang J *et al* 2015 A flexible fiber-based supercapacitor-triboelectric-nanogenerator power system for wearable electronics *Adv. Mater.* **27** 4830–6
[64] Yang J *et al* 2015 Eardrum-inspired active sensors for self-powered cardiovascular system characterization and throat-attached anti-interference voice recognition *Adv. Mater.* **27** 1316–26
[65] Lin Z-H *et al* 2013 Enhanced triboelectric nanogenerators and triboelectric nanosensor using chemically modified TiO_2 nanomaterials *ACS Nano* **7** 4554–60
[66] Fan F-R *et al* 2012 Transparent triboelectric nanogenerators and self-powered pressure sensors based on micropatterned plastic films *Nano Lett.* **12** 3109–14
[67] Hou T-C *et al* 2013 Triboelectric nanogenerator built inside shoe insole for harvesting walking energy *Nano Energy* **2** 856–62
[68] Hu Y *et al* 2013 Triboelectric nanogenerator built on suspended 3D spiral structure as vibration and positioning sensor and wave energy harvester *ACS Nano* **7** 10424–32
[69] Tang W, Meng B and Zhang H X 2013 Investigation of power generation based on stacked triboelectric nanogenerator *Nano Energy* **2** 1164–71
[70] Wang S *et al* 2014 Maximum surface charge density for triboelectric nanogenerators achieved by ionized-air injection: methodology and theoretical understanding *Adv. Mater.* **26** 6720–8

[71] Wang Z *et al* 2014 Enhancing the performance of triboelectric nanogenerator through prior-charge injection and its application on self-powered anticorrosion *Nano Energy* **10** 37–43

[72] Zhang X-S *et al* 2014 High-performance triboelectric nanogenerator with enhanced energy density based on single-step fluorocarbon plasma treatment *Nano Energy* **4** 123–31

[73] Zhang X-S *et al* 2013 Frequency-multiplication high-output triboelectric nanogenerator for sustainably powering biomedical microsystems *Nano Lett.* **13** 1168–72

[74] Yang Y *et al* 2013 Silicon-based hybrid energy cell for self-powered electrodegradation and personal electronics *ACS Nano* **7** 2808–13

[75] Li T *et al* 2016 Lightweight triboelectric nanogenerator for energy harvesting and sensing tiny mechanical motion *Adv. Funct. Mater.* **26** 4370–6

[76] Dzenis Y 2004 Spinning continuous fibers for nanotechnology *Science* **304** 1917

[77] Li D and Xia Y 2004 Electrospinning of nanofibers: reinventing the wheel? *Adv. Mater.* **16** 1151–70

[78] Li Z *et al* 2018 Multilayered fiber-based triboelectric nanogenerator with high performance for biomechanical energy harvesting *Nano Energy* **53** 726–33

[79] Yin Y *et al* 2018 Stretchable and tailorable triboelectric nanogenerator constructed by nanofibrous membrane for energy harvesting and self-powered biomechanical monitoring *Adv. Mater. Technol.* **3** 1700370

[80] Yu B *et al* 2018 A biomimetic nanofiber-based triboelectric nanogenerator with an ultrahigh transfer charge density *Nano Energy* **48** 464–70

[81] Qiu Q *et al* 2019 Highly flexible, breathable, tailorable and washable power generation fabrics for wearable electronics *Nano Energy* **58** 750–8

[82] Li Z *et al* 2017 Nanofibrous membrane constructed wearable triboelectric nanogenerator for high performance biomechanical energy harvesting *Nano Energy* **36** 341–8

[83] Chen F *et al* 2019 A novel triboelectric nanogenerator based on electrospun polyvinylidene fluoride nanofibers for effective acoustic energy harvesting and self-powered multifunctional sensing *Nano Energy* **56** 241–51

[84] Qin Z *et al* 2019 Wearable and stretchable triboelectric nanogenerator based on crumpled nanofibrous membranes *ACS Appl. Mater. Interfaces* **11** 12452–9

[85] Kim W *et al* 2019 Humidity-resistant triboelectric energy harvester using electrospun PVDF/PU nanofibers for flexibility and air permeability *Nanotechnology* **30**

[86] Kim S-R, Yoo J-H and Park J-W 2019 Using electrospun AgNW/P(VDF-TrFE) composite nanofibers to create transparent and wearable single-electrode triboelectric nanogenerators for self-powered touch panels *ACS Appl. Mater. Interfaces* **11** 15088–96

[87] Im J-S and Park I-K 2018 Mechanically robust magnetic Fe_3O_4 nanoparticle/polyvinylidene fluoride composite nanofiber and its application in a triboelectric nanogenerator *ACS Appl. Mater. Interfaces* **10** 25660–5

[88] Cheon S *et al* 2018 High-performance triboelectric nanogenerators based on electrospun polyvinylidene fluoride-silver nanowire composite nanofibers *Adv. Funct. Mater.* **28**

[89] An S, Sankaran A and Yarin A L 2018 Natural biopolymer-based triboelectric nanogenerators via fast, facile, scalable solution blowing *ACS Appl. Mater. Interfaces* **10** 37749–59

[90] Martins P, Lopes A and Lanceros-Mendez S 2014 Electroactive phases of poly (vinylidene fluoride): determination, processing and applications *Prog. Polym. Sci.* **39** 683–706

[91] Mandal D, Yoon S and Kim K J 2011 Origin of piezoelectricity in an electrospun poly (vinylidene fluoride-trifluoroethylene) nanofiber web-based nanogenerator and nano-pressure sensor *Macromol. Rapid Commun.* **32** 831–7

[92] Park S H *et al* 2010 Robust superhydrophobic mats based on electrospun crystalline nanofibers combined with a silane precursor *ACS Appl. Mater. Interfaces* **2** 658–62
[93] Zhong G *et al* 2011 Understanding polymorphism formation in electrospun fibers of immiscible poly (vinylidene fluoride) blends *Polymer* **52** 2228–37
[94] Kwak S S *et al* 2016 Triboelectrification-induced large electric power generation from a single moving droplet on graphene/polytetrafluoroethylene *ACS Nano* **10** 7297–302
[95] Lee J H *et al* 2015 Control of skin potential by triboelectrification with ferroelectric polymers *Adv. Mater.* **27** 5553–8
[96] Seung W *et al* 2017 Boosting power-generating performance of triboelectric nanogenerators via artificial control of ferroelectric polarization and dielectric properties *Adv. Energy Mater.* **7** 1600988
[97] Yu H *et al* 2013 Enhanced power output of an electrospun PVDF/MWCNTs-based nanogenerator by tuning its conductivity *Nanotechnology* **24** 405401
[98] Huang T *et al* 2015 Enhanced power output of a triboelectric nanogenerator composed of electrospun nanofiber mats doped with graphene oxide *Sci. Rep.* **5** 13942
[99] Yu Y and Wang X 2016 Chemical modification of polymer surfaces for advanced triboelectric nanogenerator development *Extreme Mech. Lett.* **9** 514–30
[100] Fang Z *et al* 2018 Surface texturing and dielectric property tuning toward boosting of triboelectric nanogenerator performance *J. Mater. Chem.* A **6** 52–7
[101] Saadatnia Z, Esmailzadeh E and Naguib H E 2019 High performance triboelectric nanogenerator by hot embossing on self-assembled micro-particles *Adv. Eng. Mater.* **21** 1700957
[102] Zhao L M *et al* 2016 A size-unlimited surface microstructure modification method for achieving high performance triboelectric nanogenerator *Nano Energy* **28** 172–8
[103] Wang Z L 2013 Triboelectric nanogenerators as new energy technology for self-powered systems and as active mechanical and chemical sensors *ACS Nano* **7** 9533–57
[104] Kanik M *et al* 2015 A motion- and sound-activated, 3D-printed, chalcogenide-based triboelectric nanogenerator *Adv. Mater.* **27** 2367–76
[105] Li Z L *et al* 2015 Beta-cyclodextrin enhanced triboelectrification for self-powered phenol detection and electrochemical degradation *Energy Environ. Sci.* **8** 887–96
[106] Huang T *et al* 2015 Human walking-driven wearable all-fiber triboelectric nanogenerator containing electrospun polyvinylidene fluoride piezoelectric nanofibers *Nano Energy* **14** 226–35
[107] Li Z B *et al* 2019 Small-sized, lightweight, and flexible triboelectric nanogenerator enhanced by PTFE/PDMS nanocomposite electret *ACS Appl. Mater. Interfaces* **11** 20370–7
[108] Park S-H *et al* 2011 Dye-sensitized solar cells based on electrospun polymer blends as electrolytes *Sol. Energy Mater. Sol. Cells* **95** 296–300
[109] Dai Y *et al* 2011 Ceramic nanofibers fabricated by electrospinning and their applications in catalysis, environmental science, and energy technology *Polym. Adv. Technol.* **22** 326–38
[110] Cavaliere S *et al* 2011 Electrospinning: designed architectures for energy conversion and storage devices *Energy Environ. Sci.* **4** 4761–85
[111] Laforgue A 2011 All-textile flexible supercapacitors using electrospun poly (3, 4-ethylenedioxythiophene) nanofibers *J. Power Sources* **196** 559–64
[112] Cao R *et al* 2018 Self-powered nanofiber-based screen-print triboelectric sensors for respiratory monitoring *Nano Res.* **11** 3771–9

ₕtning Source UK Ltd.
Keynes UK
50745040721
/4UK00003B/151